ADVANCES IN

Immunology

VOLUME 43

ADVANCES IN
Immunology

EDITED BY

FRANK J. DIXON

*Scripps Clinic and Research Foundation
La Jolla, California*

ASSOCIATE EDITORS

K. FRANK AUSTEN
LEROY E. HOOD
JONATHAN W. UHR

VOLUME 43

ACADEMIC PRESS, INC.
Harcourt Brace Jovanovich, Publishers

San Diego New York Berkeley Boston
London Sydney Tokyo Toronto

COPYRIGHT © 1988 BY ACADEMIC PRESS, INC.
ALL RIGHTS RESERVED.
NO PART OF THIS PUBLICATION MAY BE REPRODUCED OR
TRANSMITTED IN ANY FORM OR BY ANY MEANS, ELECTRONIC
OR MECHANICAL, INCLUDING PHOTOCOPY, RECORDING, OR
ANY INFORMATION STORAGE AND RETRIEVAL SYSTEM, WITHOUT
PERMISSION IN WRITING FROM THE PUBLISHER.

ACADEMIC PRESS, INC.
San Diego, California 92101

United Kingdom Edition published by
ACADEMIC PRESS, INC. (LONDON) LTD.
24-28 Oval Road, London NW1 7DX

LIBRARY OF CONGRESS CATALOG CARD NUMBER: 61-17057

ISBN 0-12-022443-7 (alk. paper)

PRINTED IN THE UNITED STATES OF AMERICA
88 89 90 91   9 8 7 6 5 4 3 2 1

CONTENTS

The Chemistry and Mechanism of Antibody Binding to Protein Antigens
ELIZABETH D. GETZOFF, JOHN A. TAINER, RICHARD A. LERNER, AND H. MARIO GEYSEN

I.	Introduction	1
II.	Criteria for Systematic Studies	5
III.	Structural Chemistry from X-ray Crystallography	8
IV.	Defining the Chemistry of Antigenic Recognition	27
V.	Defining the Mechanisms of Antigenic Recognition	39
VI.	Dominant Epitopes and Collective Response	46
VII.	Evaluation of Predictive Methods	49
VIII.	An Experimental Approach to the Design of Antipeptide Antibodies	59
IX.	Frequency Data for Individual Amino Acids in Protein Epitopes	65
X.	Overview of Chemistry and Binding Energy	72
	References	84

Structure of Antibody–Antigen Complexes: Implications for Immune Recognition
P. M. COLMAN

I.	Introduction	99
II.	Principles of Protein–Protein Interactions	100
III.	Antigen Receptor Structure	105
IV.	Antigens	117
V.	Three-Dimensional Structures of Complexes	119
VI.	The Interface Adaptor Hypothesis	125
	References	128

The γδ T Cell Receptor
MICHAEL B. BRENNER, JACK L. STROMINGER, AND MICHAEL S. KRANGEL

I.	Introduction: Immunological Antigen Receptors	133
II.	T Cell Receptor (TCR) γ Genes	136
III.	TCR γδ Proteins	146
IV.	TCR δ Genes	153
V.	Phenotype and Distribution of TCR γδ Lymphocytes	161
VI.	Ontogeny of TCR γδ	165
VII.	Functional Capabilities of TCR γδ Lymphocytes	171
VIII.	Summary and Hypothesis	174
	References	178

Specificity of the T Cell Receptor for Antigen
Stephen M. Hedrick

I.	Introduction	193
II.	Nature of the Determinant Recognized by T Cells	194
III.	Studies on the Specificity of T Cells for Peptide Antigens	197
IV.	Formation of the Antigen-MHC Determinant Recognized by the TCR: Stable Formation in the Absence of T Cells	206
V.	TCR Combining Site	215
VI.	Strategies of Immune Recognition	223
	References	225

Transcriptional Controlling Elements in the Immunoglobulin and T Cell Receptor Loci
Kathryn Calame and Suzanne Eaton

I.	Introduction	235
II.	Immunoglobulin Gene Expression	236
III.	Immunoglobulin Transcriptional Enhancer Elements	240
IV.	Immunoglobulin Promotor Elements	251
V.	Possible Mechanisms Mediating Transcriptional Regulation of Immunoglobulin Genes	258
VI.	T Cell Receptor Gene Expression	266
VII.	Perspectives	269
	References	270

Molecular Aspects of Receptors and Binding Factors for IgE
Henry Metzger

I.	Introduction	277
II.	Sites on IgE Interacting with Receptors	278
III.	The Mast-Cell Specific Receptor	282
IV.	Immunoregulatory Fc_ϵ Receptors and IgE-Binding Factors	298
V.	Surface Receptors on Monocytes, Eosinophils, and Platelets	305
VI.	Other IgE-Binding Proteins	307
	References	308

Index — 313

Contents of Recent Volumes — 325

The Chemistry and Mechanism of Antibody Binding to Protein Antigens

ELIZABETH D. GETZOFF, JOHN A. TAINER, AND RICHARD A. LERNER

*Department of Molecular Biology,
Research Institute of Scripps Clinic,
La Jolla, California 92037*

H. MARIO GEYSEN

*Department of Molecular Immunology,
Commonwealth Serum Laboratories,
Parkville 3052, Victoria, Australia*

I. Introduction

Although the science of immunology began with studies on immunochemistry, a series of important advances in biological techniques led the major focus in this field away from the chemistry of antibody binding for several decades. Among these discoveries, hybridoma technology (Kohler and Milstein, 1975) was particularly important, because it allowed the purification of individual antibody specificities from the entire set of responses resulting from presentation of a protein immunogen. While this ability to make individual antibody specificities rapidly continues to develop, advances in the definition of epitopes, and hence in the nature of the interaction chemistry, have proceeded more slowly. Within the past few years, however, the field of immunology has been returning to immunochemistry with renewed interest due, in part, to the development of peptide synthesis technologies. These methods enable one to prepare antipeptide antibodies of predetermined specificity (Lerner, 1984) and to screen monoclonal and polyclonal antibodies for the identification of the critical antigenic sequences and even individual side chains that they recognize (Geysen *et al.*, 1984, 1985a,b, 1987b,c; Houghten *et al.*, 1986; Houghten, 1985).

During the same period, crystallographic studies of antibodies, protein antigens, and most recently, complexes of antibodies with protein antigens have made the detailed stereochemistry of the antibody–antigen interaction available at the atomic level. Complementing this knowledge of the structural stereochemistry are recent results in the successful design of a substrate for antibody catalysis (Pollack *et al.*, 1986) based on the known active-site structure of an antibody, and of new catalytic antibodies generated using transition-state analogs as immunogens (Tramontano et al., 1986, 1988; Jacobs et al., 1987; Napper et al., 1987). Taken together, these new approaches and results bring to immunology a renewed interest in understanding and using fundamental principles of immunochemistry.

There are already a number of authoritative reviews on protein immunogenicity (Berzofsky, 1985; Benjamin et al., 1984; Crumpton, 1974). Here, we ignore the biological complexities associated with the production of antibodies to address the chemical nature of the recognition process. As part of this work, we will consider the known antibody combining site structures for their implications regarding the stereochemistry of interaction. Detailed crystallographic information on the structure of antibody–antigen complexes appears elsewhere in this volume (see Colman, 1988). In addition, the recent application of immunochemistry to the design of antibodies with enzymatic activities is outside the scope of this review, but has recently been described elsewhere (Lerner and Tramontano, 1987, 1988).

In order to consider the chemistry of antigenicity, we have, by necessity, had to select the fraction of antigenic sites that can be defined in sufficient stereochemical detail. Although there are many different ways to identify protein epitopes, here we focus primarily on results from X-ray crystallography and peptide mapping, because these currently provide the most suitable data base for the consideration of the chemistry and mechanism of antibody binding to protein antigens.

Many excellent studies of epitopes have been done using fine specificity mapping techniques (reviewed by Benjamin et al., 1984); however, the available sequence variability of proteins puts definite limitations on this approach for understanding the role of individual amino acids. Chemical modification studies allow assembled topographic epitopes to be identified by chemical footprinting, but currently require specific amino acids, such as lysine and threonine (Burnens et al., 1987). Like chemical modification, proteolytic digestion techniques and blocking of epitopes by other antibodies can provide a footprint of the combining region (Jemmerson and Paterson, 1986a,b; Smith-Gill et al., 1982; Berzofsky et al., 1982; East et al., 1982).

In one sense, fine specificity mapping, blocking by other antibodies, chemical modification, and proteolytic digestion approaches are superior to peptide mapping techniques, because they provide a means to study assembled topographic epitopes that may be missed using peptide methods. From a conformational point of view, X-ray crystallography provides the most complete understanding of antibody–antigen interactions, if the complex in question can be crystallized.

In another sense, methods that identify the topographic protein interface buried by the bound antibody (i.e., X-ray crystallography and, to some extent, footprinting studies) provide an incomplete picture, because they define the boundaries of the epitope without necessarily identifying those residues most critical to the binding chemistry. Both fine specificity and peptide mapping approaches aim to identify the subset of the amino acid side chains critical to specific binding; this subset of residues might appropriately be termed the antigenic determinants for a given epitope. Peptide mapping can be further applied to generate a comprehensive, self-consistent data base on the frequency of response to sequential antigenic sites, which is ideal for identifying overall chemical patterns and for probing the interaction chemistry.

Most antibodies raised against intact proteins recognize conformational sites and, consequently, do not bind peptides or proteolytic fragments with high affinity (Benjamin et al., 1984). Yet, some antibodies raised against native protein antigens do bind to peptide homologs of the protein sequence (Dorow et al., 1985; Lando et al., 1982; Lando and Reichlin, 1982; Rodda et al., 1986a), allowing the chemistry of these sites to be probed at the resolution of single amino acid side chains (Geysen et al., 1987b,c; Getzoff et al., 1987; Fieser et al., 1987). Although protein epitopes can be expected to have a conformational component (Berzofsky, 1985; Benjamin et al., 1984), we will use the term "sequential epitope" to refer to epitopes or portions of epitopes identified by peptide mapping in which complete conformational information is not immediately available. Some antibodies raised against native protein appear to recognize denatured forms of the protein only, and although these antibodies may be important for biochemical and diagnostic identifications, they are not the primary focus of this review.

For our purposes, sequential epitopes denote the subset of antigenic sites on native proteins that can be mapped by antibody binding to peptide homologs of portions of the protein sequence and competitively inhibited by native protein at reasonable concentrations (about 10^{-5} or less). Peptide mapping probes only a subset of protein epitopes, but the chemistry of these sites is relevant, provided that proper controls indicate reasonably high antibody affinities (about 10^7 or higher) for the native protein and that native protein competes with the peptide homologs for antibody binding (Geysen et al., 1987b,c; Fieser et al., 1987; Rodda et al., 1986a). As noted previously (Benjamin et al., 1984), proteins are an extremely diverse set of antigenic molecules, spanning wide variations in secondary structure, size, assembly state, and degree of foreignness from the responding host. Several apparent controversies appear to be the result of attempts to apply a technique such as peptide mapping to proteins that are poor candidates for that approach. For example, sequence differences in cytochrome c between the protein antigen and the host protein (2–10%) are limited, yet every amino acid difference in cytochrome c can elicit antibodies (Benjamin et al., 1984). Peptide mapping of sequential epitopes is unlikely to succeed on cytochrome c, because the differing amino acids are not sequential, but likely to succeed on foreign proteins (e.g., myohemerythrin or proteins from pathogens) sharing little or no homology with those of the host. All the above-mentioned approaches to identification of epitopes appear to be useful, provided the nature of the specific protein antigen is considered in the experimental design. For example, cytochrome c is a good candidate for chemical footprinting techniques due to the large number of Lys residues.

Crystallographic approaches to characterizing antibody–antigen complexes are limited by practical constraints, resulting from the experimental difficulties of protein crystallization and from the extensive time and equipment requirements. Nevertheless, the crystallographic structures of antibody complexes with protein antigens (Amit et al., 1985, 1986; Colman et al., 1987a; Sheriff et al., 1987b) provide a wealth of information with which all other biochemical and immunological experiments must be reconciled. Yet, it is currently safest to assume that the occurrence of residues in a crystallographic interface is a necessary, but not sufficient, indication that they contribute critically to the specific chemistry of binding. Moreover, for the purposes of understanding antibody binding to protein antigens and identifying underlying chemical patterns, it is desirable to obtain a large, systematic data base on the frequency of response to sets of antigenic sites for characterized proteins. It is not practical to obtain such a data base from X-ray crystallographic structures alone. The most complete understanding of antigenic recognition, therefore, requires a synthesis of data from X-ray crystallography and from other complementary experiments such as peptide mapping. Ultimately, a complete picture of the immune response will include a detailed description of the structural chemistry of the T cell epitopes (reviewed by Paterson, 1988; Berzofsky et al., 1987). In this regard, the recently solved structure of the related class I major histocompatibility antigen (Bjorkman et al., 1987a,b) represents an important breakthrough. Moreover, despite current technical difficulties in the assay of T cell epitopes, due to the ternary nature of the interaction, there is no *a priori* reason that the overall approach to defining sequential epitopes presented here cannot be applied to rigorous and systematic studies of the interaction chemistry for T cell epitopes.

This review is based on information derived from crystallographic structures of antibodies and antibody–antigen complexes, from a systematic data base on the frequency of response to sequential epitopes on the protein antigens myohemerythrin (MHr), myoglobin, and tobacco mosaic virus (TMV) coat protein, from the cross-reactivity of antipeptide antibodies with native MHr, and from the identification of the critical side chains in sequential epitopes in 14 other proteins. This synthesis of crystallographic and peptide mapping information provides a framework for posing a consistent and testable model for the chemical nature of antibody binding to protein antigens. Minimal requirements for objective, quantitative analysis of data on epitopes and related intrinsic structural parameters are proposed. Also addressed are evidence for the existence of dominant epitopes and their correlation with antigen stereochemistry, the extent of individual animal and cross-species variation in antigenic response, and the validity of a defined, collective response. As to practical aspects, we shall consider the validity of predicting antigenic sites from amino acid sequences and the selection of appropriate sites for the production of antipeptide antibodies that will react with the intact protein antigen. Finally, this information is integrated to assess the implications for the chemistry, mechanism, and binding energy of antibody–antigen interactions.

II. Criteria for Systematic Studies

The field of immunochemistry is poised for extremely rapid advances in both theoretical and practical understanding of the chemistry of antibody recognition. Applications are already evident for progress toward the design of vaccines (Getzoff *et al.*, 1988a,b; Cease *et al.*, 1987; Francis *et al.*, 1987; McCray and Werner, 1987; Good *et al.*, 1987; Sutcliffe *et al.*, 1983; Gerin *et al.*, 1983; Beachey *et al.*, 1981; Audibert *et al.*, 1981) and catalytic antibodies (Lerner and Tramontano, 1988). However, the level of productivity for the next few years may well be as limited by problems in the analysis of immunochemical data and the design of experiments as by technical restrictions. The sciences of mathematics and physics, for example, are blessed by having explicit, defined standards for the evaluation of data, including the specification of confidence limits. These standards not only provide criteria for judgement of experimental data, but also offer assurance that any given set of results can be duplicated in different laboratories within the confidence limits allowed by the experimental context. Such standards are obviously much more difficult to apply and evaluate for biological systems. However, the recent determination of X-ray crystallographic structures for antibody–antigen complexes provides a data base of specific, objective, explicitly defined information on the binding interfaces of both antibodies and protein antigens against which other experiments can be tested.

A prerequisite for combining the results of structural and immunological experiments is the establishment of appropriate criteria for the systematic analysis of these data bases (Van Regenmortel, 1987; Van Regenmortel *et al.*, 1988). As a first step, we have recently proposed three requirements for studies of the chemistry of antigenic recognition (Geysen *et al.*, 1987a): (1) compatible methodology; (2) consistent, objective, and accurate results; and (3) defined criteria for correlation. Without such guidelines, time may be lost on subjectively appealing interpretations and apparent correlations based on qualitative assessments that do not allow reliable duplication.

In this review, we suggest specific steps for increased rigor in the analysis and interpretation of immunochemical results, which should be applicable to studies of both B cell and T cell epitopes. For example, the expansion of improved systematic data bases on well-defined systems will facilitate the design of experiments to probe the basic chemistry of immune recognition. The extreme diversity of the immune response requires the proper assessment of the frequency, as well as the existence, of the immune response to a given epitope in a given species (Geysen *et al.*, 1987c). We also consider the type of variation in response across species and show how this can be considered quantitatively (Section VI). To apply these suggestions to experimental questions, we have systematically evaluated current methods for the prediction of antigenic sites from the linear sequence and have compared the results with those expected from the selection of both long and short peptides at random (Section VII). We believe that the use of a systematic and objective approach greatly simplifies a number of supposed quandaries in the study of both the interaction of antipeptide antibodies with native protein antigens and the interaction of antiprotein antibodies with peptides antigens. For example, our statistical analyses (Section VII) show that observed differences in the antibodies to longer peptides may follow directly from the statistical probabilities associated with the increased length and do not require any unique properties of longer peptides per se (although longer peptides may also have more conformational character). Finally, we have attempted to show that the detailed stereochemical information available from

TABLE I
CHARACTERIZATION OF IMMUNOGLOBULINS OF KNOWN CRYSTALLOGRAPHIC STRUCTURE

| Antibody | Antigen | Antibody | | | RMSD[a] (Å) | | Resolution Å | Residual Error (%) | Reference |
		Source	Molecule	Class	L	H			
D1.3	Hen egg-white lysozyme	Mouse	Fab	IgG1, κ	1.39	1.06	2.8	28	Amit et al. (1986)
NC41	Influenza virus neuraminidase	Mouse	Fab	IgG2a, κ	—	—	3.0	35	Colman et al. (1987a)
HyHEL5	Hen egg-white lysozyme	Mouse	Fab	IgG, κVI	0.72	0.89	2.54	24.5	Sheriff et al. (1987b)
McPC603	Phosphorylcholine	Mouse	Fab	IgA, κI	0.00	0.00	2.7	22.5	Satow et al. (1986)
J539	β(1–6)-D Galactan	Mouse	Fab	IgA, κVI	0.75	0.80	2.6	19	Suh et al. (1986)
HEd10	Single-stranded poly(dT)DNA	Mouse	Fab	IgG, κ	0.87	0.99	3.0	27.2	Cygler et al. (1987)
Newm	γ-Hydroxy vitamin K	Human	Fab'	IgG1, λI	1.69	1.29	2.0	46	Saul et al. (1978)
Kol	—	Human	Fab	IgG1, λI	1.36	0.65	1.9	26	Marquart et al. (1980)
Dob	—	Human	IgG	IgG1, κIII	—	—	6.0	—	Silverton et al. (1977)
Mcg	2,4-DNP[b] derivatives and other aromatic compounds	Human	L dimer	λV	1.29	—	2.3	37.1	Edmundson et al. (1974)
Loc	—	Human	L dimer	λI	—	—	3.0	27	Chang et al. (1985)
Rhe	—	Human	V_L dimer	λI	1.64	—	1.6	14.9	Furey et al. (1983)
REI	—	Human	V_L dimer	κI	0.48	—	2.0	—	Epp et al. (1974)

[a] RMSDs (root mean square deviations) from McPC603 were calculated separately for the heavy and light chains.
[b] DNP, dinitrophenyl.

X-ray crystallographic structures of antibody combining sites suggests a fundamental basis for the existence of chemical biases in the binding of antibodies to protein antigens.

Addressing these questions involved making certain compromises and choosing priorities in both experimental and analytical methods, and we fully recognize that later work may identify at least some problems with our approach. Our hope is to provide not only a useful presentation of current experimental results, but also a framework for the further improvement of methods to combine and analyze this complex biological, biochemical, and biophysical information.

III. Structural Chemistry from X-ray Crystallography

The X-ray crystallographic structures of antibodies, antigens, and antibody–antigen complexes provide an objective data base for defining the nature of the antibody–antigen complex at atomic resolution. The basic stereochemistry of antibody structures has been reviewed in detail elsewhere (Colman, 1988; Saul and Poljak, 1985; Davies and Metzger, 1983; Novotny et al., 1983; Marquart and Deisenhofer, 1982; Amzel and Poljak, 1979; Padlan, 1977; Davies et al., 1975), including details of the domain associations to form the combining site (Chothia et al., 1985). Recently, two X-ray structures have been determined for antibody–lysozyme complexes (Sheriff et al., 1987b; Amit et al., 1985, 1986) and one for an antibody–neuraminidase complex (Colman et al., 1987a), as reviewed elsewhere (Colman, 1988; Davies et al., 1988). Here, we consider only those structural aspects particularly relevant to the chemistry of the antibody–antigen interaction. This

---

FIG. 1. α-Carbon backbone and overall surface shape of the intact Dob antibody structure. In this **T**-shaped structure, determined by Silverton et al. (1977), the antigen combining sites are located at the ends of the two horizontal Fab arms formed by the association of the light chains (α-carbon backbone as red lines and surface as light blue dots) and heavy chains (α-carbon backbone as yellow lines and surface as blue dots). Dob has a substantial deletion in the hinge region, which presumably limits its segmental flexibility greatly. The molecular surface in this and following figures represents the area accessible to a water-sized (1.4-Å radius) probe sphere as calculated numerically with MS or analytically with AMS (Connolly, 1983a,b). This surface is composed of convex regions, formed by the solvent-accessible van der Waals' surface of individual atoms and concave regions smoothing over small gaps and crevices inaccessible to the probe sphere. Molecular surfaces are especially appropriate for examining intermolecular interactions, because they focus on surface regions accessible to interacting molecules. All computer graphics line and dot renderings (Figs. 1–4, 6, 7, and 10) were made using the programs GRAMPS (O'Donnell and Olson, 1981) and GRANNY (Connolly and Olson, 1984). This computer graphics model of Dob was rendered by A. Olson and J. Tainer.

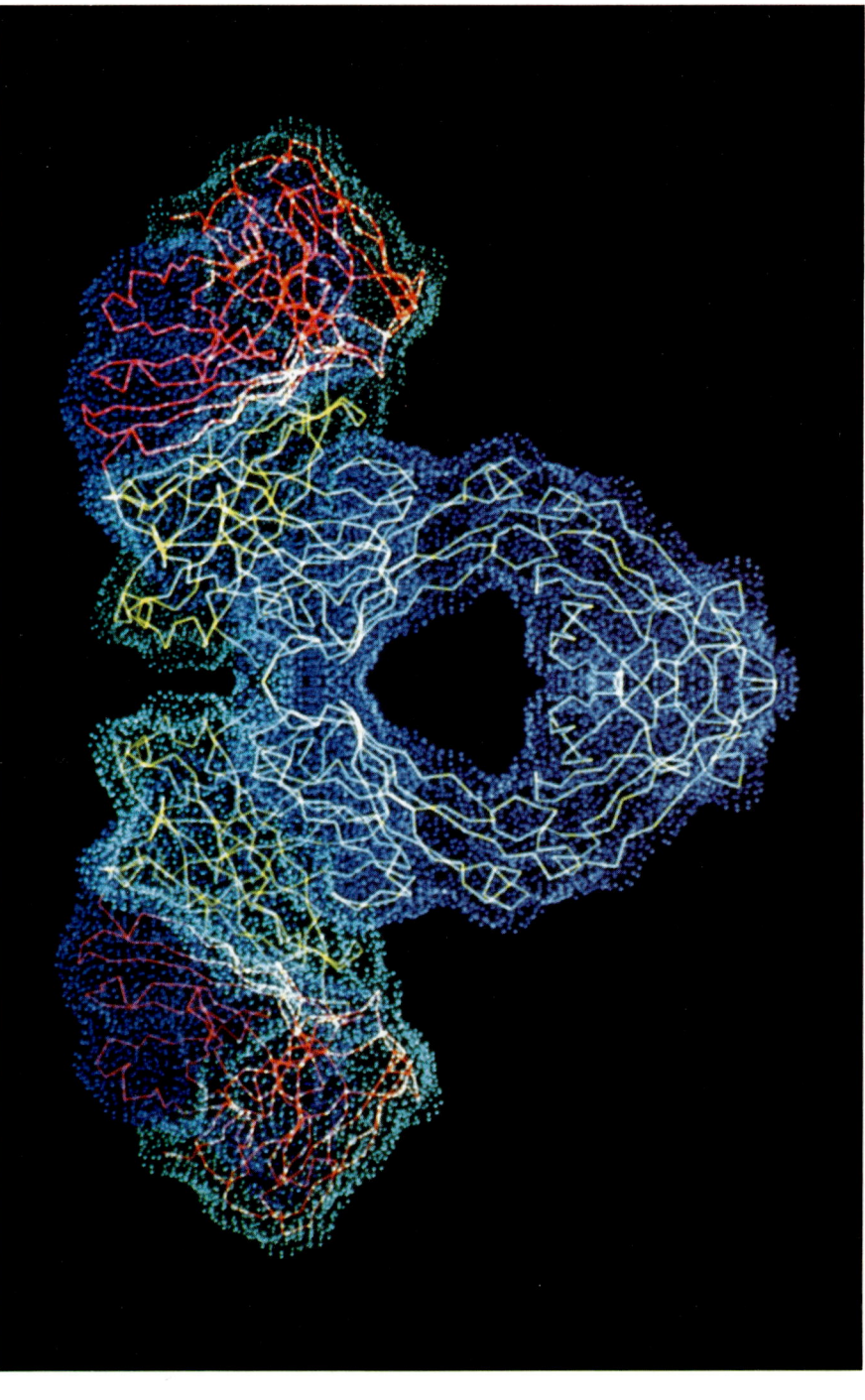

FIG. 1

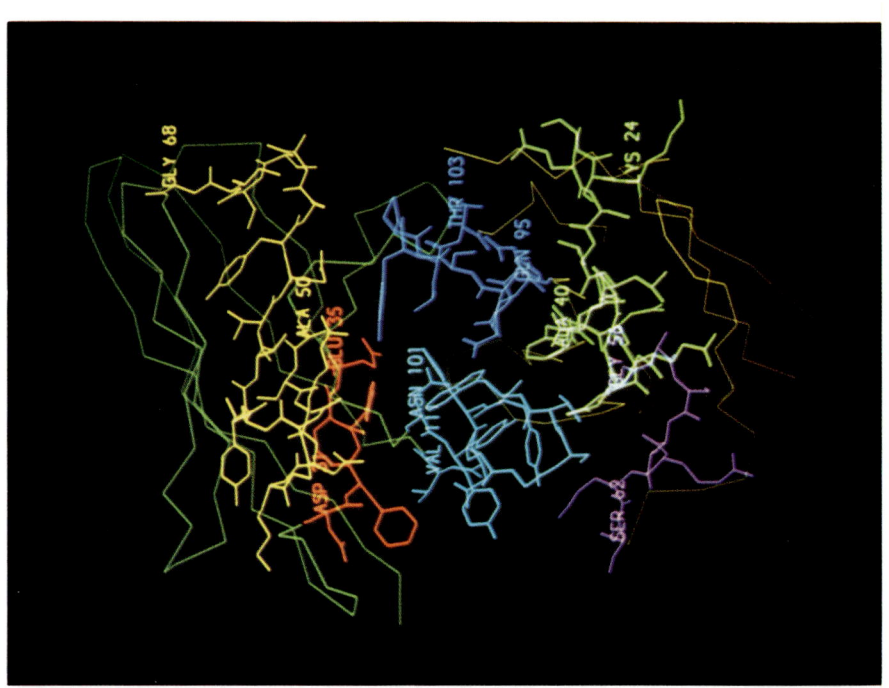
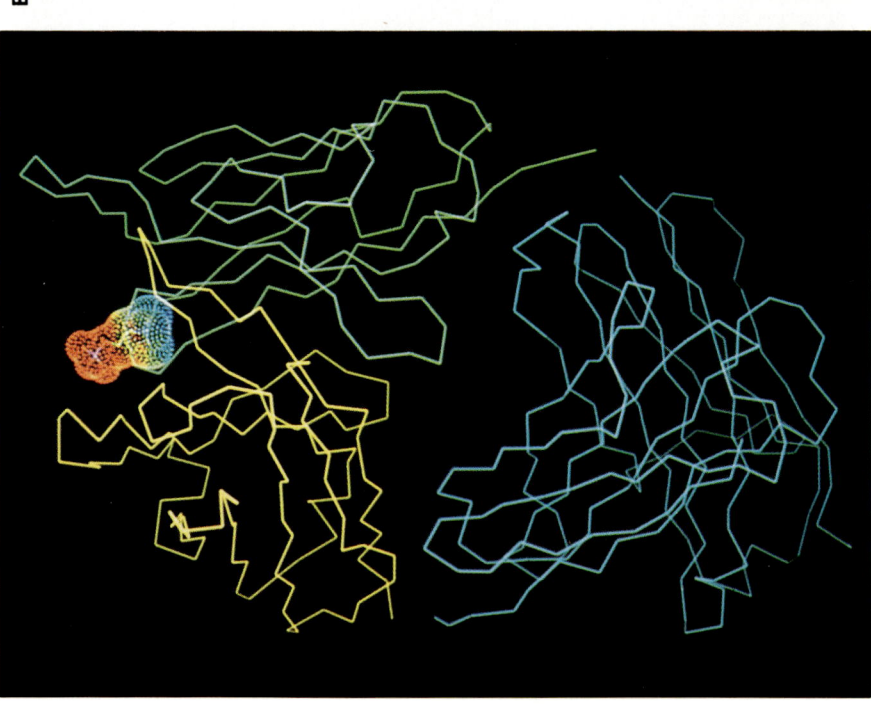

FIG. 2

analysis of antibody stereochemistry is based on the structural data from 10 variable light (V_L) and 7 variable heavy (V_H) domains (Table I, excluding antibodies NC41, Dob, and Loc; Bernstein et al., 1977).

Only one low-resolution study (Silverton et al., 1977) defines the structure of an entire antibody molecule including both the Fab and the Fc regions (Fig. 1). The Fc region, which is normally disordered due to flexibility of the hinge sequences joining it to the Fabs (Huber and Bennett, 1983; Marquart et al., 1980; Colman et al., 1976), is visible in this crystal structure (IgG Dob) because of a deletion in the hinge sequence. Other crystallographic structures characterize Fab regions (Fig. 2A); the Fab fragment contains the V_L and V_H structural domains, along with the constant light (C_L) and first constant heavy ($C_H 1$) domains. All the domains have a Greek key β barrel fold (Richardson, 1981), a common and very stable domain structure that can tolerate sequence and conformational changes in the loop regions. The antigen combining site is formed by six hypervariable loops (Fig. 2B), known as complementarity-determining regions (CDRs), contributed from pairs of one V_L and V_H domain related by a pseudo-2-fold axis. A common nomenclature for these CDRs in sequence order along the polypeptide chain is L1, L2, and L3 for the V_L domain and H1, H2, and H3 for the V_H domain. The 2-fold arrangement between V_L and V_H places the CDRs together in a continuous surface across one end of the Fab, such that these loops form part of the V_L-V_H dimer interface and can, thereby, influence each other indirectly (Fig. 2A).

Biochemical (Smith-Gill et al., 1986, 1987a) and X-ray crystallographic results (Amit et al., 1986; Sheriff et al., 1987b; Colman et al., 1987a,b) show that both the V_L and V_H CDRs are critical to antigenic recognition. Briefly, the CDRs are highly solvent-exposed, interacting loops at the end of the V_L and V_H β domains in the free antibody that become much less exposed in the antibody–antigen complex. Two basic consequences from this global stereochemistry of the antibody combining site are considered in this section. First, the highly exposed loops

FIG. 2. Binding site of the antibody McPC603. (A) The phosphorylcholine hapten (top) with its molecular surface is shown bound in the antigen combining site of McPC603 Fab, determined from the crystallographic structure of the antibody–hapten complex (Segal et al., 1974; Satow et al., 1986). The α-carbon backbone is shown for the four β barrel structural domains of the Fab: V_L (yellow), V_H (green), and constant (blue). (B) The six complementarity-determining regions (CDRs) form the antigen-binding site. The α-carbon backbone of V_L (green) and V_H (yellow), together with the complete main-chain and side-chain atomic structure for the V_L CDRs L1 (green), L2 (magenta), and L3 (blue) and V_H CDRs H1 (red), H2 (yellow), and H3 (cyan), is shown looking down into the combining site from the solvent. Computer graphics models by E. Getzoff and J. Tainer.

easily accommodate charged side chains that may aid antibody–antigen interaction kinetics in solution and may provide increased affinity in the complex due to charge pairing in the absence of damping solvent. Second, CDRs should allow inducible complementarity, because their local conformation depends on extensive solvent interactions in the free antibody and extensive protein interactions in the antigen–antibody complex, as well as on amino acid sequence, positional constraints of rejoining the β domain, and interactions with other CDRs.

Portions of the CDRs have been modeled due to the constraints from the connections to the framework regions (Smith-Gill et al., 1987b; Chothia et al., 1986; de la Paz et al., 1986; Rees and de la Paz, 1986; Fine et al., 1986; Mainhart et al., 1984; Feldmann et al., 1981; Stanford and Wu, 1981; Padlan et al., 1976b). Although some of these CDR models successfully match crystallographic data, some are, as yet, untested, and at least some predicted regions, based on structures of unbound antibodies, conflict with the experimental data on the complexed antibody. Examination of the protein structure data base indicates that the same six or seven amino acid sequences can have different conformations in different three-dimensional environments (Wilson et al., 1984). By analogy, some variability resulting from the change in the three-dimensional environment between the free (unbound) antibody and the antibody–antigen complex might be expected within the CDRs, and this is discussed in more detail below.

FIG. 3. Electrostatic complementarity between phosphorylcholine and the antibody McPC603. (A) The electrostatic potential surface of the dipolar phosphorylcholine hapten complements the electrostatic potential at the McPC603 combining site. The molecular surfaces are shown as dots colored by electrostatic potential (red, < -15 kcal mol^{-1}; yellow, -15 to -5 kcal mol^{-1}; green, -5 to $+5$ kcal mol^{-1}; cyan, $+5$ to $+15$ kcal mol^{-1}; blue, > 15 kcal mol^{-1}). The skeletal molecular model is colored by atom type (red, oxygen; blue, nitrogen; green, carbon). (B) The calculated electrostatic field of McPC603, color coded by the electrostatic potential (see Fig. 3A), shows the effect of electrostatic forces in precollision alignment of the dipolar phosphorylcholine hapten into the correct position for binding. The skeletal model of the bound phosphorylcholine hapten (lower center) is shown in its docked position. A second phosphorylcholine hapten with its molecular surface (top center) is shown 12 Å from the binding site in an orientation aligned by the field from the antibody. The electrostatic potential of the antibody's molecular surface complements the charges of the bound phosphorylcholine hapten, and the field shows significant orientation energy for the hapten 12 Å from the binding pocket. The electrostatic field is shown as arrows color coded by the potential at their origin, using the same code as for the molecular surface. These electrostatic calculations and computer graphics models (E. Getzoff, V. Roberts, A. Olson, and J. Tainer, unpublished results) are based on the atomic structure (Satow et al., 1986; Padlan et al., 1976a; Segal et al., 1974).

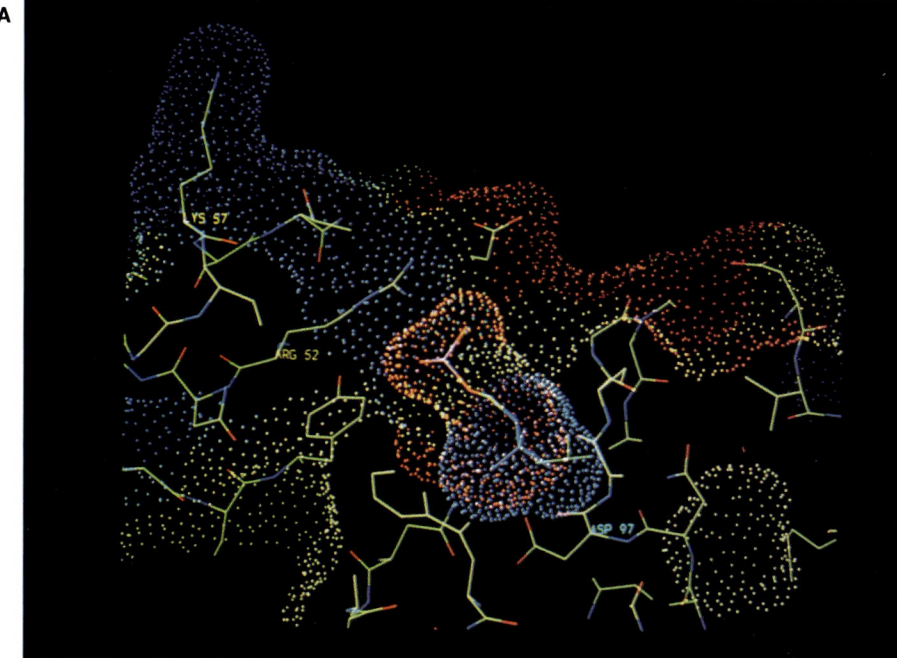
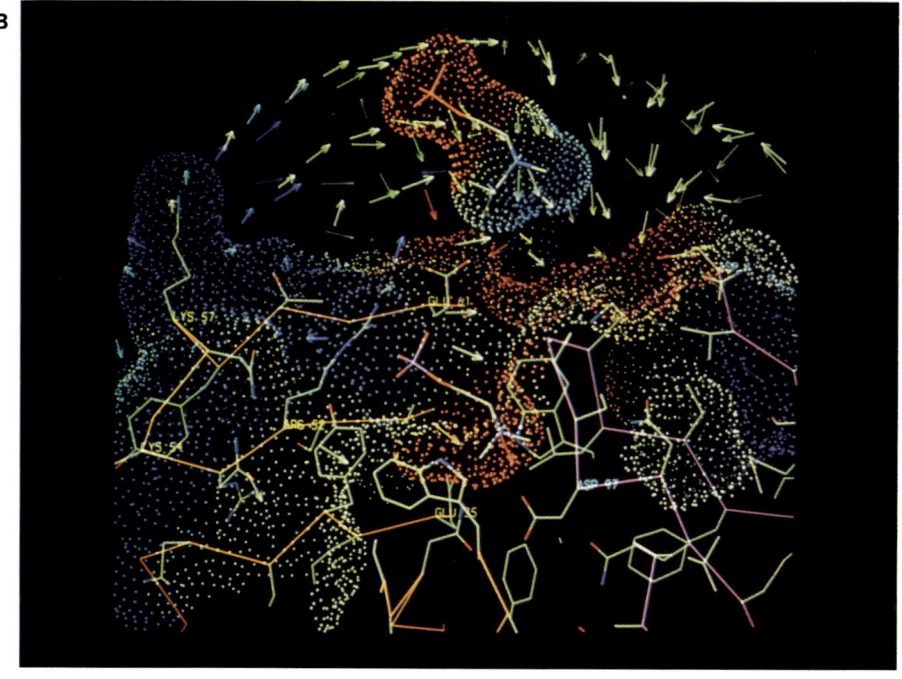

Fig. 3

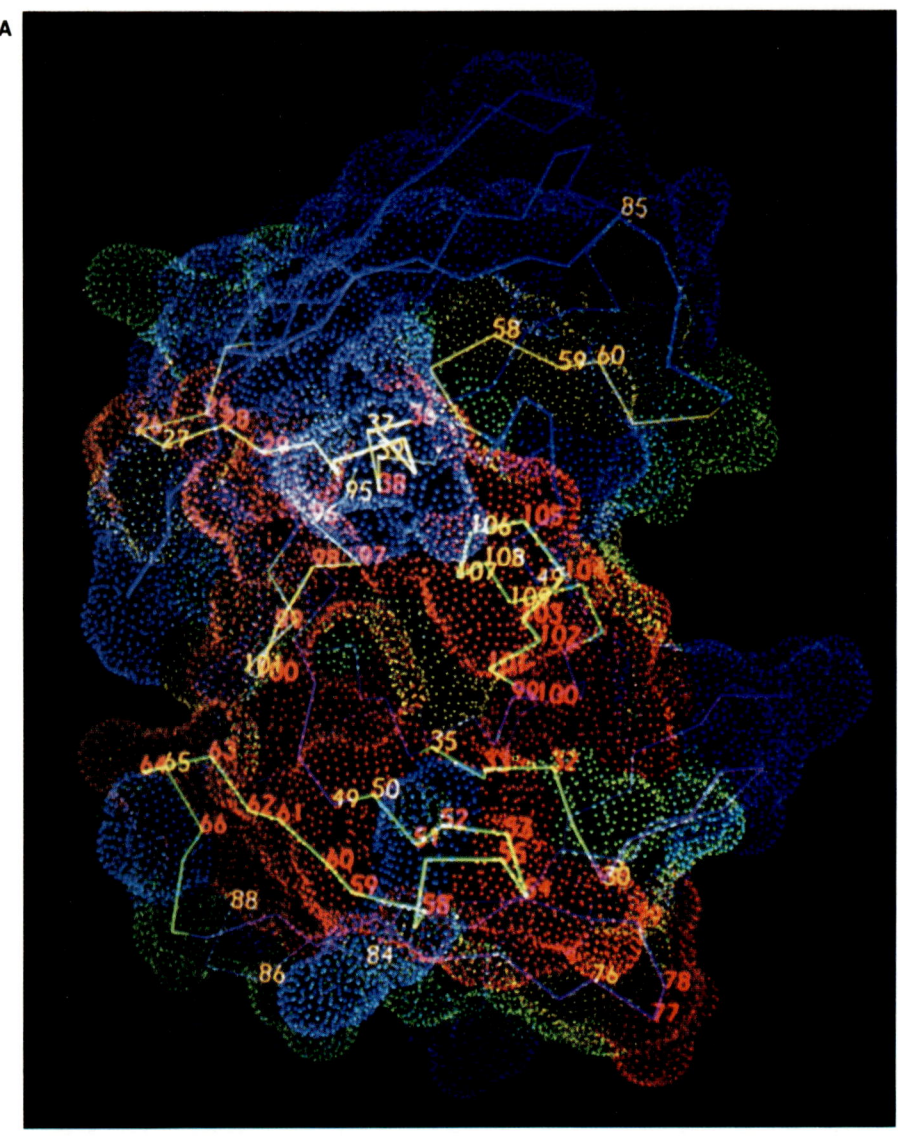

Fig. 4

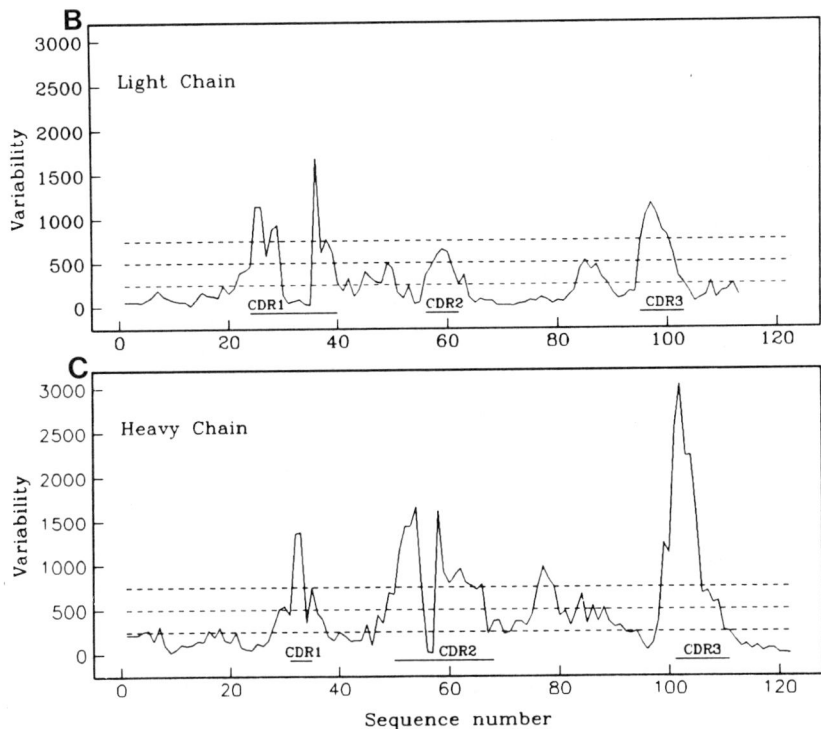

Fig. 4. Immunoglobulin sequence variability. (A) Surface variability, mapped onto the three-dimensional molecular surface topography of the V_L and V_H domains of McPC603, shows highly variable surface (red) surrounding a less-variable region (green) near the center of the combining site. Labeled α-carbon backbones (lines) and external molecular surfaces (dots) are shown, looking down into the combining site from the solvent. The α-carbon backbones are color coded for framework residues of V_L (blue, top) and V_H (magenta, bottom), with both V_L and V_H CDRs color coded red for CDR1, yellow for CDR2, and green for CDR3. In both V_L and V_H chains, residues with high-average surface variability are labeled by number at the α-carbon position in yellow (more variable) and in red (most variable). The external molecular surface dots are color coded by sequence variability classified into four categories (red, most variable; yellow, more variable; green, less variable; and blue, least variable). The highly (most and more) variable regions are defined as potential IDRs; the majority are also CDRs. The most prominent highly variable surface region outside the CDRs is formed by heavy chain residues 76–79 (bottom right), which are conformationally adjacent to H2. Other framework residues with high surface variability are light chain 49 (behind) and 85 (upper right), heavy chain 30, 49, and 99–100 (adjacent in sequence to H1, H2, and H3, respectively), and heavy chain 84, 86, and 88 (lower left). Crystallographic coordinates (Satow et al., 1986; Padlan et al., 1976a; Segal et al., 1974) are taken from the Brookhaven Protein Data Base (Bernstein et al., 1977). (B and C) The linear relationship of surface variability to sequence position is plotted for (B) V_L and (C) V_H sequences from 25 mouse and human immunoglobulins. Residues are numbered sequentially to match the sequence of McPC603. Variability is averaged over six residues and is plotted at the third position to allow appropriate mapping onto the three-dimensional structure. CDRs are shown by horizontal bars: L1 (residues 24–40), L2 (residues 56–62), and L3 (residues 95–103) and H1 (residues 31–35), H2 (residues 50–68), and H3 (residues 101–111). Long, dashed horizontal lines separate the four categories of surface variability used to color code A. [Adapted from Kieber-Emmons et al. (1987).]

A. ELECTROSTATIC RECOGNITION BETWEEN ANTIGEN AND ANTIBODY

Electrostatic forces have been implicated in a variety of biologically important intermolecular interactions, including drug orientation by DNA (Dean, 1981), macromolecular assembly (Perutz, 1978; Getzoff et al., 1986b), substrate binding and catalysis (Sheridan and Allen, 1981; Warshel and Levitt, 1976; Getzoff et al., 1983), and macromolecular complementarity with inhibitors, drugs, and hormones (Blaney et al., 1982; Weiner et al., 1982; Pullman et al., 1982). Considerable evidence now supports the importance of electrostatic forces in stabilizing macromolecular interactions (Matthew, 1985; Ohlendorf and Matthew, 1985; Matthew and Gurd, 1986). However, characterization of electrostatic forces for antibody–antigen interactions in particular has been hampered, until recently, by lack of crystallographic structural information, although their potential importance has been recognized (Karush, 1962, 1978).

Calculations of the electrostatic potential on the molecular surface of the variable region of Fab McPC603 reveal strong electrostatic complementarity between the antibody and the bound phosphorylcholine hapten (Getzoff, 1985; Getzoff et al., 1985) (Fig. 3A), which extends to the expanded spherical harmonic surface calculated from molecular dynamics snapshots (Max and Getzoff, 1988). Around the hapten binding site, the electrostatic field generated by the arrangement of charges in the antibody molecule appears to guide the dipolar hapten into the orientation found in the crystallographic coordinates of the docked complex (Fig. 3B). The electrostatic contributions of specific residues in or near the binding site can be determined from these calculations: arginine-52 and glutamic acid-61 of the V_H chain, in conjunction with aspartic acid-1 of the V_L chain, are influential in aligning the field to direct the hapten toward the binding region. Lysine residues 57 and 67 from the V_H chain and 36 from the V_L chain also contribute to a lesser extent. Electrostatic alignment should be significant, because the overall electrostatic potential energy between the antibody and the hapten exceeds thermal energy, even when the hapten is located over 12 Å from its binding site (Fig. 3B). Such calculations confirm and quantitate the local electrostatic stabilization of the hapten, as suggested in analyses of the crystal structure (Padlan et al., 1976a; Segal et al., 1974): V_H arginine-52 interacts with the phosphate and V_H glutamic acid residues 35 and 61 are located near the positively charged choline. Calculations further indicate that V_L aspartic acid-97 also influences local electrostatic interactions.

In the interaction of the antibody HyHEL5 (Table I) with lysozyme, crystallographic evidence shows that two arginine residues of lysozyme interact closely with two glutamic acid residues of the antibody (Sheriff et al., 1987b). Calculations reveal an extensive and energetically important electrostatic complementarity between the two surfaces buried in the interface; quantitation of the role of these forces in precollision guidance and binding energy is still in progress. These calculations, based on the crystal structures of antibodies McPC603 and HyHEL5, suggest that electrostatic forces can strongly influence immune recognition (E. D. Getzoff, V. A. Roberts, A. J. Olson, and J. A. Tainer, unpublished results). However, antibody affinity to another lysozyme epitope has been increased by removing charged residues at the site periphery (Roberts et al., 1987), suggesting that the net electrostatic contributions to the binding energy will vary with the antibody and will probably depend highly on the complete removal of solvent from interacting charged side chains (see Section X).

Other experiments indicate a role for electrostatic forces in the binding of some antibodies. Charge mutations in the McPC603-related phosphorylcholine-binding antibody S107 have been shown to influence hapten and DNA binding (Chien et al., 1988; Scharff et al., 1985; Teillaud et al., 1983). The replacement of glutamic acid-35 of the heavy chain (which appears to stabilize the binding of the positively charged choline to McPC603) by alanine greatly decreases binding of the hapten. The mutation of aspartic acid-101 of the heavy chain to alanine, far from the antigen binding site, appears to cause a local rearrangement in the structure of the active site region, so that phosphorylcholine no longer binds (Chien et al., 1988). For the HEd10 antibody, studies of ionic strength effects on DNA binding indicate that approximately two phosphate groups from the DNA backbone are involved directly in the interaction (Lee et al., 1982).

B. PATTERNS OF SEQUENCE VARIATION IN THE ANTIBODY VARIABLE DOMAIN

1. *Effects of Sequence on Stereochemical Variation*

Historically, complementary relationships in the recognition of antigens by antibodies have been attributed to the hypervariable nature of the CDRs (Kabat et al., 1987; Kabat, 1978, 1983; Wu and Kabat, 1970). CDRs form the antigen binding site (Kabat et al., 1976) and affect the self-association of V_L and V_H chains (Hamel et al., 1984; Stevens et al., 1980). Conversely, antibody residues that are not classically hypervariable can be complementary in the context of idiotope recognition (Victor-Korbin et al., 1985), as discussed in the following section.

A theoretical approach to the evaluation of allowed stereochemical variation in antibodies, referred to as surface variability analysis, attempts to couple both intrinsic and extrinsic factors of antigenicity by considering the evolutionary variability of protein surface regions (Kieber-Emmons and Köhler, 1986a,b). This method, which measures amino acid sequence variability as a function of hydrophilicity and evolutionary sequence variation, was developed primarily to characterize potential idiotope- (antibody epitope-) determining regions (IDRs). It also provides an objective measure of the relative stereochemical flexibility or conformational permissiveness of the CDRs. To evaluate sequence variation by this approach, hydration potentials defining the affinity of each amino acid side chain for the solvent water (Wolfenden et al., 1979; Moews et al., 1981) are averaged over six residues for each sequence and are inverted to make hydrophilic values positive. Then, the resulting hydrophilicity profiles are averaged to form a consensus value at each sequence position and are assayed for variability, according to the formula of Wu and Kabat (1970): variability thus represents the number of different hydrophilicity values divided by the frequency of the most common hydrophilicity value. The product of the consensus and variability of hydrophilicity values is used to define a surface variability index, giving maximal values for surface-exposed sequences that varied significantly during evolution, i.e., those likely to have multiple allowed conformations (Kieber-Emmons et al., 1987).

Surface variability profiles (Fig. 4) for a family of 25 mouse and human immunoglobulin variable region sequences show that the sequence variation corresponds primarily to hypervariable regions, with the majority of the variability in this immunoglobulin family residing in the V_H chain (Kieber-Emmons et al., 1987). Some framework residues show significant variability, and framework residues have been shown to participate in the antibody–antigen contact region (Amit et al., 1986; Sheriff et al., 1987b). Within the V_L chain (Fig. 4B), the relative surface variability of L3 is somewhat greater than L1, which is much greater than L2. Within the V_H chain (Fig. 4C), H3 surface variability is greater than H2, which is greater than H1. In both the V_L and V_H chains, CDR3 appears to have the most conformational freedom, based on sequence variation, but the V_H chain CDR3 is by far the most highly variable.

The effect of this primary structure variability on the three-dimensional molecular surface topography of the antibody combining site can be assessed (Kieber-Emmons et al., 1987) by mapping sequence variability values (classified into the four categories listed in Fig. 4 as follows: most variable, more variable, less variable, and least variable) onto the surface of the three-dimensional structure of McPC603. As shown in Fig. 4A, highly variable surface topography surrounds a less-variable region near the center of the combining site.

The high level of sequence variation in H3 could just reflect a large number of allowed conformations, or it could indicate that H3 has a greater capacity for inducible complementarity upon antigen binding. However, if H3 maintains a rigid conformation in binding protein antigens, then it might be supposed that the stabilization of each such conformation must come primarily from interactions within this structural element, because the antibody structure obviously tolerates significant changes. This has the practical consequence that functional chimeric antibodies can be made by grafting mouse CDRs onto the human antibody framework (Jones et al., 1986). Experimental and theoretical evidence would then suggest that a polypeptide loop, stabilized primarily by interactions within the loop, is unlikely to be a very rigid structure (except for special cases, such as disulfide-rich and metal-containing loops). Therefore, the sequence variability evidence would suggest that conformational variation in CDR3 probably includes possible inducible complementarity upon interaction with protein antigens, when the CDR3 environment changes from solvent to protein.

2. On the Antigenic Nature of Antibodies

Idiotopes, the antigenic sites on antibodies, can activate antibody clones bearing complementary binding sites through a self-recognition process. One model proposes that the response to such self-proteins is directed against sequence regions that exhibit the highest evolutionary variability (Benjamin et al., 1984). Therefore, according to this model, sequence-variable regions are antigenic, and evolutionarily conserved regions induce immunological tolerance. From a structural chemistry perspective, variable, and therefore antigenic, regions can accommodate local changes in conformation and should correlate with sequence regions that are flexible and surface exposed (Tainer et al., 1985a). Thus, the combination of intrinsic factors, such as mobility and accessibility, and extrinsic host factors, such as immunological tolerance, immune response genes, idiotype networking, and structural gene repertoire, appear to influence the antigenicity of antibodies, as well as other protein antigens (Benjamin et al., 1984; Berzofsky, 1985). Yet, any functionally important mobility will be limited, because of the fidelity of the antiidiotypic response, as evidenced, for example, in probing protein active sites (Lugwig et al., 1987).

Based on the evaluation of sequence variation (Kieber-Emmons *et al.*, 1987), the probable IDRs cover a continuum of binding sites in the variable region (see Fig. 4). The large repertoire of IDRs should allow many combinatorial possibilities for idiotope expression in three dimensions, including those formed solely by V_L chain residues, those formed solely by V_H chain residues, and those formed by residues of both chains. Topographic mapping of one idiotypic system has shown a linear idiotope map, spanning from the antigen binding site to the vicinity of the constant region (Greenspan and Monafo, 1987). Surface-variable regions including framework residues may be recognized by more cross-reactive antiidiotypic antibodies, since fewer CDR residues are involved.

Of the framework residues in the V_L chain, only 49 and 85 (numbered sequentially according to McPC603) exhibit high surface variability (Fig. 4). Both are isolated sequentially and spatially from the CDRs. In the V_H chain, framework residues with high sequence variability include those adjacent in sequence to CDRs (30, 49, 99, and 100), those conformationally adjacent to CDRs on the surface (76–79), and those distant from the CDRs (84, 86, and 88). Two clusters of V_H chain framework residues (76–79 and 84, 86 and 88) form likely IDRs; residues 76–79 form a protruding β bend made up of most variable residues, whereas residues 84, 86, and 88 (outwardly facing residues along a β strand) form a relatively small, flat surface patch of more variable residues. These surface topography and surface variability characteristics suggest that the region encompassing residues 76–79 forms the most likely IDR outside of the antibody combining site.

The structural stereochemistry of the antibody has also been used to identify possible IDRs by calculation of large probe accessibility to identify protruding areas (Novotný *et al.*, 1986a). Interestingly, this approach suggested that only about 25% of the CDRs would be strongly antigenic and identified a minimum of 40 potential IDRs. Antigenic sites on proteins have also been identified by calculation of a local dissimilarity value based on three-dimensional neighbors (Padlan, 1985), which is related to the sequence variation mapping discussed above. These structurally based methods promise to be useful for identification and understanding of IDRs and other protein epitopes.

C. Induced Complementarity in the Antibody

1. *Immunological, Biochemical, and Crystallographic Data*

Recent results have suggested a role for the surface mobility of the antigen in antibody–antigen interactions (reviewed in Tainer *et al.*,

1985a). For the known antibody–antigen complexes, the surfaces of the protein antigen buried by the antibody contact most frequently represent regions of high acessibility and local mobility. Antibody segmental flexibility is relatively well known and is discussed elsewhere (Colman, 1988; Huber and Bennett, 1983). The existence of possible functionally important, induced complementarity in the CDRs upon antibody binding to proteins is controversial due to the limited experimental data available (Huber, 1986). The binding region formed by light chain dimers can change in response to different solvent conditions (Chang et al., 1987). Circular dichroism (Holowka et al., 1972) and fluorescence polarization (Schlessinger et al., 1975) measurements, both of which are sensitive to small conformational changes, have suggested that antigen binding changes the environment of aromatic residues, many of which occur in the combining site. The existence of different conformational states for free and bound antibodies has been further suggested by kinetic studies (Zidovetzki et al., 1981; Metzger, 1978; Lancet and Pecht, 1976). In addition, combined experimental and computational evidence suggests that sequence changes in antibodies may allow local side-chain rearrangments, resulting in the positioning of new side chains in the combining site and altering binding activity (Chien et al., 1988).

Unfortunately, no pairs of crystal structures are available for a given antibody both as the free molecule and bound to its protein antigen. The only published examples of Fab structures in both bound and free forms are the Fab–hapten complexes of McPC603 and Newm, which have a central area for specificity that shows no significant conformational changes upon hapten binding (Segal et al., 1974; Amzel et al., 1974). Nuclear magnetic resonance (NMR) data on McPC603 is consistent with a single encounter step (Dower et al., 1977) and no large changes upon hapten binding (Gettins et al., 1982). Experiments with V_L chain dimers, however, indicate that adding haptens can cause changes, including movement of contact residues to improve complementarity with haptens and even prying the V_L chains apart (Edmundson and Ely, 1985, 1986; Ely et al., 1978, 1985; Edmundson et al., 1984). There is also some evidence for hapten-induced changes in the conformation of an intact IgG (Okada et al., 1985).

The expectation that an antibody–antigen interaction could affect the local antibody conformation comes from the observation that the surface area buried in the antibody–lysozyme complexes (about 700 $Å^2$) (Sheriff et al., 1987b; Amit et al., 1986) is comparable to that between the Fab variable domains (Getzoff et al., 1986b). It seems likely that these large buried interfaces, which exclude solvent, can significantly shift the local minimum energy conformation in their contact regions, particularly causing side-chain rearrangements, as has been proposed for the assembly of domains in oligomeric proteins (Richardson, 1981).

2. Superposition of Variable Domain Structures

In the absence of a data base of structures for antibodies with and without bound protein antigen, it is worth examining existing structural data for clues to functionally important antibody flexibility. High-resolution structures are known for at least nine Fabs and four V_L or light chain dimers (Table I). One way to probe possible, functionally significant mobility is to examine CDRs after their respective variable domains have been superimposed, based on their consistent features. The framework (nonhypervariable) residues of antibody variable domains are spatially closely superimposible, whereas the CDRs vary in shape and length, as well as sequence, although their ends are similarly positioned by framework residues. CDRs with consistent conformations in known structures probably represent regions that are relatively well ordered, despite changes in environment and sequence, and, therefore, are unlikely to show inducible complementarity upon antigen binding.

In Fig. 5, the structures of antibody variable domains were compared by superimposing individual V_L and V_H domains onto the corresponding domains of McPC603. A slightly modified version of the common β sheet framework residues of Chothia and Lesk (1987) was used to define the one-to-one correspondence needed for the superposition. In the numbering scheme of Kabat et al. (1987), these included V_L residues 4–6, 9, 11–13, 19–25, 33–49, 53–55, 61–76, 84–90, and 97–107 and V_H residues 3–12, 17–25, 33–52, 56–60, 68–82, 88–95, and 102–112. To use a consistent set of residues for all the superpositions, we excluded V_L 10 and 106A, because these residues are deleted in some sequences. Root

---

FIG. 5. Conformational variation of the CDRs. (A) Ten V_L and (B) seven V_H structures from available X-ray crystallographic coordinates (see text and Table I) were separately superimposed using conserved framework residues. The α-carbon backbone is colored light gray for framework residues, red for CDR1, yellow for CDR2, and green for CDR3. The differential conformational variation of the CDRs is apparent. (A) For V_L, L1 (red, right) shows the largest variation, including two structures with long insertions; L3 (green, left) shows significant variation in the open turn region of the L2 loop, and L2 (yellow, right) is largely conserved except for the deletion in Newm. (B) For V_H, H3 (green, left) is by far the most conformationally variable, H2 shows some variation at the open turn end, and H1 (red, center) is highly conserved conformationally. This and all other solid (raster) images (Figs. 5, 9, and 10) were rendered, using the programs RAMS or MCS (Connolly, 1983b, 1985). [This comparison taken from V. Roberts, J. Tainer, A. Feinstein, and E. Getzoff (unpublished results).]

Fig. 5

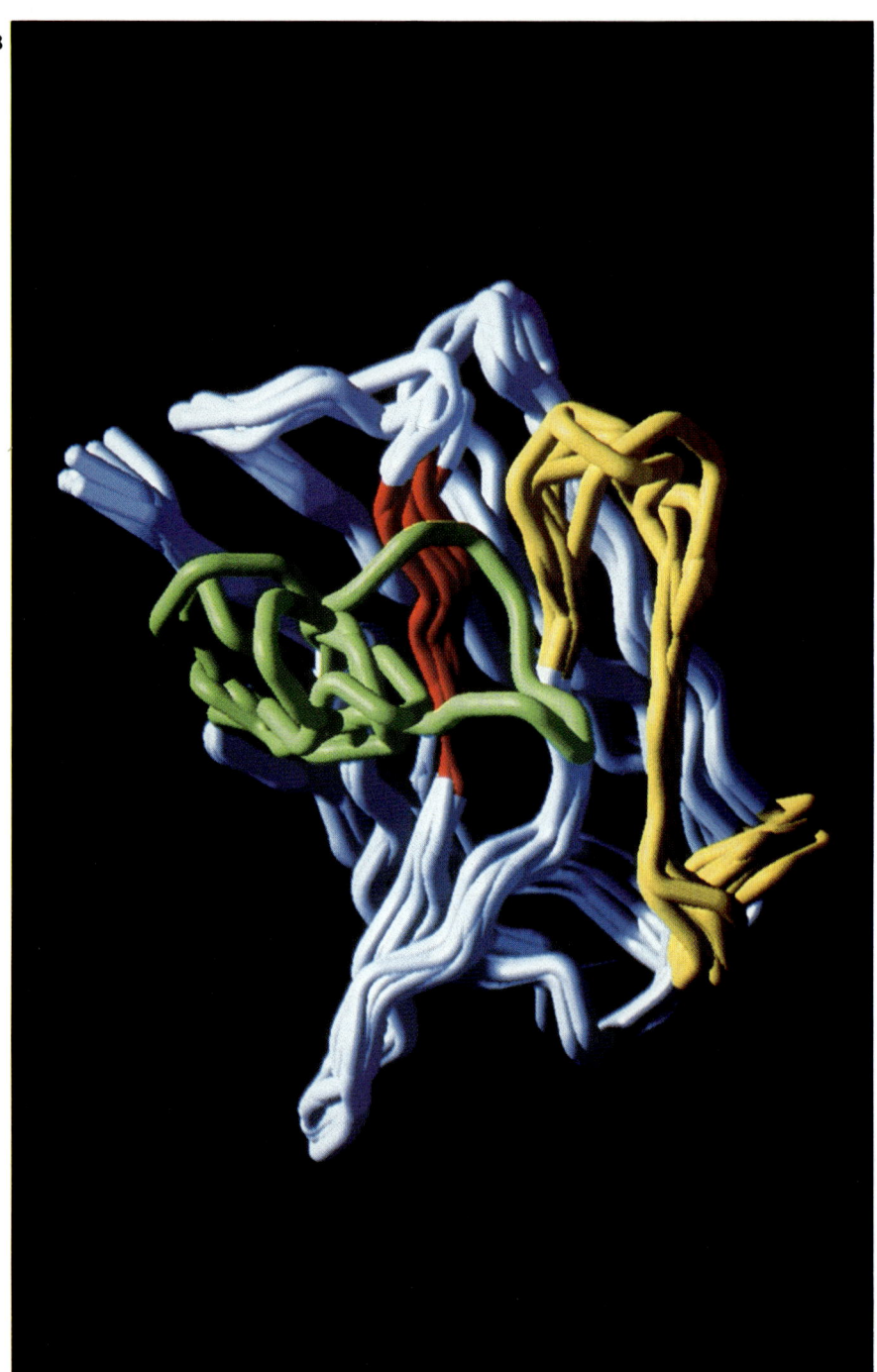

Fig. 5

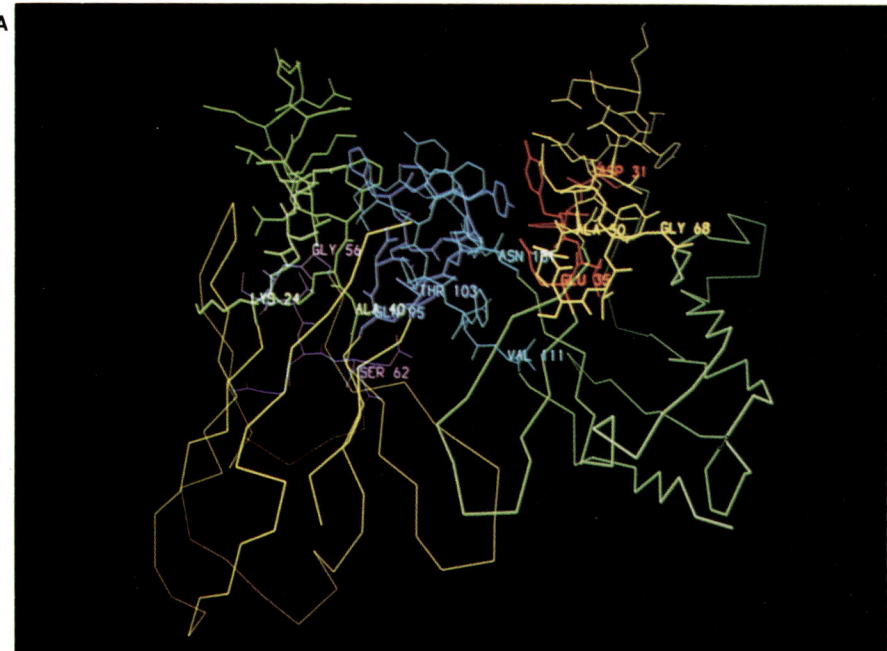
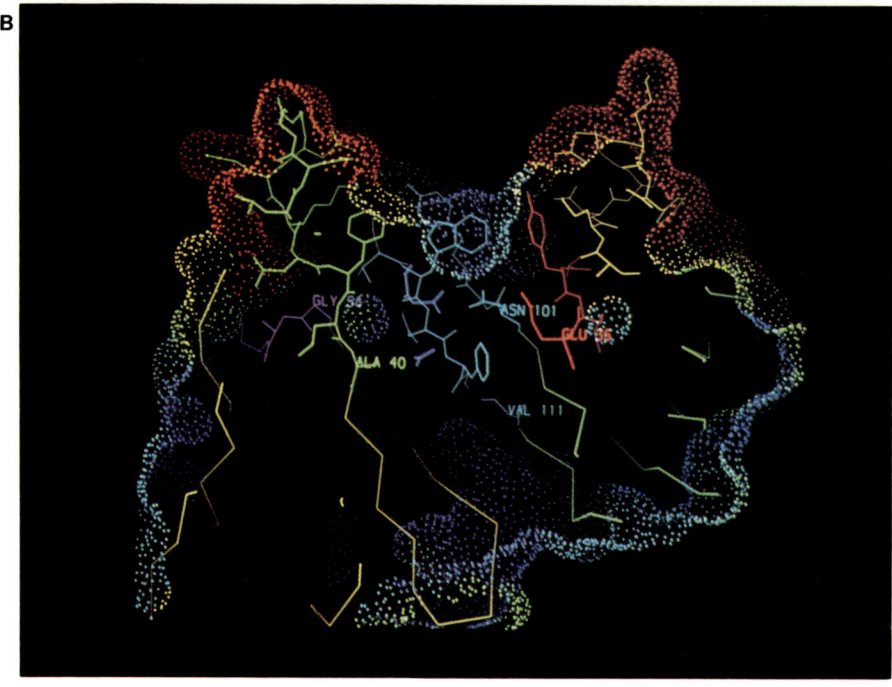

Fig. 6

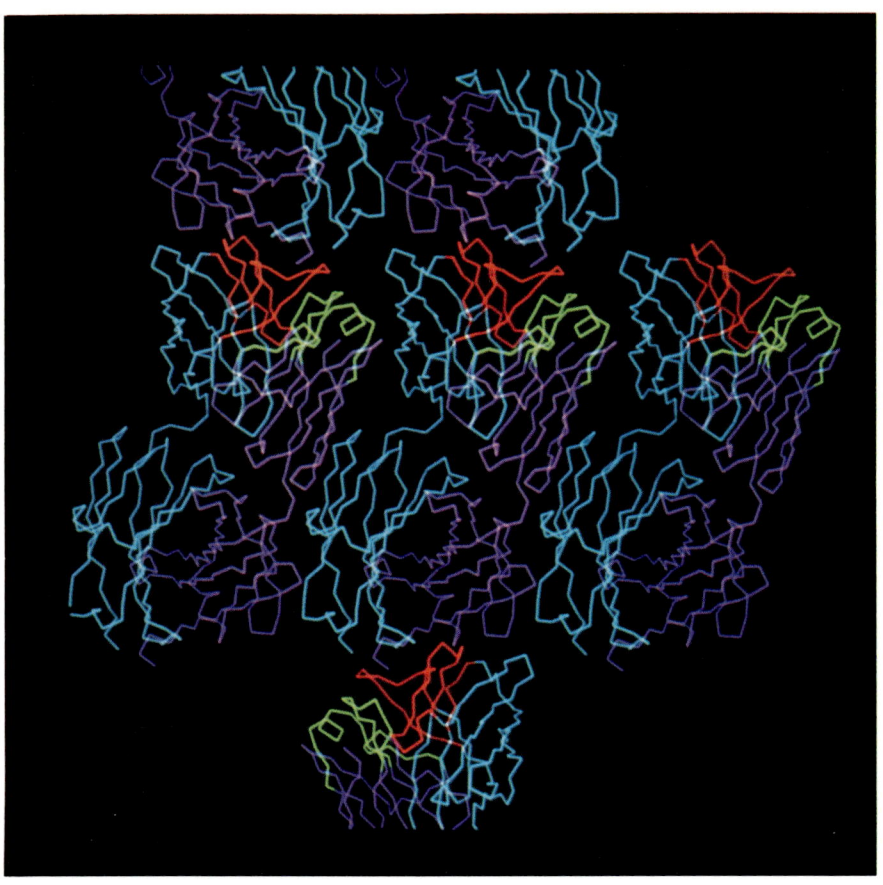

Fig. 7

mean square (RMS) deviations from the McPC603 structure varied from 0.48 Å (REI) to 1.69 Å (Newm) for V_L and from 0.80 Å (J539) to 1.29 Å (Newm) for V_H (Table I). Examination of the 10 superimposed V_L domains (D1.3, HyHEL5, McPC603, J539, HEd10, Newm, Kol, Mcg, Rhe, and REI) and 7 V_H domains (D1.3, HyHEL5, McPC603, J539, HEd10, Newm, and Kol) indicates differential conformational variability among the different CDRs (Fig. 5A and B).

In V_L, L1 is a Greek key loop (a connection across the end of the β barrel structure joining nonadjacent β strands) overlapping into β strands at the two ends. L2 is primarily an extended chain and β turn, and its conformation is highly conserved in the superpositions. Like L1, L3 is at the end of two β strands and is primarily formed by the loop between them. The largest variation in L3 occurs at the turn region at the middle of the loop. In V_H, H1 is primarily an extended β strand with minimal conformational variation at the N-terminal end of the strand. H2 is formed by two adjacent β strands, and the largest variation in this CDR is in the loop joining the two strands. H3 is at the end of two β strands, primarily formed by the loop between them.

FIG. 6. Order and flexibility in the McPC603-binding site. The stereochemical relaionship of the CDRs and their resulting surface topography and mobility are shown. (A) The arrangement of the McPC603 CDRs for V_L (left) and V_H (right) results in the ineraction of L3 and H3 to form the central portion of the combining site. For V_L, the α-carbon backbone of the framework (yellow, left) is shown with full atomic coordinates for L1 (green), L2 (magenta), and L3 (blue, front and center). For V_H (left), the α-carbon backbone of the framework (green, right) is shown with full atomic coordinates for H1 (red), H2 (yellow), and H3 (blue, front and center). (B) The patterns of local mobility based on the average main-chain temperature factors reveal a somewhat concave, central ordered region formed by the lower portions of L3 and H3, surrounded by flexible portions of L1 and L3, together with H2 and H3. The molecular dot surface is colored by averaged temperature factors of the main-chain atoms: red highest (most mobile) to blue lowest (most ordered). The more mobile regions form the "hot lips" for the combining site, while the binding pocket remains relatively well ordered. This differential order in the combining site should accommodate both specificity in the binding pocket and flexibility for functionally important inducible complementarity. (Computer graphics model by E. Getzoff, A. Feinstein, and J. Tainer.)

FIG. 7. Crystal packing for the Fab Kol. In the crystallographic structure of Fab Kol (Marquart et al., 1980), the antigen combining site shows direct interactions with neighboring molecules in the crystal. In this cross-section of the crystal lattice, the α-carbon backbones are color coded to show the location of the framework and CDR residues for the light chains (lavender framework bonds and green CDRs) and the heavy chains (light blue framework bonds and red CDRs). Crystal packing places the CDRs from one Fab against the C_L and C_H1 of the next. (Computer graphics model by E. Getzoff and J. Tainer.)

As seen in Fig. 5A, the conformational CDR variability in the V_L chain is in the order L1 greater than L3, which is much greater than L2, with the exception of Newm in which a section of L2 has been deleted (Amzel et al., 1974; Saul et al., 1978; Poljak et al., 1974). Within the V_H chain (Fig. 5B), the conformational variability of H3 is greater than H2, which is much greater than H1. The L1, L3, H2 and H3 CDRs show the most conformational variability. These data would suggest that L1, L3, H2 and H3 are the most likely candidates for functionally important inducible complementarity. This idea is borne out in the two antibody–antigen complexes for which the atomic structures of all the CDRs have been defined. For the D1.3 antibody, lysozyme contacts primarily H3 (Amit et al., 1986); for the HyHEL5 antibody, lysozyme contacts primarily L3 and H2 (Sheriff et al., 1987b).

3. Mobility Data from Crystallographic Temperature Factors

Structural implications for inducible complementarity can be derived from measures of mobility, as well as conformational variation. One measure of mobility involves patterns of average main-chain temperature factors determined from X-ray structure refinement (Tainer et al., 1984; Westhof et al., 1984). As a consequence of the structural chemistry of antibodies, the CDRs are loop structures and ends of β strands on one end of the more conformationally stable β barrel. In general, the average main-chain temperature factors show maxima and minima that correlate with loops and strands, respectively. Within the limits of available temperature factor data, the relative mobility of the CDRs matches their deviations from superposition (E. Getzoff and A. Feinstein, unpublished results). This potentially important differential mobility of the CDRs is also seen in molecular dynamics snapshots of McPC603 visualized with spherical harmonic representations of the consensus molecular surface (Max and Getzoff, 1988). The effect of stereochemistry and surface topography on the relative mobility of the CDRs, derived from the temperature factor data on McPC603, is shown in Fig. 6. Any inference of mobility from X-ray crystallographic temperature factor data must take into account possible structural error and the effects of crystal packing (Sheriff et al., 1985). For example, Fab Kol forms strong crystal contacts with the combining region, effectively recognizing itself (Fig. 7; Matsushima et al., 1978), and perhaps indicating that the stereochemical features of CDRs have evolved to make good protein–protein interactions.

D. INDUCED COMPLEMENTARITY IN THE ANTIGEN

Of the three structures of antibody–antigen complexes that have been published, both the D1.3 (Amit et al., 1986) and HyHEL5 (Sheriff *et al.*, 1987b) antibodies bind to relatively flexible regions of lysozyme, whereas the NC41 antibody binds to flexible regions of neuraminidase, including a disordered loop (Colman *et al.*, 1987a). A note in proof (Amit *et al.*, 1985) suggested that the D1.3 antibody binds to well-ordered regions of lysozyme. One sequential piece of this D1.3–lysozyme interface (residues 18–27) is in a section with tight crystal contacts (Artymiuk *et al.*, 1979), which can result in the lowering of the crystallographic temperature factors (Sheriff *et al.*, 1985), and, hence the perceived mobility of this piece of the lysozyme structure; the other sequential piece (residues 116–129) is in the very mobile C-terminal region. Lysozyme structures with fewer crystal contacts in this interface region (Rao *et al.*, 1983) do show local mobility maxima in both sequential pieces of the interface with antibody D1.3. In addition, experimental data on lysozyme mobility from hydrogen–deuterium exchange indicate that both areas of the D1.3 epitope are flexible (Williams and Moore, 1985) and both appear mobile, as based on the side-chain displacements found among different crystal structures and molecular dynamics calculations (Ichiye *et al.*, 1986). Thus, most evidence indicates that the lysozyme regions, bound by antibodies in the crystallographically characterized complexes, are relatively mobile.

In Section IV, we discuss the use of the protein MHr in systematic studies of antigen mobility (Tainer *et al.*, 1984; Geysen *et al.*, 1987c; Fieser *et al.*, 1987). The MHr crystallographic structure is unique in that its temperature factors have been corrected for crystal contacts, based on the MHr monomer and the hemerythrin octamer structures (Sheriff *et al.*, 1985). Therefore, the adjusted MHr temperature factors should reflect mobility accurately, allowing assessment of differential mobility in this antigen.

Of course, a critical role for local mobility in functionally important, inducible complementarity would not require equal mobility of all portions of the antibody or antigen interface. As discussed in Sections III, B and C, if induced fit does occur, it probably involves unequal participation of different parts of the antibody combining site. Investigations of mobility and induced complementarity in the antigen are primarily directed toward detecting intrinsic antigenic biases; such biases reflect a common chemistry of antibody recognition, whose overall pattern will span a diversity of stereochemical sites. The difficulty of identifying common chemistry from the diverse system of immune recognition should be kept in mind, when evaluating the three existing X-ray structures of antibody–antigen complexes. In Section VI, studies on the frequency of the antigenic response to the protein MHr in several animal species are considered in regard to concepts of species response, collective response, and immunodominance and their relationship to common chemical patterns for antibody–antigen interaction.

In general, analysis of the three published structures of antibody–antigen complexes suggests that the global structure of the antigen is unchanged by the binding process (Amit *et al.*, 1985, 1986), but that local concerted changes may occur (Sheriff *et al.*, 1987b; Colman *et al.*, 1987a). An important conclusion from the first crystallographic structure of an antibody–antigen complex (Amit *et al.*, 1986) is that the D1.3 antibody does not cause global unfolding of the antigen, lysozyme, upon binding. The possibility of smaller, but, nonetheless, critically important, movements of side chains remains a more controversial issue, in part due to technical problems with the analysis and interpretation of small, functionally important, conformational changes. Changes in the antigen are evident in the lysozyme–antibody complex, determined by Sheriff *et al.* (1987b), which has a high binding affinity of 2×10^9, and in the structure of the antibody–neuraminidase complex (Colman *et al.*, 1987a). For the D1.3 antibody–antigen complex (Amit *et al.*, 1986), the authors were cautious about possible overinterpretation of side-chain movements; the maximum structural changes for the complex were reported as 1.6 Å for the α-carbon positions.

Rigorously defining the existence and significance of small amounts of local structural change is a difficult problem. In one approach, the vector shifts between the free and bound positions of each atom in the interface are examined to differentiate between small shifts, resulting from error (which should be more or less random), and concerted shifts that together suggest a significant, but possibly small, local change. Using this method, we have examined the local environment of the lysozyme involved in the D1.3 antibody–antigen complex. Comparison of the coordinates from D1.3-bound lysozyme (Amit *et al.*, 1986) with coordinates of unbound lysozyme (Rao *et al.*, 1983; Phillips, 1974) suggests that concerted shifts may be occurring in the bound lysozyme structure that affect the surfaces of certain key residues buried by the antibody–antigen complex. Although the exact nature of these movements varies with the pair of lysozyme structures being superimposed, there appears to be a consensus picture: side-chain movements of exposed Arg-21, along with smaller concerted movements of other residues (Asp-18, Asn-19, Tyr-23, Ser-24, Leu-25, Gln-121, and Ile-124), increase the area of antibody interaction with these side chains. Even if the magnitudes of observed shifts are within the range of the error in the structure, the shifts would still be significant if they were concerted, i.e., errors are random, but concerted shifts would lie in the same directions. From evaluation, the differences observed between the superimposed lysozyme structures appear to indicate concerted shifts in some lysozyme side chains in the contact region, thus implying that induced complementarity may be involved.

Studies on site-directed mutants of trypsin and subtilisin suggest that functionally important changes in residue positions (smaller than those identified in crystallographic structures of antigen–antibody complexes) occur in high-resolution crystal structures (better than 2-Å resolution) of mutant enzymes. These small, local changes are often associated with large changes in K_m and k_{cat} (e.g., Wells and Fersht, 1985; Howell et al., 1986). Thus, for understanding the chemistry and mechanisms of enzymes or antibodies, it is necessary to identify and study such small, local structural changes that can affect the binding energy (Getzoff et al., 1987).

E. Reconciling Crystallographic and Peptide Mapping Data

To put this new level of structural information from X-ray crystallography of antibody–antigen complexes into the context of existing immunological data, a consistent nomenclature for discussing the antigenic data base is necessary. This is particularly important, given the wide variation in antibody specificity, affinity, and frequency of response to given protein antigens, as well as the highly diverse nature of protein antigens.

The two published lysozyme–antibody complexes (Sheriff et al., 1987b; Amit et al., 1986), together with a third lysozyme–antibody structure currently being refined, encompass almost one-half of the lysozyme surface (Davies et al., 1988). As additional antibody–lysozyme structures become known, the percentage of the lysozyme surface identified as antibody contact surface should increase, and we may find that lysozyme and many other proteins induce a set of antibodies that together bury all or almost all of the exposed surface. At first consideration, this would suggest that all the exposed surface is antigenic, i.e., able to interact with an antibody combining site, as previously suggested (Benjamin et al., 1984).

Before accepting a definition of antigenic residues as all those found in the interface with an antibody, it is worth considering that the essence of antigenicity involves binding energy. In the broadest sense, two contrasting possibilities could account for the interaction energy of protein antigens: (1) the total binding energy encompasses numerous, small contributions from some 15 (the average number of residues suggested to constitute the interface in the 3 determined structures) or more interacting residues, such that any single amino acid contributes a relatively small portion of the energy and specificity to the chemistry of interaction; or (2) the binding energy is primarily a function of a few critical amino acid residues, such that the other residues forming the interface are brought into proximity by the interactions of the critical residues, and may contribute slightly to or even detract from the overall binding energy.

Given the great diversity of the antibody response, it is likely that both possibilities occur, at least in some instances. If about 15 residues in an epitope contributed more or less equally to the binding energy, then conservative changes in any one residue should have a relatively small effect on the binding affinity, provided the change did not have significant effects on the remaining 14 residues. As already shown from the detailed work on lysozyme–antibody complexes (Amit *et al.*, 1986; Sheriff *et al.*, 1987b), changes in a single amino acid side chain of the antigen can greatly affect binding even without apparent steric hindrance. For example, replacement of a lysozyme glutamine residue by a histidine greatly reduced binding by the D1.3 antibody, and replacement of a lysozyme arginine residue by a lysine similarly affected binding by the HyHEL5 antibody. Thus, to understand the interaction chemistry, it is necessary to define the amino acids critical to the binding energy (the antigenic determinants in a chemical sense) and not simply to identify the residues buried in the interface. The existence of heteroclitic antibodies, those raised against one protein antigen that have a higher affinity for another related protein antigen (Smith-Gill *et al.*, 1984), suggest that the contributions of some residues to binding affinity can be increased, i.e., that not all interacting residues are contributing fully to the binding affinity.

One technical point should be appreciated: a protein interface does not have a unique quantitative definition. In fact, from the published articles, it is apparent that various crystallographic groups use different definitions for residues involved in the interface. Even if one were to assume that a list of all residues in the interface could be defined as the epitope, one would then have to ask what definition to use for "interface" (e.g., the residues buried to a certain radius probe sphere, the residues within van der Waals' contact distance, only the side chains interacting directly, or those interacting directly and through water molecules, etc.). Ultimately, molecular dynamics and energy calculations may contribute more information regarding residue interactions and, consequently, even more complex possibilities regarding a definition of the interface.

The existing crystal structures appear to concur with peptide mapping data, discussed in Section IV. Crystallographically defined epitopes on proteins consist of conformationally adjacent, continuous pieces of sequence (similar to the clusters observed from peptide mapping data on MHr discussed in Section IV). The advantage of analyzing crystallographic structures is that all residues in the interface area can be identified. Other approaches, such as fine structure specificity analysis (Benjamin *et al.*, 1984; Mehra *et al.*, 1986; Smith-Gill *et al.*, 1984), peptide release by proteolysis (Jemmerson and Paterson, 1986a,b), chemical modification of bound and free antigen (Burnens *et al.*, 1987; Cooper *et al.*, 1987), and peptide mapping (Geysen *et al.*, 1984, 1987b,c, 1988b; Houghten, 1985; Houghten *et al.*, 1986), identify only some of the side chains that are critical for competitive binding or are near the center of the binding site. Thus, fine specificity studies can identify distinct overlapping epitopes (Darsley and Rees, 1985), which would together presumably only be a portion of the buried interface, if the crystal structures were known. Similarly, monoclonal antipeptide antibodies appear to recognize distinct, sometimes even overlapping, sets of residues within a single peptide, rather than the entire sequence (Schoofs *et al.*, 1988; Fieser *et al.*, 1987).

For the purpose of understanding binding mechanisms, distinguishing the side chains that contribute the bulk of the binding energy is important. Thus, the combination of recent crystallographic data with other experimental results on protein antigenic sites indicates that the existence of an antigen residue in the antibody–antigen interface may be a necessary, but not sufficient, indicator of its importance to the interaction chemistry. From the practical point of view, if epitopes are defined only as the interface residues identified by X-ray crystal structures of antibody–antigen complexes, relatively few epitopes will ever be known.

As noted earlier, we use the term sequential epitope for sites identified using peptides, because peptide methods detect sequential pieces of epitopes. All epitopes, however they are identified, are topographic (whether or not this information is directly available), and most are assemblies of residues from separate portions of the amino acid sequence brought together by the three-dimensional fold (Benjamin *et al.*, 1984). However, instead of adopting a new definition of protein epitopes, i.e., all (approximately 15) antigen residues buried in the crystallographic structures of antibody–antigen interfaces, we would argue for the more classical definition, i.e., all (approximately 7) antigen residues apparently critical to antibody binding, as indicated by various experimental techniques. Epitopes, as defined in this article, usually represent epicenters for the antibody–antigen interaction and not the total interface. Besides being compatible with most experimental methods of identifying epitopes (except X-ray crystallography and some footprinting techniques, which define epitopes by their boundaries), this definition explains many apparent observations of autoimmunity provoked by molecular mimicry (Oldstone, 1987). Such autoimmunity depends on sufficient similarity for cross-reactivity and sufficient variation to break immunological tolerance. If one applies the crystallographic definition of epitope as equal to the entire molecular interface, it is difficult to understand this cross-reactivity, although it could occur due to the accommodation of topographic variability by induced fit. However, this accommodation would presumably require more extensive inducible complementarity than implied by current crystallographic results (Sheriff et al., 1987b; Colman et al., 1987a; Amit et al., 1986).

In an effort to test experimentally the agreement of results from X-ray crystallography and peptide mapping, monoclonal antisera HyHEL5 and HyHEL10 were used in a blind peptide mapping study, and the outcome was compared with the published (Sheriff et al., 1987b; Davies et al., 1988) and unpublished results (E. A. Padlan, E. W. Silverton, S. Sheriff, G. H. Cohen, S. Smith-Gill, and D. R. Davies) from X-ray diffraction studies of these antibodies in antibody–lysozyme complexes (H. M. Geysen and S. J. Smith-Gill, unpublished results). Although preliminary, the peptide mapping results identified the unpublished HyHEL10 site and correctly suggested the involvement of Trp and Arg in the complex. Previously, fine specificity studies have similarly identified critical residues for HyHEL5 (Smith-Gill et al., 1982). In principle, the characterization of critical residues, originally studied by peptide mapping and comprehensive single-residue substitution in peptides (see Sections IV and V), can be done in the context of the native protein fold, using cloning and site-directed mutation methods. Toward this end, MHr has been cloned and expressed in native and mutant forms (Alexander et al., 1988).

A complete understanding of the efficiency and specificity of antibody–antigen interaction will require a synthesis of the structural description at atomic resolution with other biochemical and biophysical data. For this synthesis to proceed efficiently, appropriate and consistent descriptive terminology should be developed such that agreement is reached among the results of different experimental techniques within their respective limitations. We suggest the following nomenclature and conditions: for residues in the *interface* of antibody–antigen complexes, as defined crystallographically, the term *contact residue* is appropriate, as the antibody–antigen interface on a protein antigen represents an antigenic contact. Probably the most generally accepted objective, quantitative description of a protein–protein interface is the area buried to a probe with a 1.4-Å radius, the size of the oxygen atom in a solvent water molecule (Getzoff et al.,

1986b; Connolly, 1983a,b; Richards, 1977; Lee and Richards, 1971). This interface includes the complete topographic structure of an epitope. Residues identified by biochemical competition or other experimental means as contributing to the binding energy might appropriately be called *critical residues*. The set of critical residues might truly be considered to form the *determinant* or epicenter for an epitope. These distinctions may help to improve the current practice of defining antigenic sites operationally, resulting in confusion in the integration of results from different experimental techniques (as discussed in detail by Van Regenmortel *et al.*, 1988).

IV. Defining the Chemistry of Antigenic Recognition

A. GENERATION OF A SYSTEMATIC DATA BASE

The vertebrate immune system represents an extremely diverse set of binding proteins. Whereas a great deal of information is available on the generation of this diversity (Ephrussi *et al.*, 1985; Honjo, 1983; Tonegawa, 1983; Wall and Kuehl, 1983), relatively little is known about the detailed chemistry and mechanisms of antigen–antibody interaction for protein antigens. In most cases, antigenic sites in proteins are likely to be topographic and assembled, i.e., have interaction surfaces dependent on the tertiary folding of the protein chain (Colman *et al.*, 1987a; Sheriff *et al.*, 1987b; Amit *et al.*, 1986; Benjamin *et al.*, 1984). Except for the few known X-ray crystal structures of antibody–antigen complexes and those cases in which sequenced mutational or evolutionary variants of the protein are available, the exact amino acid residues to which an antibody binds have been difficult to determine. Moreover, antigenic amino acid residues are normally operationally, and thus inconsistently, defined by the nature of the crystallographic, footprinting, or sequence-based experiments employed (Van Regenmortel, 1987; Van Regenmortel *et al.*, 1988). From crystallographic studies of antibody–antigen complexes, some residues identified in the contact surface with the antibody will probably be unimportant to the binding energy; whereas, from sequence-based, fine specificity studies, some residues identified to be in epitopes may instead only indirectly influence antibody binding through conformational changes in the protein antigen. In addition, proteins with both solved three-dimensional structures and defined epitopes are often highly conserved among different species and may thus produce limited immunological responses in experimental animals.

It is axiomatic that an understanding of complex systems, whether the interactions of atoms or macromolecules, depends on the ability to ask the specific experimental questions that address a single, unknown variable. In much of immunology, experiments involve complex systems containing many different variables, which can be difficult to define and impossible to eliminate. To obtain a systematic data base on a single protein antigen, a series of studies (Tainer et al., 1984; Geysen et al., 1987a,c; Getzoff et al., 1987; Fieser et al., 1987) was initiated on the immune response to MHr. MHr was selected for the following features: a well-refined atomic structure with a simple fold, providing surface areas composed mostly of only two or three separate pieces of amino acid sequence, accurate information on local mobility (from X-ray diffraction data), nonmammalian origin to reduce tolerance in experimental animals (Nossal, 1983), an iron center, providing an internal spectroscopic check for native structure, a disaggregated monomeric state at high protein concentrations, and apparent structural similarity with the biologically important bacterial pilus antigen (Getzoff et al., 1988a,b; Parge et al., 1987).

MHr functions as an oxygen carrier in certain invertebrates (Klippenstein et al., 1972). The MHr protein fold consists of an antiparallel bundle of four α-helices (named A, B, C, and D in sequence order) that surround a two-iron center at the active site, an N-terminal loop region, and shorter loops between the helices and at the C-terminus. The X-ray crystal structure has been solved and refined at high resolution for MHr from the marine wcrm *Themiste zostericola* (Sheriff et al., 1987a). In addition, the temperature factors have been corrected for crystal contacts, based on a comparative analysis with the octameric hemerythrin structure from *Themiste dyscritum* (Sheriff et al., 1985). An accurate indication of local mobility is useful for evaluating questions regarding possible inducible complementarity in the antibody and antigen (Section III, C and D) and for questions regarding the reactivity of antipeptide antibodies with native protein antigens (Tainer et al., 1985a). Examination of the exposed MHr molecular surface and corrected average main-chain temperature factors by residue indicates that a large portion (over 80%) of the molecular surface is relatively mobile. This corrected pattern of mobility is consistent with a qualitative analysis of the local and underlying stereochemical interactions (Tainer et al., 1984). Mobility is highly correlated with surface topography (Tainer et al., 1985a,b); the most highly protruding convex regions have the fewest intramolecular packing interactions, hence, the most flexibility and the most chemical interactions immediately available to an antibody.

To analyze the complex immune response to intricate antigens such as proteins, one can use either defined antibodies or defined antigens to select individual specificities. In a complementary approach to the use of selective antibodies (monoclonal technology), selective antigens (peptide homologs of the protein sequence) have been used to define the individual specificities within polyclonal anti-MHr antisera. The use of polyclonal antisera allows a study of the total antibody response of each animal to the whole protein and avoids unusual monoclonal specificities not favored *in vivo* (Metzger *et al.*, 1984). Moreover, the specificity of each antiserum can be delineated with defined peptide antigens, so that the chemistry of antigenic sites can be examined.

In the past, many researchers have disagreed on which sites of a protein antigen are immunologically reactive, in part, because the diversity of the immune response allows most sites to induce at least some antibodies (Benjamin *et al.*, 1984; Berzofsky, 1985). MHr has been used for systematically sampling the consistent, defined portion of the total immunogenicity of a protein (the ability of sites to trigger antibody production) that could be identified by competitive binding to sequential peptide homologs of the protein's sequence (Geysen *et al.*, 1987c). This work focused on the major trends, i.e., the identification of epitopes that induced strong, frequent responses. Of course, because this approach requires using peptide antigens for the assay of protein site antigenicity, some nonsequential assembled epitopes may have escaped detection.

B. Defining Antigenicity in Terms of Recognition Frequency

Given the enormous variability of the immune response, it is desirable to identify patterns resulting from frequently recognized sites. This approach is also relevant for practical applications, requiring a high frequency of response for a given species and antigen, as in the case of vaccines. Frequently recognized sites illustrate the common aspects of the immune response, which are most important for understanding the chemistry of the interaction. To locate the immunologically reactive sites of MHr, Geysen *et al.* (1987c) screened polyclonal antiprotein antisera against peptide homologs of the protein sequence. In the first set of studies, antiprotein antibodies against the intact MHr were raised in seven outbred rabbits. The antiprotein antisera were reacted with comprehensive sets of all five- and six-residue overlapping peptide homologs of the MHr sequence, as well as selected longer synthetic peptides. A profile of the reactivity of MHr sites assessed against hexapeptides is shown in Fig. 8. In the sites selected, the native protein (usually at 10^{-5} M or less) absorbed or competed for antiprotein antibody binding to peptides covering all of the MHr sequence, except for three regions of very low immunological reactivity (residues 21–28, 47–53, and 96–108), which were not tested due to the poor signal-to-background ratio. The results obtained in different laboratories with three different assays (Geysen *et al.*, 1987c) agreed and confirmed that the ability of the antiprotein antibodies to react with peptides related to their ability to bind the native protein. The overall correctness of

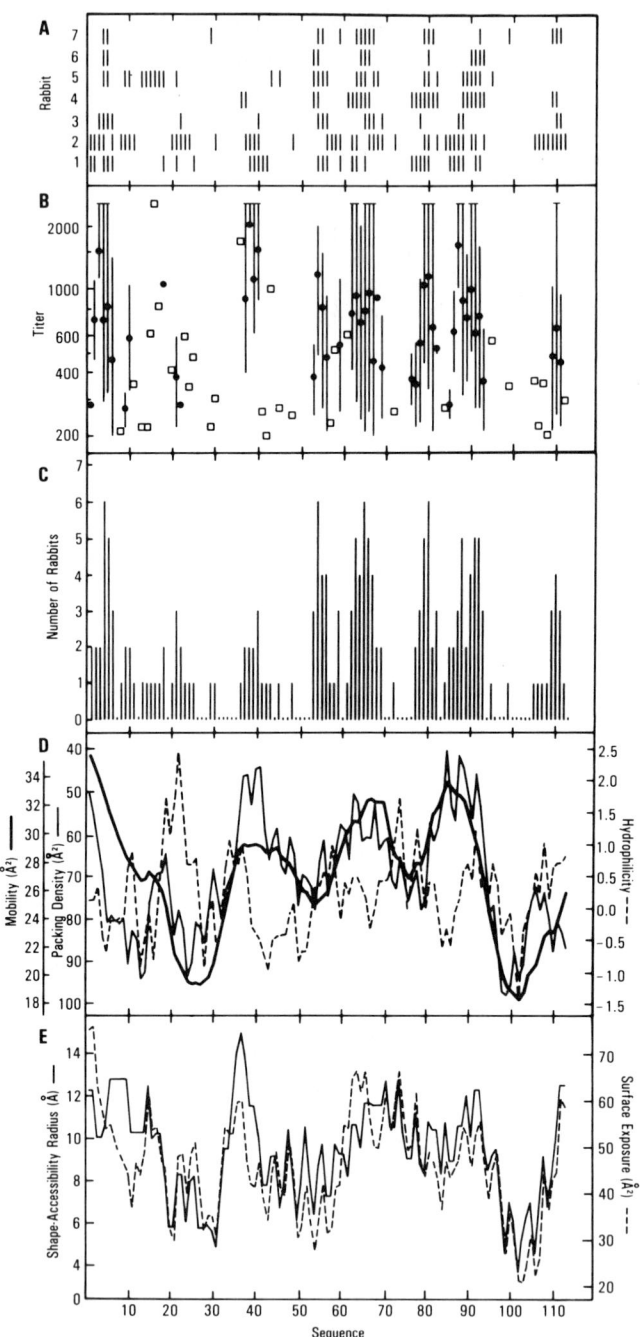

site identification was supported by the consistency of the data and by the clustering of the sites in the folded protein structure. Immunoprecipitation assays on the reactivity of longer peptides confirmed the corresponding hexapeptide reactivities measured by enzyme-linked immunosorbant assay (ELISA).

The studies on MHr resolved apparent inconsistencies in the current immunological literature. Collectively, the antibody populations from the 7 antisera reacted with hexapeptides encompassing all 118 residues

---

FIG. 8. MHr antigenic response and stereochemistry. Profiles of antigenic response (A–C) and stereochemical properties (D–E) are plotted as a function of all possible MHr hexapeptides. Antibody recognition of MHr is measured by the reactivity of anti-MHr to hexapeptide analogs of the MHr sequence, which compete for antibody binding by the native protein. Each parameter is plotted at the sequence number corresponding to the first residue of the relevant hexapeptide. (A) The reactivity pattern specific to each of the seven rabbit antisera. Responses to any hexapeptide are indicated with a short vertical bar. Immunological reactivity of the first 52 hexapeptides of the sequence was highly variable, with 3 antisera reacting with only 2 or 3 hexapeptides and 1 antiserum reacting with 20. Hexapeptides 7 through 52 (as numbered by the first residue) also showed infrequent reactivity, with 18 or 46 hexapeptides unrecognized by an antiserum, 19 recognized by a single antiserum, and none seen by more than 3 antisera. For hexapeptides 53 through 93, the immune response was more consistent and more frequent; 21 of the 41 hexapeptides reacted with at least 3 of the antisera, and each of the 7 antisera reacted with between 10 and 22 hexapeptides. The least frequently reactive region of the sequence encompassed hexapeptides 94 through 104; 9 of these 11 hexapeptides were not recognized by any of the 7 antisera. (B) The individual titer (square) or, when more than one sera reacted, the geometric mean titer circle) and range of titer (vertical bar) of the antisera. Titers less than 200 (twice the slope of the background) were ignored, and those greater than 2000 were not further resolved. (C) The frequency of the antigenic response given by the number of rabbit antisera that react with each hexapeptide. High frequency and high titer are often associated. However, peptides corresponding to protein residues 3–8, 16–21, and 38–43 were exceptions for which only 1 or 2 antisera reacted, but all had high titers (1120– >2000). In contrast, titers ranged widely in regions where 6 of the 7 antisera reacted: the narrowest range of titers was more than 4-fold (480–1970). (D) Mobility (average main-chain temperature factors corrected for crystal contacts, inverted packing density (surface area buried in noncovalently bonded close packing), and hydrophilicity [using values from Hopp and Woods (1983)]. Packing density, sphere accessibility radius, and surface exposure calculations used individual van der Waals' radii. (E) Shape-accessibility radius (radius of the largest sphere that can access a side chain without collision or interpenetration of any other residue) and surface exposure [the solvent-exposed molecular surface area (Connolly, 1983a; Lee and Richards, 1971)]. Shape-accessibility radii greater than 15 Å are not distinguished. Except for hydrophilicity, all plotted stereochemical parameters depend on the three-dimensional structure, rather than being defined identically for a given amino acid type. Structural parameters are averaged over six residues, so as to correspond with immunological data based on hexapeptides. Hexapeptides are defined from the MHr sequence. [Adapted from Geysen et al. (1987c).]

of the MHr sequence, although not every hexapeptide was antigenic in these tests. Each residue (except those at the termini) was included in six different overlapping synthetic hexapeptides and thus was assayed in six different environments, at least one of which showed immunological reactivity. The result—that each MHr residue appears in a region of the protein with antigenic potential—is consistent with the interpretation that the entire protein surface can be recognized by antibodies (Benjamin *et al.*, 1984; Berzofsky, 1985). The degree of antigenicity, however, appears to depend on the region of the protein involved, and the likelihood of obtaining antibodies with any given specificity from an individual animal varies greatly.

Of the 113 hexapeptide homologs of the MHr sequence, 4 peptides were recognized by 6 of the 7 antisera, whereas 31 reacted with only a single antiserum, and 35 failed to bind antibody present in any antiserum (Fig. 8A). No antibody specificity was common to all seven rabbits. This variability of responses among members from a single species has been observed before (Rodda *et al.*, 1986a) and highlights the need to evaluate a consensus response from numerous animals (see Section VI below).

C. The Stereochemistry of the Most and Least Frequently Reactive Sites

The first step in defining the chemistry of antibody binding to protein antigens is to identify intrinsic features of a protein that distinguish the immunologically most frequently reactive and least frequently reactive sites. Several stereochemical parameters have been proposed, including molecular shape and surface topography, physical access to interacting antibody molecules, hydrophilicity or hydrophobicity, secondary structure, amino acid specificity, and local flexibility or mobility of the peptide main chain (reviewed in Berzofsky, 1985). A systematic data base for immunological reactivity developed for MHr allows examination of these and other factors for their role in immunogenicity and antigenicity (Geysen *et al.*, 1987a,c). Each factor may contribute by enhancing or decreasing the reactivity of a site, since preference for one epitope implies discrimination against others.

When chemical and structural properties (mobility, packing density, hydrophilicity, shape accessibility, and surface exposure) for each hexapeptide of MHr were plotted versus sequence number and aligned with immunological frequency data (Fig. 8), certain trends became apparent (Geysen *et al.*, 1987a,c). Plots of mobility, packing density (inverted), shape accessibility and, to a lesser extent, surface exposure showed overall patterns that matched those for the experimentally determined immunological reactivity. This was confirmed statistically by a comparison of less antigenic (those recognized by one or no rabbit antisera) and more antigenic hexapeptides (two or more antisera). These two groups differed significantly (t test, $df = 111, p \leq 0.001$) with regard to structure-associated properties: antigenic frequency correlated best with mobility ($t = 5.75$), followed by packing density ($t = 4.66$), shape accessibility ($t = 3.87$), and surface exposure ($t = 3.23$). Hydrophilicity differences were less significant ($t = 1.86, p > 0.05$). All of the plotted stereochemical and antigenic properties shared one feature: the deepest minimum occurred between hexapeptides 99 and 106. This mapped to the only truly buried part of the MHr structure, the residues of the D-helix located underneath the N-terminal loop. Small shifts in the alignment of critical points between the structural and immunological plots highlighted the three-dimensional nature of protein epitopes. Local maxima and minima of chemical, structural, or immunological properties traced along the one-dimensional amino acid sequence do not necessarily identify true maxima and minima for a three-dimensional protein.

To identify correlations of immunological reactivity with structure-associated parameters in a three-dimensional way (Table II), the residues of MHr were divided into three categories, as assessed by reactivity with hexapeptides: (1) most reactive (41%), occurring in at least one hexapeptide recognized by four or more rabbit antisera; (2) least reactive (26%), not occurring in any hexapeptide recognized by more than one antiserum; (3) average (33%), all others (Geysen et al., 1987a,c). The most reactive positions include 4–10, 54–61, 63–72, 79–85, 88–97, and 110–115. The least reactive positions are 16–17, 28–36, 46–52, 75, 99–108, and 117–118. Ten- and fourteen-residue peptides, encompassing the longer stretches of least reactive positions (26–35, 42–51, 96–109, and 100–109), were also unreactive with anti-MHr antibodies, when coupled to protein carriers and assayed by ELISA.

1. Mobility and Packing Density

Qualitatively, the trend across the three categories of immunological reactivity in MHr coincides with changes in mobility (Table II). Decreasing mobility corresponds to decreasing immunological reactivity from most reactive, to average, to least reactive positions. The mobility of the most reactive and of the least reactive residues differs significantly from the mean (Monte Carlo method, $p < 0.001$). The major maxima and minima of the immunological profiles align best with those from plots of mobility and inverted packing density (Fig. 8D). Although the mobility curve is smoother, all 5 of its peaks overlap with maxima in the frequency (Fig. 8C) and mean titer (Fig. 8B) plots; the 2 prominent

TABLE II
MOLECULAR SURFACE CHEMISTRY AND IMMUNOLOGICAL REACTIVITY[a]

Class of immunological reactivity	Number of residues	Mobility (Å²)	Percentage total exposed surface area							Surface area (Å²)		Percentage exposed side chain		
			Potential			Hydro-phobic	Hydro-philic	Side chain	Main chain	Total exposed	Per residue	Per residue	Hydro-phobic	Hydro-philic
			−	0	+									
Most	48	29.0	48	34	19	66	34	75	25	2337	49	32	29	41
Average	39	27.6	37	40	23	61	39	73	27	2005	51	34	31	42
Least	31	23.2	23	42	35	65	35	83	17	1209	39	23	17	48
MHr	118	27.0	39	38	24	64	36	76	24	5552	47	30	26	43

[a] Individual residues are divided into three classes of antigenic reactivity as defined in the text. All stereochemical parameters listed are functions of the three-dimensional structure, not the linear sequence, and are derived from X-ray crystallographic results (Sheriff et al., 1987a). Mobility is calculated from the average main-chain temperature factors, corrected for crystal contacts (Sheriff et al., 1985). The percentage total exposed surface area is calculated from the molecular surface area (Connolly, 1983a; Lee and Richards, 1971; Richards, 1985), summed for each category as follows: negative, neutral, or positive areas based on the electrostatic potential calculated from partial charges on all atoms (Getzoff et al., 1986b), hydrophobic (carbon) or hydrophilic (noncarbon) atoms, and side-chain or main-chain atoms. The total exposed surface area and surface area per residue are calculated from the molecular surface area accessible to a water sized (1.4 Å) probe sphere. Exposed side-chain area is expressed as a percentage for hydrophobic (carbon), hydrophilic (noncarbon), or all side-chain atom surface area, exposed in the context of the protein structure, calculated from the total side-chain surface area in the absence of neighboring protein residues (Shrake and Rupley, 1973). [Adapted from Geysen et al. (1987c).]

temperature factor minima correspond to reactivities with at most one antiserum and titers of less than 500. The peak in reactivity near hexapeptide 54 falls in a local mobility minimum, but this region remains more mobile than the global minima near hexapeptides 27 and 102. The packing density profile (inverted), although more serrated, resembles the overall mobility profile; the major differences are an additional minimum near hexapeptide 13, and a shift in C-terminal peak.

2. Electrostatic Potential

Intriguingly, the most frequently recognized sites in MHr generally correspond to regions of negative electrostatic potential (Table II). The percentage of the total surface area associated with negative electrostatic potentials (−) is greatest for the most reactive positions (Monte Carlo method, $p = 0.02$) and least for least reactive positions ($p = 0.005$), while the converse is true, but less significant, for positive electrostatic potentials (+). In each case, the percentages of total exposed surface area associated with the most and the least reactive positions differ by about a factor of two. The percentages of negative and positive surface area associated with regions of average reactivity closely match those found for the whole protein.

3. Hydrophilicity

Although the percentage of the total exposed surface area contributed by hydrophilic (noncarbon) atoms is nominally largest for the residues of average reactivity in MHr (Table II), all three categories of antigenic reactivity have percentages of hydrophobic and hydrophilic surface area similar to that for the whole protein. Furthermore, the hydrophilicity profile (Fig. 8D) clearly does not match the pattern of immunological reactivity as well as the mobility and inverted packing density profiles. The highly reactive site at hexapeptide 4 is a hydrophilicity minimum, and the highest hydrophilicity peaks at hexapeptides 22 and 74, which correspond to valleys in the mobility and inverted packing density plots, show only average and low-reactivity values, respectively, for both titer and frequency.

4. Shape Accessibility and Exposed Surface Area

Half (59) of the residues of MHr are accessible to a probe sphere of 15-Å radius, and another 7 residues are accessible to spheres of 11- to 14-Å radius, suggesting that 56% of the residues are directly available to the ends of a fairly flat antibody binding region. In contrast, the shape accessibility of 34 residues (29%) is restricted to spheres of 2-Å radius or smaller. The most and least antigenic positions are not directly distinguished by their shape-accessibility radius: all contiguous stretches of most antigenic positions include one or more residues accessible only to a solvent-sized sphere with a radius of ≤ 2 Å, and all contiguous stretches of least antigenic positions include one or more residues accessible to a sphere with a radius of ≥ 15 Å. Whereas the least antigenic positions in MHr are thus accessible to antibody binding, they are also closely associated with narrow grooves containing tightly bound water molecules identified in the X-ray crystallographic structure.

The plot of shape accessibility versus sequence (Fig. 8E) shows similarities with mean titer and frequency profiles, as well as with plots of average main-chain temperature factors, inverted packing density, and surface exposure. However, compared to the mobility and inverted packing density profiles (Fig. 8D), the shape-accessibility plot distinguishes less well between the peak of antigenicity near hexapeptide 54 and the minimum near hexapeptide 30. Continuing this trend, the plot of solvent-exposed surface area (Fig. 8) shows more of a minimum near hexapeptide 54 (a reactivity peak) than hexapeptide 30 (a reactivity minimum). The exposed surface area per residue is the lowest for the least immunoreactive residues (Table II), supporting a relationship between surface exposure and antigenicity (Novotný et al., 1986b; Fanning et al., 1986). However, this trend does not continue: the most antigenic residues do not have more exposed surface area per residue than those with average antigenicity.

5. Side-Chain Contributions

When total solvent-exposed surface area is divided into side-chain and main-chain contributions (Table II), the least reactive residues of MHr are again distinguished (Monte Carlo method, $p = 0.04$). The ratio of main-chain to side-chain surface area is similar for positions with most and average reactivity and lower for the least reactive positions. In the folded protein, the side-chain surface exposure can also be measured as a percentage of the total surface exposure of the same side-chain conformation in the absence of the other protein residues. This percentage of exposed side chain (Table II) is marginally lower for the

FIG. 9. Superassemblies of MHr sequential epitopes. Three categories of immunological reactivity (red, most reactive; yellow, average; blue, least reactive) are mapped by color coding onto the solvent-exposed molecular surface of MHr, shown in front and back views. Raster color graphics images of opposing sides of the solid external molecular surface of MHr show that the most highly reactive surface regions form superassemblies (three-dimensional clusters) that wrap around one side (left view) of the protein and almost connect on the other side (right view). The four α-helices are oriented vertically, with the two-iron active site near the top. [Adapted from Geysen et al. (1987c).]

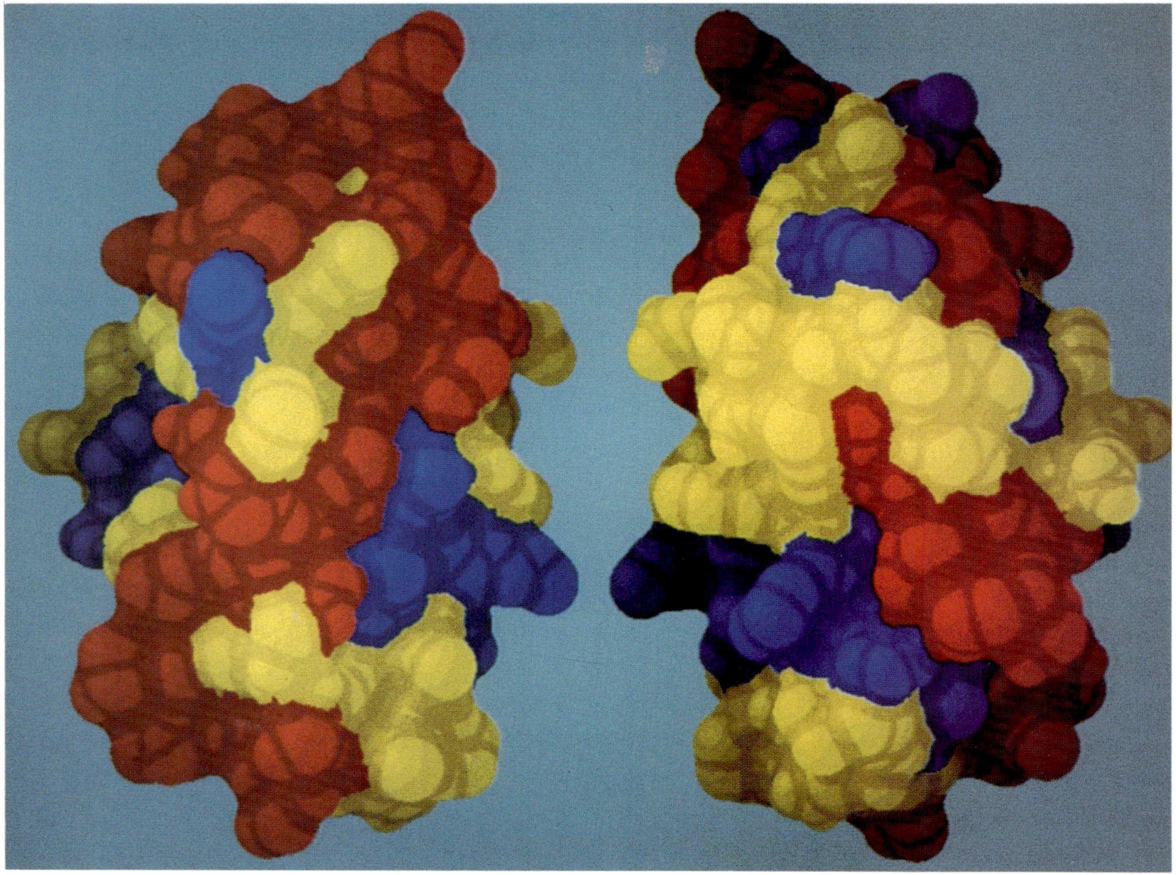

FIG. 9

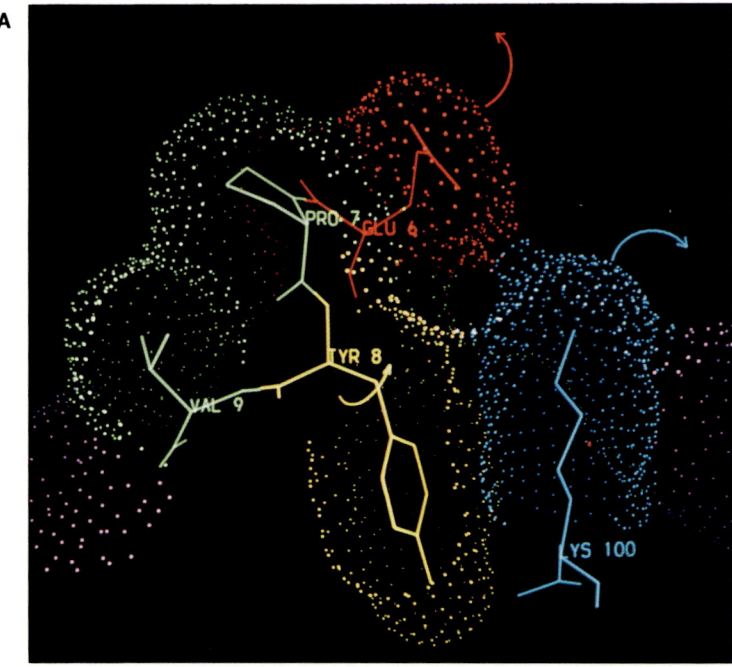
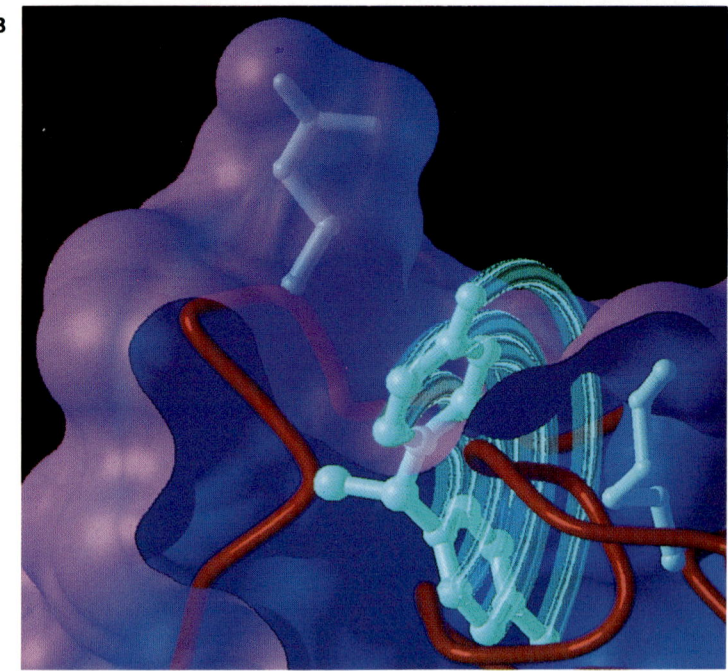

Fig. 10

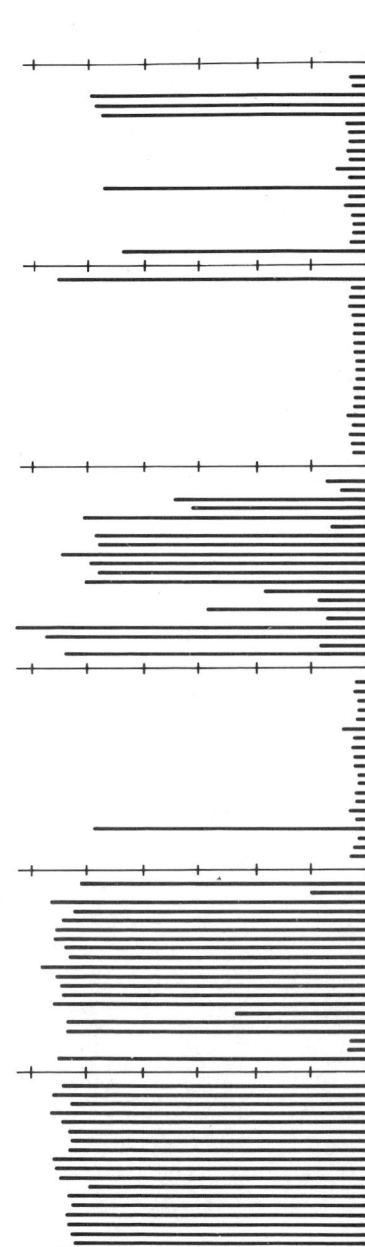

FIG. 10. Stereochemistry, mechanism, and critical residues of MHr site 4–9. (A) The spatial relationship and interactions of critical side chains for MHr site 4–9, Ile-Pro-Glu-Tyr-Val, of the N-terminal loop are viewed in a slice through the MHr atomic model (isolated by clipping planes) as displayed on computer graphics. Outside the site, the molecular surface is colored purple. Within the antigenic site, bonds and surface are colored blue for exposed, critical, positively charged side chains, red for exposed, critical, negatively charged side chains, yellow for bonds and individually surfaced side chains of critical buried residues, and green for noncritical residues. Curved arrows indicate plausible side-chain movements, when antibody binding to critical exposed polar side chains results in the breaking of one or more charged and/or hydrogen-binding interactions. Binding of the critical exposed Glu-6 side chain (red, top center) could disrupt the interaction with the Lys-100 side chain (blue, right), which packs against and forms a gate to the surface for the buried Tyr-8 side chain (yellow, center). (B) A postulated side-chain movement of Tyr leads to a resulting change in the antigenic surface topography. The MHr exposed molecular surface (shown transparent purple) is sliced away to reveal its inner surface (deep blue) and α-helical backbone (red tube). The stereochemistry of three critical residues suggests a mechanism of induced fit involving side-chain movements that break a charged hydrogen bond between surface-exposed Glu-6 (light blue bonds, upper left) and Lys-100 (light blue bonds, right), allowing previously buried Tyr-6 side chain (center, blue bonds) to rotate out from the buried native position (lower) and interact with the antibody molecule (not shown). Thus, in the proposed mechanism of interaction, antigenic recognition causes bond rotations that change the surface topography of the antigenic site. Rendered by M. Pique on CONVEX and CRAY computers using MCS (Connolly 1983b, 1985) and the Dicomed film writer at the San Diego Supercomputer Center. (C) A replacement net identifies MHr residues essential for antibody binding to peptide homologs for MHr site 4–9, Ile-Pro-Glu-Tyr-Val, antiserum 7. ELISA absorbances for each hexapeptide, measuring the reactivity with antibodies raised against MHr, are represented as a percentage of the mean absorbance of the six replicates of the parent peptide (thicker bars). Each block of bars represents 20 hexapeptides that differ from the parent peptide by replacement of the amino acid residue indicated underneath. The identity of each substituted residue (1 of the 20 genetically coded amino acids) is indicated by the position of the individual bar within each block, based on alphabetical order of the single letter code for the amino acids: 1A (Ala), 2C (Cys), 3D (Asp), 4E (Glu), 5F (Phe), 6G (Gly), 7H (His), 8I (Ile), 9K (Lys), 10L (Leu), 11M (Met), 12N (Asn), 13P (Pro), 14Q (Gln), 15R (Arg), 16S (Ser), 17T (Thr), 18V (Val), 19W (Trp), and 20Y (Tyr). Thus, the leftmost bar of A represents peptide APEPYV, and the rightmost bar represents peptide IPEPYY. In site 4–9, Glu-6 and Tyr-8 are essential and Val-9 selected (selectively replaceable). [Adapted from Getzoff et al. (1987).]

least reactive positions ($p = 0.10$), largely as a result of a decreased hydrophobic (carbon atom) contribution ($p = 0.02$). Similarly, the immunological reactivity profiles match carbon exposure better than noncarbon exposure; the least reactive positions seem to associate with areas of concave surface shape in which the most exposed atoms are frequently the ends of polar side chains.

6. Secondary Structure

Although secondary structure is not a dominant factor in the degree of antigenicity, the least reactive residues in MHr are predominantly found in α-helices. Twenty-seven least reactive residues are distributed among the four helices, whereas the remaining four residues occur in the N- or C-terminal loops. This bias may be caused in part by the peptide assay system, since the 3.6-residue helical repeat (visible in most of the structural plots) may reduce the ability of a hexapeptide to mimic a surface region or conformational determinant of a protein antigen. Except for the lack of most reactive residues in the A-helix, however, each of the four helices, as well as the longest loop, contains residues in all three categories of immunological reactivity. For example, the N-terminal loop includes residues with most (4–10), average (1–3 and 11–15), and least (16–17) reactivity. The shorter loops together also include all three categories of reactivity. Areas of average reactivity occur in every helix and connecting or terminal loop region of MHr secondary structure.

7. Sites Mapped by Peptides Cluster into Conformational Epitopes

In general, peptide mapping identifies sequential epitopes (which are primarily portions of assembled epitopes) where antibody binding to peptide homologs is competitively inhibited by native protein. However, where peptide mapping results are examined in terms of the three-dimensional topography of the MHr atomic structure, these sequential epitopes cluster into groups of superassemblies that may represent assembled, conformational epitopes. Although all epitopes will not be identified by this approach, the immunological reactivity probed by peptides varies relatively smoothly over the three-dimensional surface of the protein, as well as the linear sequence (Fig. 9). Along the polypeptide chain, the most and least reactive residues are always separated by residues with average reactivity. On the molecular surface, the most reactive residues cluster into patches, as do the least reactive residues, and the two sets of patches are usually separated by surface area from residues of average reactivity. The most frequently recognized positions form superassemblies, wrapping around the MHr molecule, to encompass the more flexible and less tightly packed areas of the protein. Residues from least reactive positions cluster into a cylindrical core within the most tightly packed portion of the protein, extending through the molecule to include residues at two surface regions. The clustering of residues with like antigenic properties and the relatively smooth variation of immunological reactivity at the molecular surface suggest that conformational or assembled determinants may often consist of residues contributed by parts of two or more linear sequences, as others have also shown (Sheriff *et al.*, 1987b; Amit *et al.*, 1986; Jemmerson and Paterson, 1986a; Thornton *et al.*, 1986; Barlow *et al.*, 1986).

V. Defining the Mechanisms of Antigenic Recognition

A. IDENTIFICATION OF CRITICAL RESIDUES

Chemical interactions are described in terms of reaction mechanisms. Thus, the modern interpretation of molecular interactions details electron and/or bond movements and conformational rearrangements in terms of the specific stereochemistry of the molecular species involved. By contrast, the binding of antibodies to their antigens, originally formulated by Ehrlich (1900) as a hard-body interaction represented by a lock-and-key model, remains widely accepted (Mariuzza *et al.*, 1987), despite experimental evidence to the contrary (Celada and Strom, 1972; Lubeck and Gerhard, 1982; Parham, 1984; Tainer *et al.*, 1984; Westhof *et al.*, 1984; Paterson, 1985; Geysen *et al.*, 1987c). Advances in peptide synthesis and methods of immunoassay now enable one to map the antigenicity of a protein at the resolution of single amino acid residues (Geysen *et al.*, 1984, 1985a; Rodda *et al.*, 1986a, b; Houghten, 1985; Houghten *et al.*, 1986; Fieser *et al.*, 1987). This degree of precision, coupled with present high-resolution atomic structures of protein antigens, allows mechanistic models for antibody binding to be postulated.

In a study of MHr sequential epitopes, the role of individual amino acids was examined in the binding of rabbit antiprotein antisera to more than 1500 distinct peptide analogs, differing from the protein sequence by single amino acid replacements (Getzoff *et al.*, 1987). These data were then combined with information from the three-dimensional crystallographic structure of MHr for evaluation of probable mechanisms of antibody binding. One important question in this study was whether the chemistry of the antibody–antigen interaction arises simply from the chemistry of the antigen's static exposed molecular surface, or whether the observed biases for mobility and electrostatic forces may reflect a common concerted or multistep interaction mechanism, allowing inducible complementarity. The consensus picture suggests an interaction mechanism in which initial antibody binding to solvent-exposed amino acid side chains can promote local conformational changes that alter the interaction of other antigen side chains with the antibody.

The roles of individual side chains in each of 7 sequential epitopes were characterized by measuring the reactivities of appropriate antisera with sets of 120 peptides that included all possible analogs (replacement nets) differing from the parent peptide by replacement of a single amino acid residue (Fig. 10; Getzoff et al., 1987). Those residues for which replacement by alternative amino acids resulted in loss of reactivity with the antibody were postulated to be essential residues for the interaction. By using such replacement net data, each residue of a peptide was assigned to one of four descriptive categories: essential (irreplaceable), selected (selectively replaceable), partially replaceable, or generally replaceable. For example, in MHr site 4–9 (Fig. 10C), Ile-4 can be replaced by any of the other 19 commonly occurring amino acids, without affecting peptide binding by the antisera (generally replaceable). Similarly, Pro-5 can be replaced by most other amino acids without influencing antibody binding. In contrast, antigenic reactivity is lost when Glu-6 or Tyr-8 is replaced with any of the other amino acids; these two residues are irreplaceable and thus essential to the interaction with antibody. Pro-7 is partially replaceable; Val-9 is selected, allowing only Ala, Ile, Ser, or Thr as replacements. The term *critical* is used to encompass both essential and selected residues.

Based on the studies of MHr (Getzoff et al., 1987), the specificity requirements for the essential residues at a given antigenic region appear to be less variable than the responses of individual antisera to a given epitope. Residues critical to the interaction of the induced antibodies from one rabbit are usually critical to the interaction of antibodies from other responding rabbits. This is generally true both within the same peptide epitope (e.g. Phe-80 and Lys-83 in MHr site 80–85) and between overlapping peptide sites (e.g., Tyr-67, Glu-69, and Val-70 in MHr sites 63–68, 65–70, and 68–72). Thus, intrinsic stereochemical properties of specific residues may bias the selection of the critical residues recognized by the immune system. Individual animals may respond to different sites, but once a site is recognized, the chemical basis for interaction appears to be similar.

Although all MHr residues critical for antigenic interactions with the seven rabbit antisera could not be unambiguously determined, those that were identified (Table III) clustered together in three dimensions to form microassemblies. The critical residues of eight of the nine defined sites (including three overlapping sites) formed clusters composed of closely interacting side chains (Getzoff et al., 1987). Thus, the antigenic

TABLE III
PROPERTIES OF CRITICAL RESIDUES IN SEQUENTIAL MHr EPITOPES[a]

Residue identity	Secondary structure	Percentage exposed side-chain area			Packing density	Sphere accessibility
		Hydrophobic	Hydrophilic	All		
6 Glu	Loop	81	70	77	32	15
8 Tyr	Loop	0	0	0	140	2
9 Val	Loop	49	0	49	67	15
16 Val	Loop	0	0	0	96	0
17 Phe	Loop	56	0	56	63	15
18 Tyr	Loop	22	10	19	98	15
19 Glu	Loop	81	32	63	33	15
20 Gln	A-helix	46	89	70	35	15
54 His	B-helix	0	0	0	92	1
55 Phe	B-helix	2	0	2	120	2
56 Thr	B-helix	75	57	70	36	15
57 His	B-helix	26	36	29	78	15
58 Glu	B-helix	0	0	0	65	0
63 Asp	Turn	54	25	39	49	15
64 Ala	Turn	66	0	66	21	15
66 Lys	Turn	79	95	85	33	15
67 Tyr	Turn	5	3	5	132	2
69 Glu	Loop	34	60	44	61	15
70 Val	Loop	16	0	16	77	8
71 Val	C-helix	60	0	60	39	15
72 Pro	C-helix	44	0	44	35	15
80 Phe	C-helix	0	0	0	122	0
81 Leu	C-helix	26	0	26	78	8
82 Glu	C-helix	61	57	59	46	15
83 Lys	C-helix	25	35	29	80	15
84 Ile	C-helix	0	0	0	105	0
85 Gly	C-helix	—	—	—	23	13
91 Val	Loop	0	0	0	102	1
92 Asp	Loop	83	28	54	37	15
93 Ala	Loop	94	0	94	7	15
94 Lys	Loop	68	59	65	46	15
95 Asn	D-helix	0	16	11	72	11
111 Asp	D-helix	0	0	0	46	0
112 Phe	D-helix	19	0	19	106	15
113 Lys	Loop	50	87	64	71	15

[a] Critical residues are labeled by residue number and three-letter amino acid code. Secondary structure is taken from the crystallographic structure of MHr (Sheriff et al., 1987a). Exposed side-chain area for hydrophobic (carbon), hydrophilic (noncarbon), and all side-chain surface area exposed in the context of the protein structure is expressed as a percentage of the total side-chain surface area in the absence of neighboring protein residues. Packing density is determined numerically, as the surface area ($Å^2$) buried in noncovalently bonded close packing, using a modified version of the program MS (Connolly, 1983a). Sphere accessibility is determined analytically as the radius (Å) of the largest spherical probe that can access a given side chain without collision or interpenetration with the surface from any other residue (E. Getzoff and J. Tainer, unpublished method). Sphere radii of zero represent side chains inaccessible to a water-sized (1.4-Å radius) probe sphere, and sphere radii ≥ 15 were evaluated as 15 Å (roughly the radius of an immunoglobulin binding domain. [Adapted from Getzoff et al. (1987).]

superassemblies are constructed from a connected set of microassemblies, which themselves cluster in three dimensions. These antigenic microassemblies are not restricted to clusters formed by sequence-local neighbors. Critical residues interacting spatially within one epitope often adjoin critical residues of nearby epitopes, and together form a network of closely interacting side chains that represent the backbone of the three-dimensional superassemblies formed by the most reactive sites. For MHr, the structural architecture of the critical residues includes interlocking van der Waals' interactions (as identified by contact distances and individual atomic radii) and hydrogen bonds to form a connected network, involving all of the defined sites except for site 4–9 in the N-terminal loop.

B. Evidence for Induced Complementarity

Structural properties were identified for 35 critical residues in MHr (Table III). The antigenic specificity conferred by residues shown to be highly exposed at the protein surface (defined from the X-ray crystallographic structure) is readily appreciated. What is significant, however, is that residues whose side chains are at least partially buried can apparently confer the same degree of specificity (Getzoff et al., 1987). As shown in Table III, each antigenic site for which critical residues were determined includes one or more highly exposed critical side chains (with > 55% of the potential surface area of the side chain exposed) that are accessible to an antibody binding domain (as modeled by a 15-Å radius sphere) and one or more largely buried critical side chains (<5% exposed surface) that are nearly or totally inaccessible to a solvent-sized probe (1.4-Å radius). These data show that antibody–antigen interactions for MHr are dependent on the presence of both highly exposed, predominantly hydrophilic residues and mostly buried, predominantly hydrophobic residues. Antibody interaction with buried critical residues in each antigenic site may affect the protein fold and thus its conformational stability by reducing key interactions between MHr secondary structural elements.

The finding that at least partially buried hydrophobic side chains of the protein are frequently critical for the binding of antiprotein antibodies to peptides raises two questions. (1) Are these residues also critical for antibody binding to the native protein antigen? (2) Are buried residues directly implicated in the binding process or do they simply act to bias the peptide conformation toward one suitable for antibody binding? Competition data measured in solution-phase assay show that native MHr inhibits the binding of antiprotein antibodies to peptides corresponding to the sites for which critical residues were defined (Geysen *et al.*, 1987c). Thus, antibody binding to these sites is not limited to the binding of protein fragments, but is general to both protein and peptides. The data base for MHr also suggests that buried side chains of the protein are directly involved in binding to the antibody. First, only one amino acid type often permits binding, rather than a set of acceptable amino acid replacements with similar properties, as would be expected for side chains performing a conformational role. Second, buried critical residues are not randomly located, but show a specific positional relationship to the solvent-exposed critical residues, which is most frequently seen as a gating phenomenon (see specific examples below). Third, the clustering of critical buried side chains from nearby sites (identified by using different noncontiguous peptide sequences) is logical for side chains involved in antibody binding, but not for those with only conformational roles. Finally, monoclonal antiprotein and antipeptide antibodies for sequential epitopes within the MHr C-helix also show the same sites and patterns of critical residues (Fieser *et al.*, 1987).

The stereochemical relationships of the critical residues identified in the MHr sequential epitopes suggested that the mechanisms of interaction may often involve a process (local induced complementarity or concerted conformational change) in which the specific recognition of highly exposed side chains is coupled to movements allowing the interaction with additional, sometimes partially or completely buried, usually hydrophobic side chains (Getzoff *et al.*, 1987). The degree of motion (induced fit), implied by combining the antigenic and structural data, varies from site to site and ranges from small local changes in a few side chains to a combination of side-chain rotations with some movement of the secondary structural elements. The structural implications for a proposed multistep interaction mechanism for two sites (Getzoff *et al.*, 1987) are detailed below.

MHr site 4–9 includes essential residues Glu-6, Tyr-8, and selected residue Val-9. As shown in Fig. 10A and B, Glu-6 forms a salt-bridge (charge–charge interaction) through a bound water molecule to Lys-100. Tyr-8 is buried from solvent exposure (0-$Å^2$ exposed area) and is inaccessible to a 1.4-Å radius probe sphere (Table III). The direct interaction of the incoming antibody with Glu-6 can be envisioned to break this salt-bridge, allowing the exposure of buried Tyr-8, which lies underneath the salt-bridge in the crystal structure. The hydrophobic portion of the Tyr-8 side chain packs against the hydrophobic aliphatic section of the side chain of Lys-100, and the hydroxyl oxygen atom of Tyr-8 hydrogen bonds to the Lys main-chain nitrogen atom. Thus, changes in the salt-bridge could disrupt both van der Waals' contacts

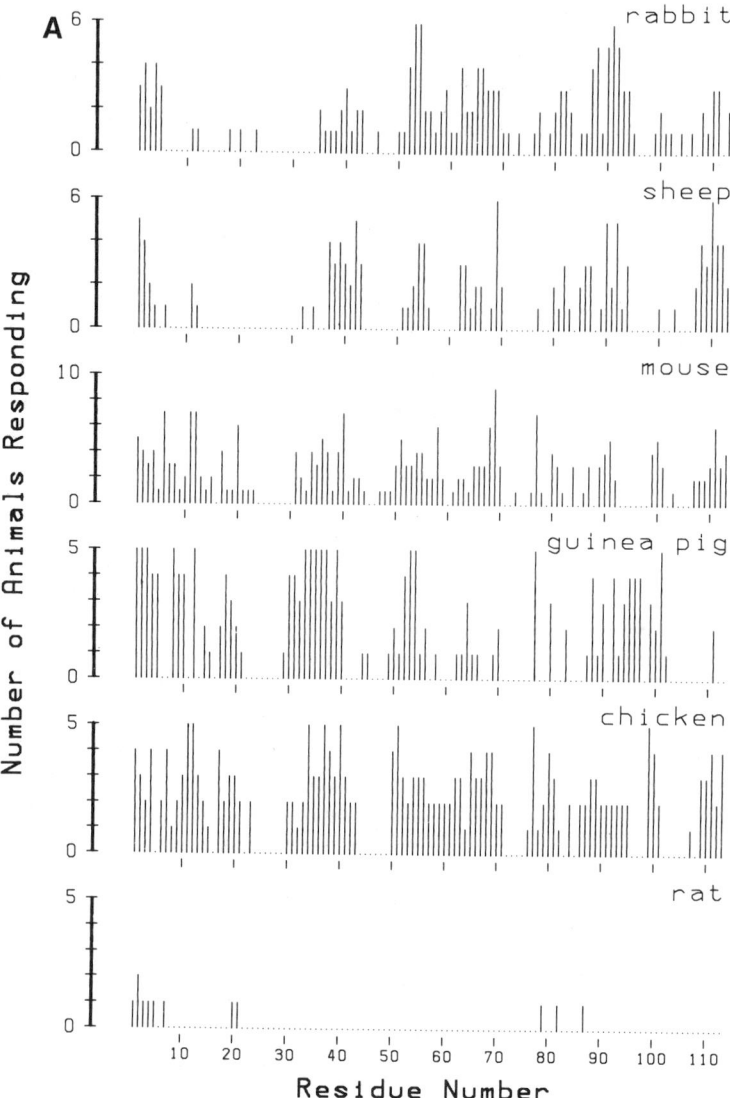

FIG. 11. Immunological responses to MHr in different species. (A) The frequency of the antigenic response is given by the number of antisera from each species that react (respond with a titer of >200) with each hexapeptide (numbered according to the position of the amino-terminal residue in the MHr sequence). Species are indicated at the top of each panel. (B) The collective responses to MHr in six species (chicken, guinea pig, rabbit, mouse, sheep, and rat) are compared. The area of each square is proportional to the summed, normalized response for each indicated species. Upper squares (rows) correspond to the species identified to the left, and lower squares (columns) correspond to the species identified above. Area of overlap indicates the magnitude of the shared response, and nonoverlapping areas indicate the proportion of the response unique to the corresponding species. [Adapted from Geysen et al. (1988a).]

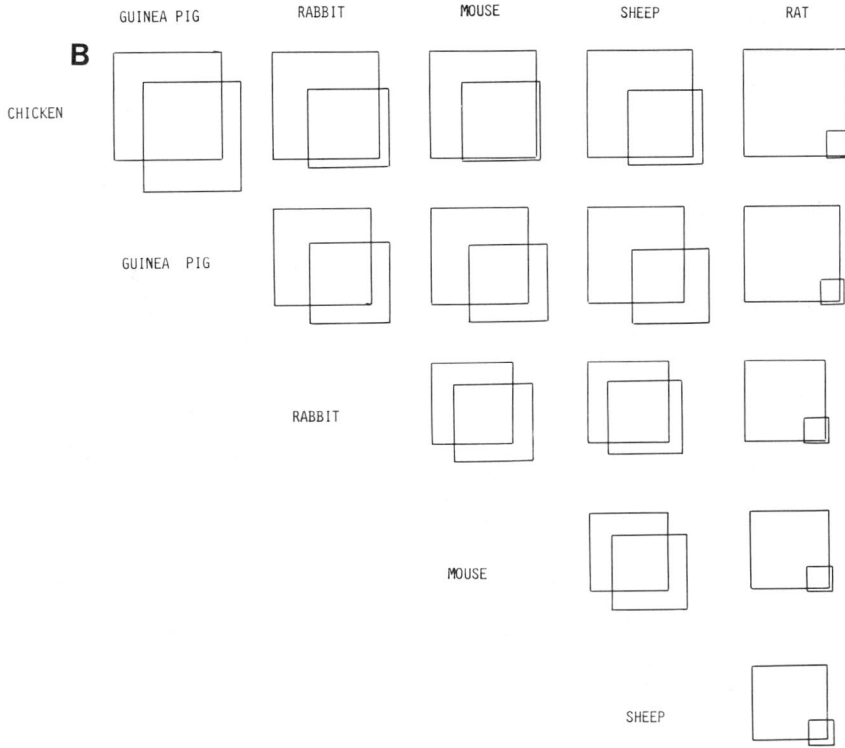

Fig. 11. (Continued)

and a hydrogen bond stabilizing the position of Tyr-8. An antibody positioned to simultaneously recognize the side chains of Glu-6, Val-9, and Tyr-8 (after rotation about the C_α–C_β bond to allow accessibility) would also contact Pro-7. This side chain is mostly replaceable, but substitution with large aromatic side chains (Phe, His, Trp, or Tyr) eliminates antibody binding, presumably by steric hindrance. This proposed mechanism of binding and induced fit for site 4–9 can proceed naturally from the process of antibody–antigen recognition of the highly exposed Glu-6, through disruption of the indirect salt-bridge between Glu-6 and Lys-100 and concomitant release of the bound water molecule, to expose Tyr-8 for antibody binding.

In site 90–95, solvent-exposed Asp-92 and Ala-93 and buried Asn-95 are critical to antibody recognition in all antisera examined. Solvent-exposed Lys-94 and buried Val-91 are also selected in specific antisera. The side chain of Asn-95 makes charged hydrogen bonds to both Asp-92 and Lys-94 within the same epitope and to Lys-83 in adjoining site

80–85. Antibody binding to Asn-95 (or to Lys-83) could thus destabilize both site 80–85 and site 90–95. The side chain of Asn-95 is buried underneath Asp-92 and Lys-94 and could become exposed upon antibody binding to exposed, essential Asp-92 and/or upon conformational changes in the Lys-94 side chain, which is highly mobile (based on its high-temperature factors) in the crystallographic structure. Thus, only relatively minor, predominantly side-chain movements of the antigen appear to be required to allow interaction with the antibody.

In the seven MHr sites analyzed (Getzoff et al., 1987), the environments and general structural properties associated with the microassemblies formed by critical residues suggested a consensus mechanism in which antibody binding to one or more solvent-exposed amino acid side chains may cause rearrangements that, at least partially, uncover previously buried side-chain atoms of the antigen. If such hypothesized rearrangements occur only for antibodies of low affinity identified by binding to peptides, then they do not represent the general, biologically important mode of antibody action. To test for this, antibody affinities for native MHr were measured by using antiMHr antibodies purified against peptides 3–16, 37–46, 63–72, and 73–82. The affinity constants for native MHr in solution fell in the range of 10^7–10^8 (Getzoff et al., 1987), which is about average for antiprotein antibodies. Based on these measured affinities and competition by native MHr in solution, the reactivities of the antibodies for the identified sites appear to be biologically relevant.

VI. Dominant Epitopes and Collective Response

Sections IV and V defined a systematic data base of the frequency of response in rabbits to sequential epitopes on MHr and characterized some of its implications for the chemistry of antibody–antigen interaction. To be sure that the observed frequency of response to MHr sites was not dictated only by the choice of animal, a comparable study (Tribbick et al., 1988) was done in five other species, as well as another group of rabbits. Figure 11A shows a summary of the responses for each group of animals: 6 rabbits, 6 sheep, 10 mice, 5 guinea pigs, 5 chickens, and 5 rats. For each species, the number of animals that responded to each hexapeptide homolog of the MHr sequence was determined.

Comparison of the overall response from the new set of rabbits with the original set (Geysen et al., 1987c) provides a basis for determining the relatedness of the responses obtained for two different species, i.e., are the differences among species greater than or comparable to the differences between two groups of the same species? The frequency response for each group of rabbits is normalized to give the fraction of the group responding to each hexapeptide. Clearly, if both groups

TABLE IV
COMPARISONS OF ANTIGENIC RESPONSES TO MHr IN DIFFERENT SPECIES

Species	In group	SNR[a]	Correlation coefficient				
			Guinea pig	Rabbit	Mouse	Sheep	Rat
Chicken	5	46.0	0.373	0.363	0.738	0.367	0.094
Guinea pig	5	38.0		0.244	0.285	0.114	0.166
Rabbit	6	27.0			0.358	0.562	0.200
Mouse	10	25.2				0.421	0.105
Sheep	6	22.5					0.125
Rat	5	2.4					

[a] Summed normalized response (SNR), the sum of the number of animals responding at each hexapeptide divided by the total number of animals tested.

produce an overall identical response to the antigen MHr, then (1) regression analysis will give a slope of 1.0, with a regression coefficient of 1.0, and (2) the summed normalized response (SNR) (average number of responses per animal; see Table IV) for each peptide for each group will be the same. What is observed, not surprisingly, is that the actual values obtained deviate from the ideal with a slope of 0.703 and a correlation coefficient of 0.654. This correlation coefficient value indicates that the two groups are related ($p \ll 0.01$), but are not identical. The SNRs are very close at 26.3 and 27 responses/animal, respectively. Although these two groups of rabbits were neither immunized with the same antigen preparation nor subjected to the same dose regimen, this level of difference can be used as the base line against which to compare groups of animals from two different species.

Table IV summarizes the correlation coefficients obtained from the comparison of each combination of the responses of two species and also indicates the number of animals tested and the SNR for each species. The magnitude of the overall response to sequential epitopes for each species varies over a wide range; chickens respond most often, averaging 46.0 responses/animal, and rats respond least often, averaging 2.4 responses/animal. Two species, mouse and sheep, each give a SNR comparable to that of the rabbit. Responses of neither mice nor sheep, however, correlate as well with the rabbits' responses as responses of the previously tested rabbits (correlation coefficient = 0.654). This suggests, as expected, that the differences between species are greater than the differences between groups of animals within the same species. Two species (chicken and guinea pig) give a significantly greater SNR (Table IV) to the MHr antigen than rabbits do, and rats give a significantly poorer response than any other species tested. The low values of the correlation coefficients of the rats' response with those of other species indicate that this difference is not just a matter of magnitude, i.e., the same normalized profile, but at a higher level. In contrast, the correlation coefficient (0.738) between the responses of chickens and mice indicates that these are at least as similar as those for the two groups of rabbits, although the SNR for chickens (46.0) is almost twice that for mice (25.2).

Another way of comparing the responses for the different species is to determine the degree of overlap in the SNR for each peptide and to treat the nonoverlap as belonging to the species giving the higher response. Figure 11B shows the comparisons between species in terms of the degree of relatedness between the responses of pairs of species. This again highlights the observation that the responses of mice correlate well with those of chickens (despite the latter having a greater magnitude), as the mouse response is almost a total subset of the chicken response.

The collective responses of different species to sequential epitopes differ both in absolute magnitude and in the extent to which well-recognized epitopes by one species are "silent" in another. These differences present difficulties in defining and understanding the intrinsic properties of the antigen that bias the immune response. Whatever the origin of the differences observed in the responses between species to the same antigen, they are not all antigen derived. Thus, the criterion of the nonmammalian origin of the antigen is insufficient to obtain a completely unbiased response to a protein antigen from a single species. Nor does the use of outbred animals result in a collective response limited solely by the intrinsic properties of the protein antigen. Therefore, relationships between these intrinsic properties and antigenicity, determined from an analysis of data restricted in its species diversity, may not define a species-independent relationship. Thus, it may be important to define a collective response to a biologically important antigen, as for example in the design of vaccines (see Section VIII).

The picture is further complicated when the response is obtained from groups of animals for which the same antigen is given in conjunction with different adjuvants, e.g., aluminum hydroxide, Freund's complete, etc., or different carriers, e.g., keyhole limpet hemocyanin, tetanus toxoid, diphtheria toxoid, etc. Although it is difficult to identify with confidence *all* of the antigenic sites of a protein antigen, it is possible to address the reverse issue, that is, are some regions of the protein antigen uniformly negative with respect to immune recognition?

For the five species with significant responses (thus excluding rats), the proportion of hexapeptides that failed to bind antibody was as follows: rabbit 36%, chicken 27%, guinea pig 43%, mouse 26%, and sheep 53%. If these "negative" peptides had occurred at random, an estimated maximum of one hexapeptide would be negative for all five species simultaneously. The data, in fact, indicate 11 peptides that failed to bind antibodies from any species: this is highly significant ($p < 0.01$) in a 2×2 contingency test. This suggests that the minima in each profile (centered on residues 26, 74, and 105) result from properties intrinsic to the structural chemistry of MHr, and it appears to validate the concept of immunodominant regions.

VII. Evaluation of Predictive Methods

With the ready accessibility of protein sequences via direct translation from DNA, a reliable method for predicting the most likely sites of antigenic epitopes would be of immense value. Clearly, from the primary structure (sequence) alone, prediction of antigenic sites is essentially limited to sequential determinants (as defined by Benjamin *et al.*, 1984) that are now recognized as often being primarily portions of conformational sites (Amit *et al.*, 1986; Sheriff *et al.*, 1987b; Colman *et al.*, 1987a; Geysen *et al.*, 1987c). The first step toward any prediction, based on intrinsic parameters of the antigen, is to identify those parameters that can bias the immune response. Despite progress in defining the chemistry of antigenic recognition, the application of various parameters to prediction from amino acid sequence data has been difficult, in part due to the three-dimensional nature of protein epitopes. Thus, although in widespread use, predictive methods are generally viewed with some skepticism. In this section, we apply objective, statistical analysis to evaluate three methods commonly in use for predicting sequential epitopes from amino acid sequence data, as based on experimental results from the responses of rabbits to MHr, TMV coat protein, and sperm-whale myoglobin.

A. Proposed Predictions from Sequence Data

Intrinsic hydrophilicity parameters for individual amino acids form the basis for the most popular method of prediction to date. This method, proposed by Hopp and Woods (1981, 1983), is essentially based on three postulates: (1) antibodies bind to any combination of amino acids numbering in the range of 5–8 residues, (2) epitopes are physically placed at the surface of protein antigens, and (3) the more hydrophilic stretches of amino acids are likely to be located at the surface of a folded protein. By assigning a hydrophilicity parameter to each amino acid and by averaging the individual values over all linear stretches of, say, six residues homologous with the primary sequence of a protein antigen, the highest values were suggested to correspond to the site of sequential epitopes.

Using algorithms similar to that of Hopp and Woods (1981, 1983), Karplus and Schulz (1985) and Welling et al. (1985) proposed alternative predictive procedures based, respectively, on predicted main-chain flexibility parameters and calculated antigenicity values for each amino acid. Main-chain flexibility parameters (Karplus and Schulz, 1985) were calculated as the average, normalized crystallographic temperatures factors (B values) of the α-carbon atoms for each amino acid derived from 31 proteins, whose structures had been determined at sufficient resolution. The predicted relative flexiibility at residue position n of a given amino acid sequence was taken as the weighted sum of the neighbor-correlated normalized B values for the amino acids at positions $n - 3$ to $n + 3$, using the weights 0.25, 0.50, 0.75, 1.00, 0.75, 0.50, and 0.025, respectively. Antigenicity values (Welling et al., 1985) were derived by analyzing antigenic determinants from a number of proteins with respect to their amino acid compositions and by comparing this with the overall amino acid compositions of the proteins themselves. For each amino acid, an antigenicity value was defined as the logarithm of the relative occurrence, which is the ratio of the percentage of occurrence in the set of antigenic determinants to the percentage of occurrence in the protein antigens.

B. Criteria for Assessing Prediction Success

To evaluate the effectiveness of an algorithm in predicting the position of sequential epitopes within the polypeptide chain, it is necessary to set defined criteria against which the success or failure of the prediction can be assessed. First, because of the inherent variation in responses of outbred members of a single species to the same antigen, a minimum proportion of animals necessary for a successful response must be defined. Second, the length (number of residues) of the peptide must be chosen prior to the evaluation of success. A trivial, but completely successful, prediction can be made in choosing a length equal to the number of residues of the protein itself. The chosen length becomes a compromise between the minimal number of residues that equate to a sequential epitope and the greater likelihood of success as the length is increased. In addition, the comparative ease of synthesizing shorter peptides, rather than longer peptides, is a consideration. Ideally, a predictive algorithm should be able to predict the actual number of residues constituting the sequential portion of that epitope, i.e., about seven residues. In practice, a more realistic length is probably about 15 residues. A peptide of this length is not too difficult to synthesize and allows for some "buffer" residues adjacent to the epitope, which minimizes the end effects introduced by extracting a short stretch of residues from a much longer sequence. Longer peptides may have more chance of having a conformation similar to the corresponding portion of the protein antigen (Van Regenmortel et al., 1988; Van Regenmortel, 1987; Moudallal et al., 1985). However, even shorter peptides may have a conformational component when coupled to a carrier (Fieser et al., 1987), and there is also the possibility that longer peptides and proteolytic fragments will assume a conformational preference distinct from that of the native protein antigen.

Table V summarizes the proportion of peptides 6–20 residues in length that encompass 1 or more sequential MHr epitopes and are recognized by a designated minimal number of the 7 rabbit antisera used in the evaluation. Clearly, with a low success threshold (in terms of the number of animals that are expected to respond) and a longer peptide length, the chance occurrence of an epitope is already high. All peptides 11–20 residues in length contain a sequential MHr epitope recognized by at least one of the 7 rabbits. Furthermore, a 50% chance of finding an epitope is achieved for more than 1 responding rabbit at a peptide length of 7 residues, more than 2 rabbits at 9 residues, more than 3 rabbits at 14 residues, and so on. We have opted to assess the three predictive algorithms using a higher threshold of greater than or equal to half of the experimental animal population. Epitopes of the antigen myoglobin, which are recognized so infrequently that this threshold would rarely be attained, are treated differently. The threshold for

TABLE V
PREDICTION OF MHr EPITOPES AS A FUNCTION OF
PEPTIDE LENGTH AND PROPORTION OF RESPONDING RABBITS

Peptide length	Number of responding rabbits (of seven tested)					
	1–6	2–6	3–6	4–6	5–6	6
6	68.1	40.7	25.6	15.0	9.7	3.5
7	82.1	50.0	35.7	21.4	16.0	7.1
8	89.1	59.4	44.1	27.0	21.6	10.8
9	94.5	63.6	50.0	32.7	26.3	14.5
10	98.1	67.8	55.0	36.6	30.2	17.4
11	100.0	72.2	59.2	40.7	34.2	20.3
12	100.0	76.6	63.5	44.8	38.3	23.3
13	100.0	81.1	67.9	48.1	42.4	26.4
14	100.0	83.8	72.3	50.4	45.7	29.5
15	100.0	86.5	76.9	52.8	48.0	32.6
16	100.0	89.3	81.5	55.3	50.4	35.9
17	100.0	92.1	85.2	57.8	52.9	38.2
18	100.0	95.0	89.1	59.4	55.4	40.5
19	100.0	97.0	92.0	61.0	57.0	43.0
20	100.0	98.9	94.9	62.6	58.6	45.4

myoglobin is set to greater than or equal to two animals, half the maximal number of four responding animals observed for any epitope. The success of the prediction algorithms was assessed at two peptide lengths, 7 and 15 residues. The former approximates the mean number of residues, comprising a sequential epitope (5–8), and the latter approximates the maximal peptide length that is easily and routinely synthesized and purified in the average laboratory.

Using antisera from rabbits immunized with one of three different antigens, MHr (Fig. 12), TMV coat protein (Fig. 13), and sperm-whale myoglobin (Fig. 14), and systematically testing all possible hexapeptides as described above (Geysen et al., 1987c; Mason and Geysen, 1988), each of the three algorithms (predicted hydrophilicity, predicted flexibility, and predicted antigenicity) was applied (Figs. 12–14), and its success evaluated (Figs. 15–17). Predictive algorithms were run on the sequence of each antigen, and the peaks (highest values) were rank ordered, highest to lowest. Peaks within three residues of each other were reduced to the single highest peak, and peaks within four residues of the carboxy-terminal end of the protein were deleted (a consequence of using hexapeptides to assess antibody response). A further reduction in the number of peaks was made by ignoring local maxima that had prediction values less than the mean. This resulted in between 5 and 14 peaks for each algorithm applied to each protein.

The proportion of rabbit antisera responding to each hexapeptide of each protein was then used as the basis for evaluation of Monte Carlo trials, determining the probability of finding an epitope for a given protein antigen in a randomly chosen peptide of a given length. The mean chance of success and the corresponding standard deviation determined for 1000 trials were compared with predictions made using each algorithm (Figs. 15–17). Because the success rates of the algorithmic predictions are low (see details below), the figures show the random prediction space within plus or minus one standard deviation, rather than within two standard deviations. Thus, the stringency of the test for success is rather low, corresponding to about two of three cases.

Figures 15–17 show that none of the three algorithms predicts better than random at a peptide length of seven residues. In each case, the first prediction (corresponding to the peak with the highest value) fails to find an epitope to which four or more of seven rabbits respond. At a peptide length of 15 residues, the predictions are somewhat better, but are still, in general, within the one standard deviation boundaries of a random selection procedure. The lack of correlation between the factors assigned to each amino acid in the three methods also indicates that one or more of the algorithms must fail to predict the location of sequential epitopes.

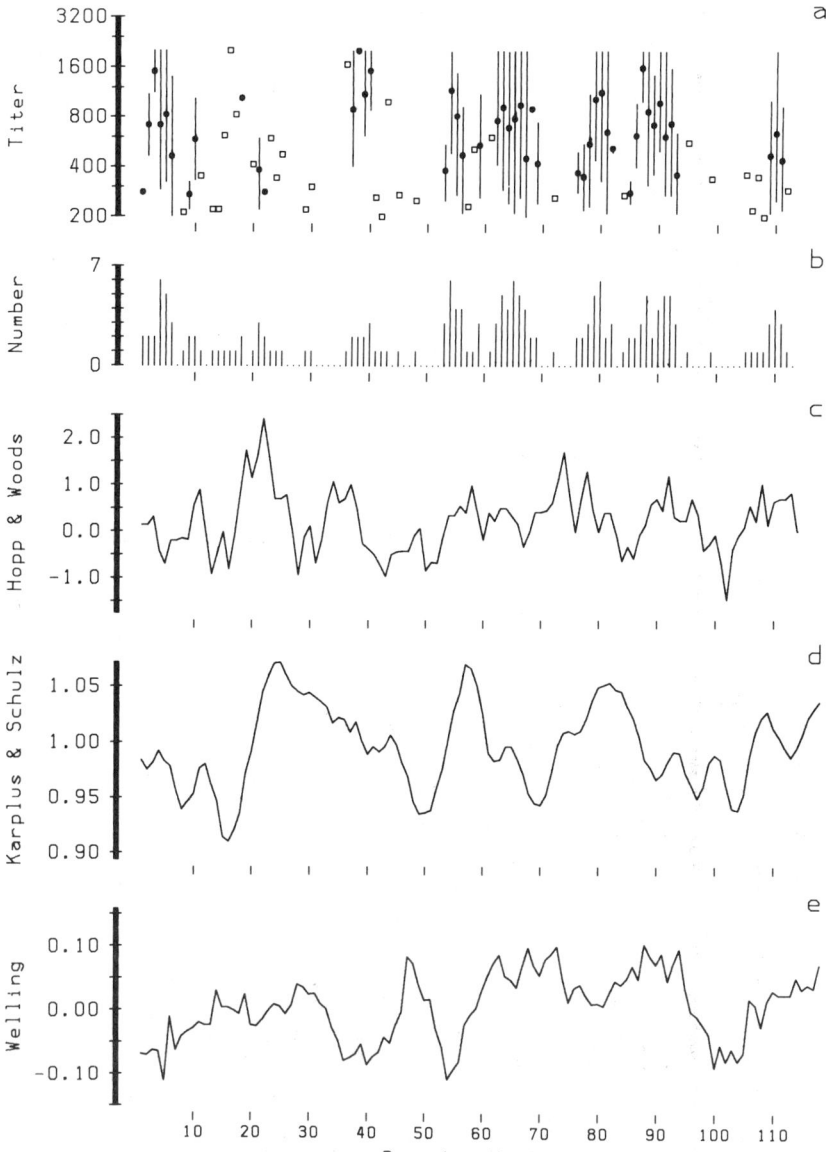

FIG. 12. Observed and predicted epitopes on MHr. (a) Titers for individual rabbit sera (open squares), geometric mean titers (closed circles), and range of individual titers (vertical lines) are plotted at the N-terminal residue of each hexapeptide. (b) Number of animals found with serum titers > 200 for each MHr hexapeptide. (c) Hydrophilicity profile as per Hopp and Woods (1983) with the average hydrophilicity calculated using a window of six residues and plotted at the N-terminal residue. (d) Flexibility profile as per Karplus and Schulz (1985) derived using a window of seven residues (reduced to 6, 5, and 4 for the three N- and C-terminal residues) and plotted above the center residue. (e) Antigenicity profile as per Welling et al. (1985) derived as for the flexibility profile.

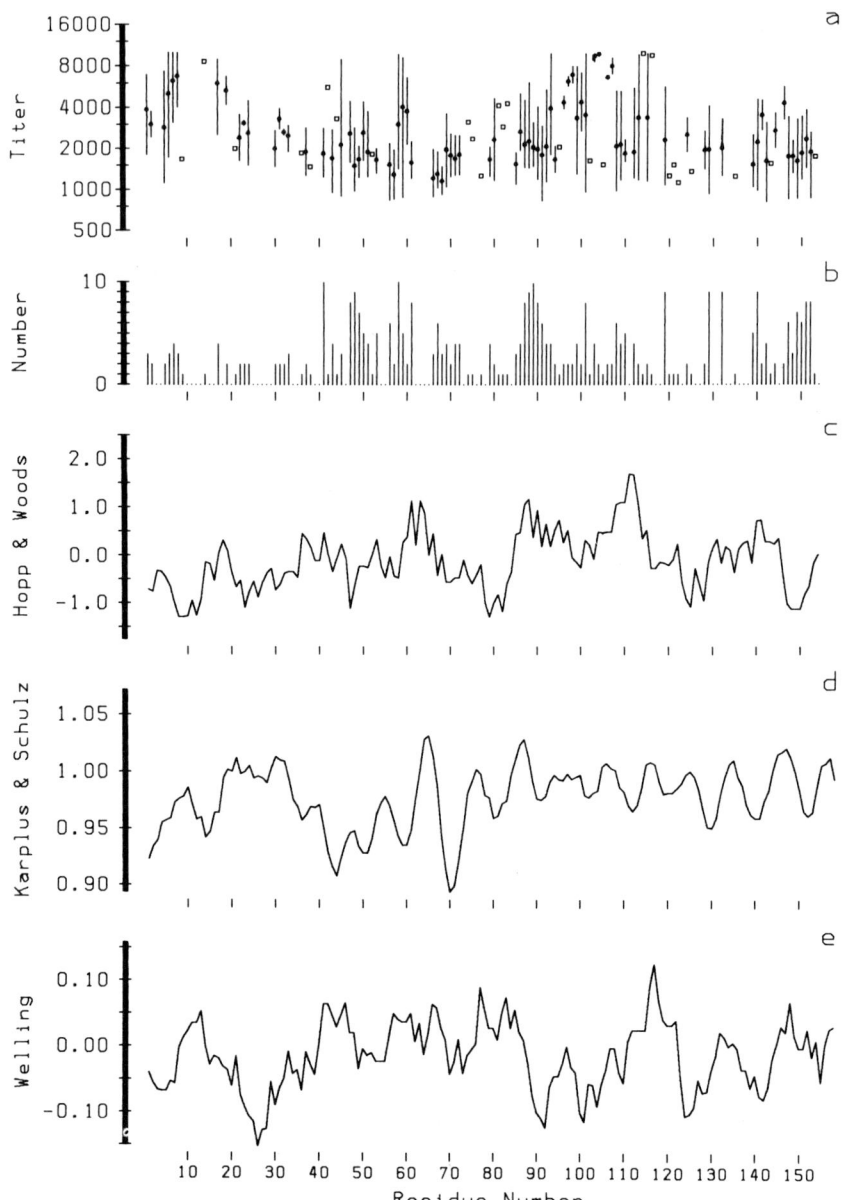

FIG. 13. Observed and predicted epitopes on TMV coat protein. (a) Titers for individual rabbit sera (open squares), geometric mean titers (closed circles), and range of individual titers (vertical lines) are plotted at the N-terminal residue of each hexapeptide. (b) Number of animals found with serum titers > 200 for each hexapeptide. (c) Hydrophilicity profile as per Hopp and Woods (1983) with the average hydrophilicity calculated using a window of six residues and plotted at the N-terminal residue. (d) Flexibility profile as per Karplus and Schulz (1985) derived using a window of seven residues (reduced to 6, 5, and 4 for the three N- and C-terminal residues) and plotted above the center residue. (e) Antigenicity profile as per Welling et al. (1985) derived as for the flexibility profile.

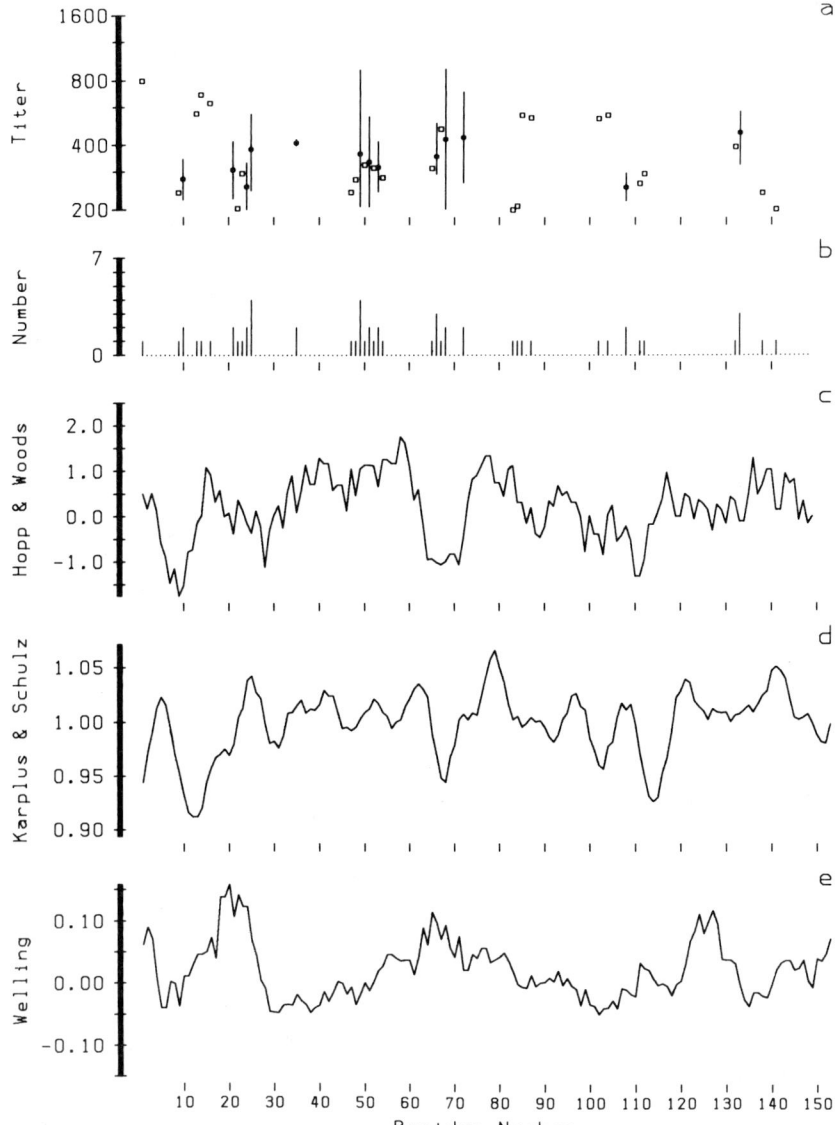

FIG. 14. Observed and predicted epitopes on sperm-whale myoglobin. (a) Titers for individual rabbit sera (open squares), geometric mean titers (closed circles), and range of individual titers (vertical lines) are plotted at the N-terminal residue of each hexapeptide. (b) Number of animals found with serum titers > 200 for each hexapeptide. (c) Hydrophilicity profile as per Hopp and Woods (1983) with the average hydrophilicity calculated using a window of six residues and plotted at the N-terminal residue. (d) Flexibility profile as per Karplus and Schulz (1985) derived using a window of seven residues (reduced to 6, 5, and 4 for the three N- and C-terminal residues) and plotted above the center residue. (e) Antigenicity profile as per Welling et al. (1985) derived as for the flexibility profile.

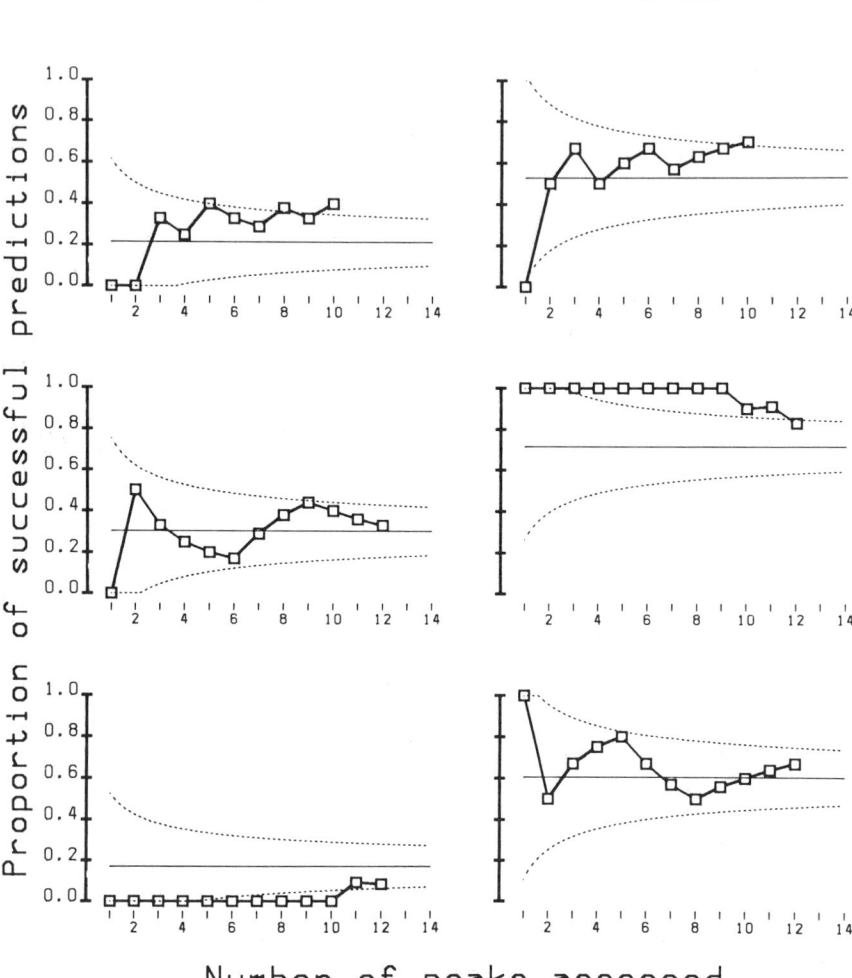

FIG. 15. Evaluation of epitope prediction from hydrophilicity values. The success with which the hydrophilicity algorithm of Hopp and Woods (1983) identifies peptides containing one or more sequential epitopes is evaluated for 7-residue (left) and 15-residue peptides of MHr (top panel), TMV coat protein (middle panel), and myoglobin (bottom panel). The horizontal axis gives the number of hydrophilicity peaks assessed in rank order, according to the values obtained from the prediction algorithm. The vertical axis gives the proportion of predicted peptides found to contain one or more of the experimentally identified sequential epitopes. The horizontal line indicates the number of peptides of the specified length determined to contain one or more sequential epitopes. Broken lines indicate the one standard deviation boundaries for a random selection of peptides, as determined from 1000 Monte Carlo trials (see the text).

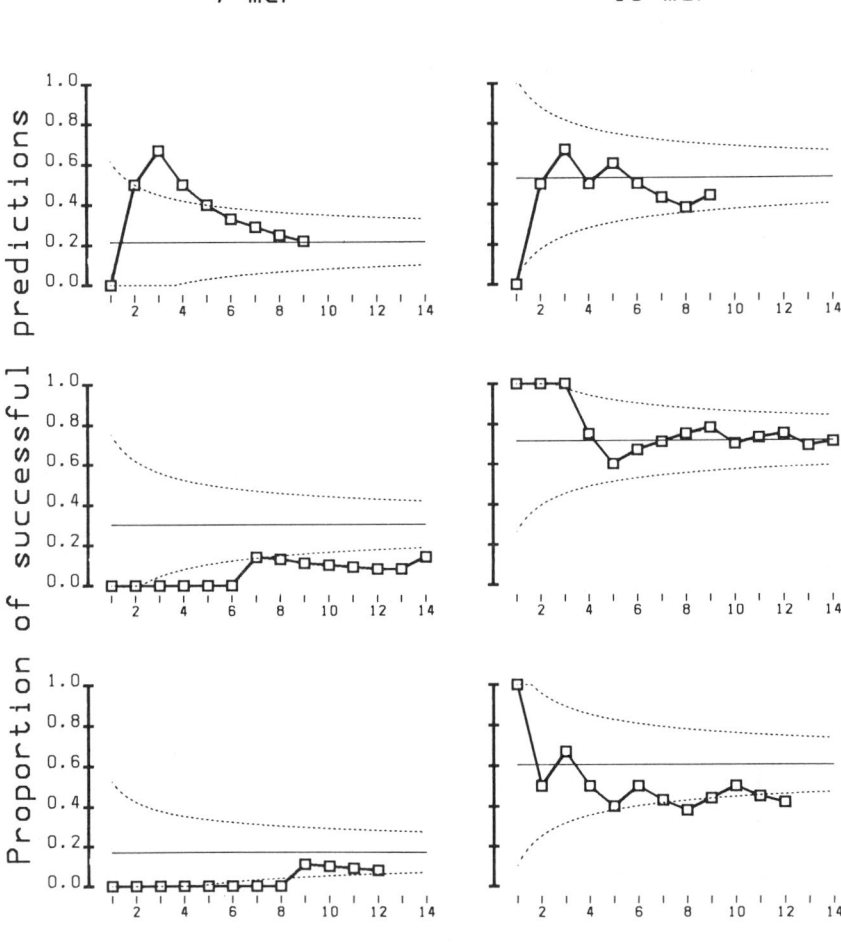

FIG. 16. Evaluation of epitope prediction from flexibility values. The success with which the flexibility algorithm of Karplus and Schulz (1985) identifies peptides containing one or more sequential epitopes is evaluated for 7-residue (left) and 15-residue peptides of MHr (top panel), TMV coat protein (middle panel), and myoglobin (bottom panel). The horizontal axis gives the number of flexibility peaks assessed in rank order, according to the values obtained from the prediction algorithm. The vertical axis gives the proportion of predicted peptides found to contain one or more of the experimentally identified sequential epitopes. The horizontal line indicates the number of peptides of the specified length determined to contain one or more sequential epitopes. Broken lines indicate the one standard deviation boundaries for a random selection of peptides, as determined from 1000 Monte Carlo trials (see the text).

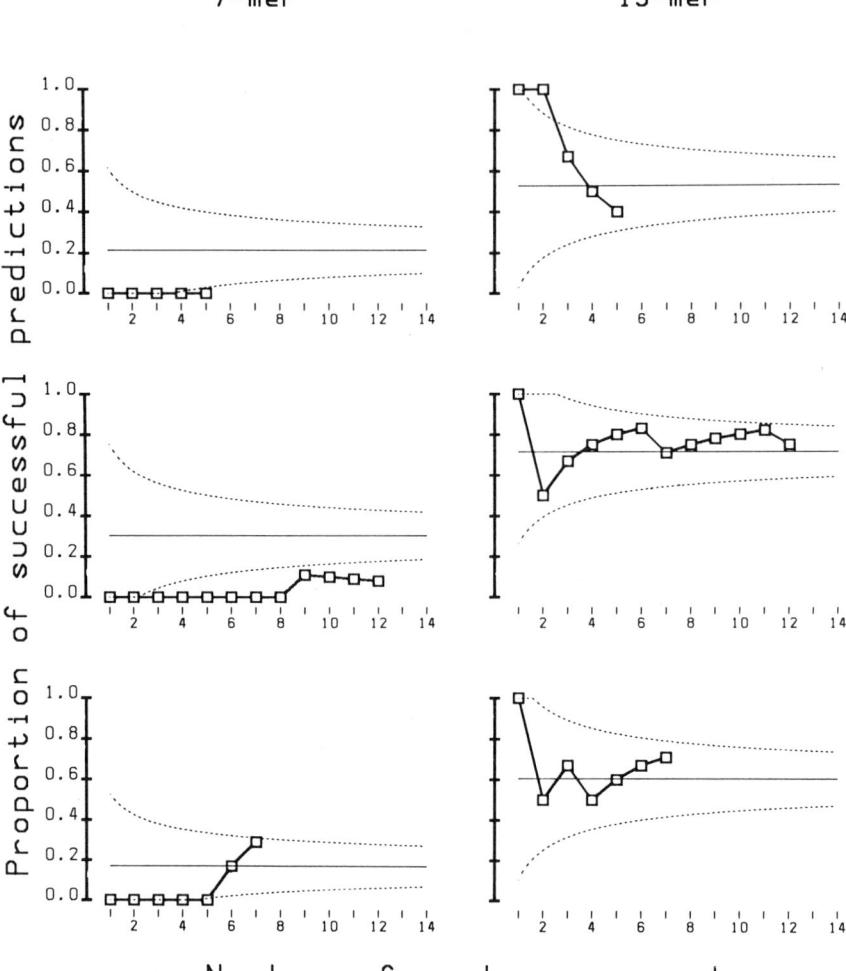

FIG. 17. Evaluation of epitope prediction from antigenicity values. The success with which the antigenicity algorithm of Welling et al. (1985) identifies peptides containing one or more sequential epitopes is evaluated for 7-residue (left) and 15-residue peptides of MHr (top panel), TMV coat protein (middle panel), and myoglobin (bottom panel). The horizontal axis gives the number of antigenicity peaks assessed in rank order, according to the values obtained from the prediction algorithm. The vertical axis gives the proportion of predicted peptides found to contain one or more of the experimentally identified sequential epitopes. The horizontal line indicates the number of peptides of the specified length determined to contain one or more sequential epitopes. Broken lines indicate the one standard deviation boundaries for a random selection of peptides, as determined from 1000 Monte Carlo trials (see the text).

From the foregoing analysis, it is apparent that the intrinsic parameters of the antigen, chosen for the predictive antigenicity algorithms described above, are insufficient to predict the immune response, at least when evaluated based on a single value for each amino acid type regardless of its environment. In the future, it may be possible to predict parameters, such as shape or mobility, from the amino acid sequence. Currently, such predictions, like those for the prediction of secondary and tertiary structure, are not reliable. The complete experimental technology exists for identifying the sequential epitopes in a chosen protein using peptide probes, thus bypassing the requirement for prediction (see Section VIII). For the determination of antigenic sites on proteins that are critical to vaccine design, the amount of work required is not unreasonable, given the social and economic importance of safe vaccines. Finally, despite the inaccuracy of current methods of prediction, there is a continuing need to refine predictive tools based on experimental results.

VIII. An Experimental Approach to the Design of Antipeptide Antibodies

One potential, practical application, resulting from the definition of intrinsic stereochemical influences on antigenic reactivity, would be the design of antipeptide antibodies for protein detection (Sutcliffe et al., 1980; Lerner et al., 1981b), diagnostics, and vaccines. Protein detection and diagnostics using peptides are now well established, with recent applications to the autoimmune deficiency syndrome (AIDS) viruses serving as an example (Gnann et al., 1987; Norrby et al., 1987). On the other hand, despite enormous advances in methodology and theoretical understanding in biochemistry, molecular biology, protein structure, and immunology, surprisingly little fundamental progress in vaccine design has been made since the time of Pasteur. Further advances will require an informed approach based on experimental results and a detailed knowledge of the stereochemical basis for protein antigenicity at a molecular level.

Prominent among the promising applications of synthetic peptide technology in biology and medicine (Lerner, 1982) is the use of peptide epitopes as vaccines (Shinnick et al., 1983; Lerner, 1983). The chosen peptide epitopes are homologs of a protein antigen from a pathogenic organism against which a vaccine is desired. These peptide epitopes have potential for vaccines, because they can generate antibodies that act to neutralize viral antigens (such as rhinovirus, foot-and-mouth disease, influenza A, hepatitis B, rabies), bacteria (such as *Streptococcus pyogenes* and *Escherichia coli*), and bacterial toxins (such as cholera and diptheria toxins) (Schmidt *et al.*, 1988; McCray and Werner, 1987; Sutcliffe *et al.*, 1983; Jacob *et al.*, 1983; Gerin *et al.*, 1983; Müller *et al.*, 1982; Prince *et al.*, 1982; Bittle *et al.*, 1982; Audibert *et al.*, 1981; Beachey *et al.*, 1981; Lerner *et al.*, 1981a). Because T cell epitopes are all peptides whose structure, possibly often amphipathic helix, is induced by interactions with helper T cell receptors and Ia antigens (Berzofsky *et al.*, 1987), synthetic peptides can represent T cell epitopes in combination with B cell epitopes. Thus, the design of peptide epitope vaccines is now being improved by identifying and including peptide sequences corresponding to defined helper T cell epitopes (Francis *et al.*, 1987; Cease *et al.*, 1987; Good *et al.*, 1987).

Of fundamental importance in designing new peptide epitopes for antibody production is the faithfulness or fidelity of the molecular mimicry of the protein antigenic site. Antibody–antigen interactions are modulated by their large surface areas, and a complete antigenic site often entails contact points that are close in space, but remote in sequence. Depending on the antigenic site, two components may contribute to the degree of fidelity: critical residues that provide dominant contributions to the binding energy, and contact residues whose mildly complementary interactions support the overall association constant for the formation of a particular immune complex. This latter component may also play an important secondary role in stabilizing the structural environment required for complete antigenic mimicry. To disentangle these possible effects and ultimately develop a functional peptide vaccine, the structural chemistry of antigenic recognition and the mechanisms of interaction, including any inducible complementarity between antibody and antigen, must be fully understood. Systematic studies on protein antigens such as MHr (Fieser *et al.*, 1987; Getzoff *et al.*, 1987; Geysen *et al.*, 1987c) have produced a data base concerning the relationship between antigenicity and protein structure that may ultimately increase our understanding of the antigenicity of pathogens and thus aid the design of effective vaccines. Currently, the design of antipeptide antibodies benefits from observed stereochemical biases of protein antigenicity, but remains a complex problem (Van Eldik and Lukas, 1987).

In a systematic evaluation of the influence of macromolecular stereochemistry on the reactivity of antipeptide antibodies with the native protein antigen (Tainer *et al.*, 1984), peptide homologs of MHr were selected based on their mobility in the parent protein, as determined from X-ray crystallographic studies (Sheriff *et al.*, 1987a). The 12 peptides synthesized included 83 of MHrs 118 residues (70% of the sequence) and involved both helical and nonhelical regions. All the peptides were chosen to have large areas of exposed molecular surface available in the context of the protein for interaction with antibodies. This study provided specific experimental evidence that the reactivity of antipeptide antibodies with the intact protein antigen was biased by local mobility. All antipeptide antisera raised against peptides representing relatively ordered surface areas (corresponding to MHr amino acid sequence ranges 22–35, 26–35, 96–109, and 100–109) showed negative or low reactivity with the protein antigen, whereas antipeptide antisera raised against peptides from relatively mobile areas showed higher reactivity against the protein antigen (Tainer et al., 1984). These results were interpreted to include the existence of induced fit for interactions of protein antigens with antipeptide antibodies. In addition, corresponding studies of antiprotein antibodies (Geysen et al., 1987c; Tainer et al., 1985a; Westhof et al., 1984; Moore and Williams, 1980) suggested that local mobility and, by implication, induced fit might be a general bias for frequent antigenic recognition. As discussed in Section III, the data from X-ray diffraction studies appear to agree with results of peptide mapping, and both are compatible with this inference, as are the more comprehensive studies on polyclonal antiprotein antibodies for MHr, myoglobin, and TMV coat protein.

The relationship between local mobility and antigenicity should apply to the design of monoclonal antipeptide antibodies that react with reasonable affinity with the native protein antigen. The studies on MHr led to the supposition that peptide homologs should be selected to be individual structural elements with average or above-average local mobility. Thus, the MHr C-helix, which represents the most mobile helical structural element, was selected. Monoclonal antibodies against MHr were isolated and screened for MHr and C-helix peptide binding, following immunization of mice with either the whole protein antigen or a peptide homolog of the C-helix (Fieser et al., 1987). The epitope specificities of one anti-MHr and six antipeptide monoclonal antibodies were examined in detail. Three monoclonals all recognized the amino acid sequence EVVPH (Table VI) at the amino end of the C-helix. Two antipeptide and one antiprotein monoclonals recognized the amino acid sequence DFLEKI at the carboxy end of the C-helix, showing that antipeptide and antiprotein antibodies could recognize the same site. This result is consistent with the observed correspondence of sites for antipeptide (Tainer et al., 1984) and antiprotein (Geysen et al., 1987c) polyclonal antibodies.

The critical residues for antigen binding were then identified, and the relative affinities of the monoclonals for intact MHr, apo-MHr, and

TABLE VI

RELATIVE AFFINITIES OF PROTEIN AND PEPTIDE-INDUCED MONOCLONAL ANTI-MHr ANTIBODIES FOR MHr, Apo-MHr, AND MHr PEPTIDES[a]

mAb	Fusion	Immunogen	Site recognized	$K_a \times 10^{-6}$ (versus MHr)	Picomoles of inhibitor at 50% inhibition			
					MHr	Apo-MHr	Peptide 1[b]	Peptide 2[c]
C	B13	Peptide–KLH	69–73[d]	0.71	70.84	3.22	340.50	160.33
I	B13	Peptide–KLH	69–73	5.40	11.61	0.14	65.75	79.10
L	B13	Peptide–KLH	69–73	1.65	21.95	1.26	39.48	13.62
A	B13	Peptide–KLH	79–84[e]	20.4	6.75	1.41	1.12	ND
F	B13	Peptide–KLH	79–84	18.6	7.83	0.20	8.85	ND
G	G4	Peptide–KLH	86–87[f]	0.25	10.28	1.40	2.47	ND
M	M3	Protein	79–84	1.58	28.50	6.73	2.06	ND

[a] The antigenic sites recognized by the mAbs were determined by omission-series and replacement-series ELISAs. Association constants (K_a) for MHr were determined by an inhibition radioimmunoassay (RIA) in which radiolabeled antibody was incubated with solid-phase MHr and various concentrations of solution-phase MHr. The ratio of free to bound antibody was plotted against concentration of soluble MHr, and the association constant was estimated as the negative inverse of this plot. ND, Not determined. [Table adapted from Fieser et al. (1987).]
[b] Residues 69–87: Glu-Val-Val-Pro-His-Lys-Lys-Met-His-Lys-Asp-Phe-Leu-Glu-Lys-Ile-Gly-Gly-Leu (in one letter code: EVVPHKKMHKDFLEKIGGL).
[c] Residues 69–85 plus an amino-terminal glycine added as a linker (the native sequence has a serine at position 68).
[d] Glu-Val-Val-Pro-His (EVVPH).
[e] Asp-Phe-Leu-Glu-Lys-Ile (DFLEKI).
[f] Gly-Leu (GL).

peptides determined (Fieser *et al.*, 1987). These three antigens differ in side-chain mobility and exposure. Loss of the iron atoms in apo-MHr allows relaxation of the native structure around the iron positions, but not complete unfolding, as indicated by circular dichroism, which showed retention of some α-helical secondary structure in the apoprotein. Peptide homologs of the C-helix provide the most conformational freedom of the epitopes and the greatest solvent exposure of residues normally buried in the protein. All of the antibodies bound apoprotein better than native protein, indicating that relaxation of the native structure by removal of the iron center increased antibody affinity for MHr (Table VI).

Whereas increased epitope flexibility may enhance antibody binding to protein antigens, not all of the antibodies examined had maximal affinities for peptide antigens, which are the most flexible form of each epitope. This suggests that increased antigen flexibility is not sufficient to maximize antibody binding. Four antipeptide monoclonals bound the MHr antigen equally well or better than the peptide. Thus, no preference for peptide over protein was observed for these antipeptide antibodies. Only one antipeptide antibody had a strong preference for peptide over native protein, and this antibody recognized the charged N-terminus of the peptide immunogen.

One antiprotein and two antipeptide antibodies (reactive with site DFLEKI) bound peptide with affinities greater than or equal to those for intact protein. However, two other antibodies (sequential epitope EVVPH) bound peptide with lower affinity than they bound MHr protein, even though these antibodies were induced with peptide. This difference may relate in part to secondary structural differences between the two epitopes; NMR spectroscopy studies have shown that the sequential epitope DFLEKI has a propensity for nascent helix formation in the free peptide in aqueous solution, but no particular structure was indicated for sequential epitope EVVPH (Dyson *et al.*, 1988).

The range of conformations adopted by immunogenic peptides may be restricted, possibly by interactions between the peptide and its protein carrier (Bahraoui *et al.*, 1986; Dyrberg and Oldstone, 1986). Also, certain free peptides can adopt conformational preferences in aqueous solution (Dyson *et al.*, 1985; Kim and Baldwin, 1984; Schulze-Gahmen *et al.*, 1985). Thus, the data on MHr C-helix monoclonals suggest that antibody binding to peptides can be influenced by the peptides' ability to fold into a particular conformation at the epitopes, either alone or via carrier influences, and supports the idea that the selection of individual structural elements may be preferable to randomly chosen sequence regions.

Critical residues were identified for distinct sequential epitopes within the MHr C-helix (Fieser et al., 1987). Sequential epitopes mapped at positions 69–73 (EVVPH) and 79–84 (DFLEKI) included residues from both the hydrophilic and the hydrophobic faces of the helix. The role of each amino acid in formation of the antigen–antibody interface was evaluated for epitopes 69–73 and 79–84, by measuring antibody binding to a set of peptide analogs in which each amino acid of the sequential epitope was replaced by each of the other 19 amino acids. Due to the amphipathic nature of the C-helix in native MHr, both these antigenic sequences contained buried hydrophobic residues. For each of these sequential epitopes, both exposed and buried residues were irreplaceable by any other amino acid without abrogating antibody binding, suggesting their direct involvement in formation of the antigen–antibody interface. One can argue that certain buried residues are seen as critical, because they are required for maintenance of an appropriate peptide structure, rather than for antibody contact. Although this explanation cannot be ruled out in all cases, it is unlikely to hold true when conservative changes still disallow antibody binding. For example, replacement of Phe-80 with tyrosine or Leu-81 with isoleucine is not allowed for binding of one monoclonal antibody to sequential epitope DFLEKI (Table VI).

Taken together, the results from studies on MHr suggest that the immunological cross-reactivity between protein antigens and homologous peptides depends on a balance between local mobility in the protein and the propensity toward particular secondary structure(s) in the peptide at the epitope location. The secondary structure(s) of the peptide that is recognized by the antibodies need not necessarily be the same as these secondary structures in the native protein and would, more likely, be intermediate to the native conformation and a disordered peptide. The apo form of MHr could represent such an intermediate structure, leading to higher reactivity of the monoclonals with this form of the antigen. A notable implication of these findings is that the protein-reactive antipeptide antibodies that recognize these intermediate structures could be important in identifying protein folding nuclei (Dyson et al., 1985; 1988; Wetlaufer, 1981; Anfinsen and Scheraga, 1975; Shrake and Rupley, 1973). The definitive test of these theories, however, will require determination, by X-ray crystallography and/or NMR spectroscopy, of the extent to which the antibodies can affect the structure of the protein and peptide antigens.

The fact that antiprotein and antipeptide antibodies recognize similar MHr sequential epitopes (Tainer et al., 1984; Geysen et al., 1987c; Fieser et al., 1987) suggests that peptide mapping of antiprotein antibodies identifies sites that are also appropriate for inducing antipeptide antibody reactivity with the native protein antigen. A reasonable conclusion, then, is that prime peptide candidates for the induction of antipeptide antibodies reactive with the native protein are individual structural elements with relatively high mobility, yet sufficient integrity, to have some local folding capability. A common example of such structural elements would be loop regions in which local side-chain interactions provide the primary folding interactions, and the shape and location allow local mobility.

IX. Frequency Data for Individual Amino Acids in Protein Epitopes

A. Epitope Level

The potential diversity of the immune repertoire, in terms of the number of different antibody molecules that can be generated, has been estimated at about 10^7 (Max, 1984). This in itself, however, does not imply that each possible antibody is functionally distinct. Nor is it axiomatic that the inherent likelihood of inducing any particular member of the repertoire is the same as that for any other member. It is well-recognized that antibody production in response to a protein antigen is not equally distributed over all the possible epitopes, but is "focused" on a subset of these. Epitopes in this latter category are usually designated as the immunodominant determinants of the particular protein antigen.

It is useful to imagine that induction of each particular antibody is associated with a probability factor that is a function of several variables. In Section IV, we have dealt with several possible stereochemical variables intrinsic to the protein antigen, for example, local mobility of the antigen at the site of the epitope, protrusion of the epitope, complementary charges of the antibody CDRs and epitopes, and individual characteristics of the antibody, etc. A further possibility exists, namely, that the residue composition of individual epitopes influences the likelihood of inducing a corresponding antibody. This influence could arise either from a chemical bias for individual amino acids inherent in the conserved structure–framework motif of all antibodies (see Section III) or, alternatively, from the necessity to include residue side chains that make special contributions to the energetics of binding between the antibody and the antigen.

To test for a possible bias toward individual amino acids, the frequency with which each residue is found in known epitopes can be compared with the expected frequency, based on the overall amino acid composition of all the protein antigens, calculated on the assumption that no bias is present. The ratio of the observed frequency and the expected frequency is then a measure of the propensity (or tendency) for a given residue to be a part of an epitope (Table VII).

Two such studies have been reported; in each, propensity factors were calculated based on a set of identified epitopes. In one study (Geysen et al., 1985b), hexapeptides homologous with the primary sequences of 12 proteins were assessed for binding to antibodies in polyclonal antisera raised to each of the respective antigens. Altogether, 310 antigenic hexapeptides were identified, and their residue composition

TABLE VII
PROPENSITY FACTORS FOR AMINO ACIDS IN SEQUENTIAL EPITOPES[a]

Amino acid	Set A	Set B	Set C
Tryptophan	0.00	0.58	0.77
Arginine	0.24	0.76	1.14
Methionine	0.26	0.85	0.41
Alanine	0.39	1.05	1.30
Cysteine	0.43	0.59	0.76
Serine	0.59	1.10	0.94
Tyrosine	0.67	0.74	1.03
Glycine	0.71	0.79	0.66
Isoleucine	0.94	1.09	0.51
Valine	0.97	1.30	0.97
Aspartic acid	1.07	1.14	1.16
Leucine	1.15	1.02	1.19
Lysine	1.20	0.95	1.61
Threonine	1.22	1.23	0.90
Histidine	1.26	0.73	2.05
Phenylalanine	1.34	0.74	0.72
Asparagine	1.41	1.12	0.84
Glutamine	1.57	0.94	0.97
Proline	1.77	1.11	0.89
Glutamic acid	1.98	1.09	0.85

[a] Propensity factors for the occurrence of each of the 20 common amino acids in sequential epitopes. Set A, from Geysen et al. (1988a), is specific for critical residues. Set B, from Geysen et al. (1985b), and set C, from Welling et al. (1985), include all (critical and noncritical) residues. Propensity factors are calculated as the ratio of the observed frequency of occurrence in epitopes to the frequency of occurrence in proteins.

analyzed to provide a table of propensity factors for each of the 20 common amino acids. In a study by Welling et al. (1985), epitopes identified in 20 proteins by other researchers were analyzed in a similar manner. The authors suggested that these factors would provide a basis for the prediction of antigenic regions in a protein and the synthesis of peptides (corresponding to sequential epitopes) that might elicit antibodies reactive with the intact protein antigen.

In neither study was every analyzed residue found to be directly involved in binding antibody. In fact, for any sequential epitope comprising a substantial number of amino acids (greater than five to seven), it is unlikely that each residue contributes to the binding interface. The noncontributing residues thus dilute the significance of the propensity factors, dependent on the relative proportion of critical residues to noncritical residues. A further restriction on the usefulness of propensity factors for understanding and predicting antigenic regions in a protein is that no account is taken of any need to provide a complementary set of residues, rather than a continuous stretch of the same or similar amino acids with the highest individual factors. For example, proteins rich in lysine (propensity factor = 1.61), e.g., histones or even polylysine, would indicate regions with relatively high predicted antigenicity values. In fact, polylysine is considered nonimmunogenic and histones are among the least antigenic proteins known.

B. Critical Residue Level

More recently, using a method for the simultaneous synthesis of large numbers of peptides (Geysen et al., 1987b), the systematic assessment of the effects of amino acid substitution on recognition between peptide epitopes and the corresponding antibody allowed the identification of all critical residues within a sequential epitope. This approach also provided a rationale for deciding the length (in terms of number of residues) of sequential epitopes, i.e., all residues between and including the outermost critical residues. Using this definition to decide on the boundaries of a sequential epitope effectively removed the "surplus" residues included in longer peptides from any consideration of the composition of that sequential epitope.

Sixty-three sequential epitopes from seventeen protein antigens were analyzed, and propensity factors were calculated for both the critical and noncritical categories of residues (Geysen et al., 1988a). Comparison of the propensity factors calculated from residues determined to be critical to antibody binding with values for hydrophilicity, hydropathicity, or the tendency for each amino acid to be exposed at the protein surface failed to show any correlation. This outcome suggested that the bias for residues facilitating antigen recognition is inherent in the overall structure of the antigen combining site of antibodies and not a consequence of antigen structure.

The high propensity factors for the hydrophobic residues proline, phenylalanine, and leucine are consistent with the suggestion that they play a dominant role in the binding energetics by way of "hydrophobic" (entropic), as well as van der Waals' interactions (see Section X). The particularly low propensity factor for arginine to occur as a critical residue may partially account for the poor immunogenicity of histones, proteins rich in this amino acid. It is also consistent with the finding that closely related proteins can often be distinguished serologically, when they differ in an epitope that includes arginine in the unreactive protein and another amino acid residue in the reactive protein. Some examples are as follows: (1) human IgG, Oz locus, in which Oz^+ equates to a lysine at position 193 and Oz^- to arginine, and (2) Thy I–I and Thy I–II, in which the latter has an arginine instead of a glutamine at position 89 (Alexander et al., 1983).

C. Antigenic Equivalence

Evolutionary conservation of function in proteins results in conservation of amino acids with an irreplaceable role in function or structure. Allowed changes at either the site of biological function or at structure-determining points can be used to indicate amino acids similar in the respective context. During the initial immune response to foreign antigens, higher animals can produce antibodies that bind specifically to the stimulating antigen. The minimal necessary repertoire of antibodies depends on the degeneracy in the specificity of each antibody for its epitope. The ability of individual antibodies to bind to similar epitopes would indicate the presence of amino acids of antigenic equivalence. Furthermore, a high frequency of interchangeability between amino acids without loss of antibody binding points to residues contributing to overall degeneracy in antibody specificity. An additional consequence of antigenic equivalence between amino acids is an increase in the likelihood of cross-reactivity occurring between an antibody and more than one antigen (Oldstone, 1987).

Sequence comparisons among closely related antigens, i.e., equivalent proteins from different species or different isolates of the same infectious agent, are often used to indicate areas of greatest sequence divergence. This variation is usually assumed to result from selective pressures of the immune response and, therefore, to indicate antigenic regions of the protein (e.g., Hagblom et al., 1985). Often, in using this form of analysis, some changes in amino acid are ignored on the basis that the substitution observed is by a "like" amino acid. For example, pairs considered equivalent are lysine and arginine, tyrosine and phenylalanine, serine and threonine, and valine and isoleucine. These implied equivalences derive from some shared physical or chemical property: for the above-named pairs, respectively, positive charge, aromatic ring, hydroxyl group, and like aliphatic structure. Important differences, however, also distinguish between the above "paired" amino acids (in the same order): size of the structure carrying the positive charge (guanyl versus amino group), ability to form hydrogen bonds (tyrosine), hydrophobicity (threonine > serine), and size (isoleucine > valine).

In a recent comprehensive study (Geysen et al., 1988a), 103 replaceability patterns (each representing an individual recognition event of antigen by an antibody) were analyzed for sequential epitopes. This data base comprised an assessment of the replaceability of 640 amino acids in 103 peptides ranging from 5 to 8 residues in length. Although derived from a data base of sequential epitopes, these values may be generally applicable: sequential epitopes identified in a systematic study of MHr often clustered to form superassemblies, apparently representing sets of assembled epitopes (see Section IV).

The replaceability of each residue in the data base was determined as a function of the number of alternative amino acids (0–19) that could be substituted without loss of antibody binding. Frequency distributions for the replaceability of each of the 20 common amino acids are shown in Fig. 18. As expected, residues within an epitope showed either a limited replaceability by alternative amino acids with the retention of antibody binding or a general replaceability by any of the other common amino acids. This is consistent with the concept that epitopes comprise both essential or critical residues (those with limited replaceability) and less critical or nonessential residues (those with a general replaceability), when assessed for contributions to antibody binding. Residues were divided into two groups based on the number of allowed replacement amino acids, with the dividing line chosen as the nadir of the frequency of replaceability distribution (Fig. 18). Thus, critical (Group I) residues are those with 0–9 allowed replacements, and noncritical (Group II) residues are those with 10–19 allowed replacements. Critical residues accounted for 435 of the 640 total residues assessed. For 171 critical residues, antibody binding was precluded by all amino acid replacements.

Data for critical residues were tabulated in the form of a replaceability matrix, which indicates the acceptability of replacements of individual critical residues within an epitope (Table VIII; Geysen et al.,

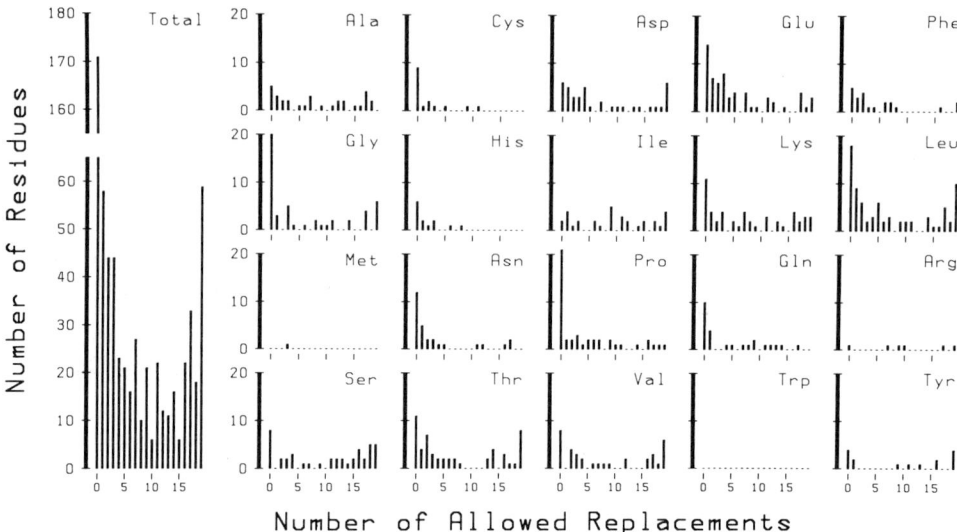

FIG. 18. Antigenic replaceability of single amino acids. Frequency distributions of the degree of antigenic replaceability for the total data base (640 residues) and for each individual amino acid, ordered according to the single-letter code. The abscissa gives the number of alternative amino acids acceptable as replacements, up to the maximum of 19 tested. [Adapted from Geysen et al. (1988a).]

1988a). When replacements were allowed, the alternative amino acids bore a physicochemical relationship to the parent residue. However, of the relationships determined to be statistically significant, traditionally considered conservative changes ranged from 29% acceptability (methionine for leucine) to 65% acceptability (valine for isoleucine, respectively). As expected, obvious clashes, such as change of charge [lysine to glutamic acid (0%) or aspartic acid (10%) and glutamic acid to lysine (6%) or arginine (4%)], aromatic hydrophobic to aliphatic hydrophobic [phenylalanine to isoleucine (16%), leucine (37%), and valine (16%)], and small residue to large [glycine to alanine (24%), valine (3%), isoleucine (6%), or leucine (6%)] were infrequently observed.

The converse size change (from large to small residue), also disallowed, is interesting in a special way. Replaceability values for this category of change evaluate the consequences of creating a void in the clustered residues of an epitope in contact with antibody. Assuming that compensatory adjustments of side chains do not occur, replacement of a residue by, say, glycine in effect creates a void between the interacting molecules the size of the replaced residue side chain.

TABLE VIII

REPLACEABILITY MATRIX FOR GROUP I AMINO ACIDS IN EPITOPES[a]

Parent amino acid	number	A	C	D	E	F	G	H	I	K	L	M	N	P	Q	R	S	T	V	W	Y	Average replaceability
A	18	100		6	17	11		50		11	6	11		22	17	17		50	22	28			3.00
C	15	7	100	7				13								27	7	7		7		1.47
D	26	12	38	100	50	4	13	23			4	12	42	4	12	4	27	15		19		2.62
E	48	21	15	42	100	(2)		8	6	6	8	21	13	13	33	4	15	10	6	8	6	2.50
F	19	11	16	16	11	100		5	16	5	37	11	11	11	11	5	32	26	16	16	21	2.74
G	33	24	3	6	3	6	100	12	6	12	6	15	3	23	9	18	24	3	3	3	3	1.61
H	13				8		23	100		15		8	15			23	8	15	8	8	8	1.85
I	17	47	29	18	29	35	(0)	(0)	100	24	59	18	12	6	29	24	24	24	65	6	(0)	4.47
K	31	16	(0)	10	(0)	10	(0)	26	13	100	10	32	23	3	26	39	29	23	13	6	(0)	2.87
L	51	12	10	8	(2)	22	(2)	8	49	18	100	29	6	4	12	10	4	4	22	6	10	2.35
M	1								100	100	100	100	100	100	100							3.00
N	23		4	4	4	4	4	9	11	9	4	13	100		13	17	9	4		4		1.04
P	37	24	8	3	5	8	14	11		16	3	16	11	100	16	5	14	11	14	8	(0)	1.97
Q	20	20	15	5	15	5	20	10	5	5	5	25	25	10	100	10	15	10	10	10	10	2.30
R	3	67				33		67	33	67	33	67		33	33	100			33	33	33	5.33
S	18	33	11	17	17	11	6	39	6	11	6	22		6		28	100	22	6	17	6	2.44
T	34	12	6	6	15	6	3	9	15	12	9	24	12	6	12	3	44	100	21	15	6	2.32
V	21	24	5	6		14		19	52	10	33	5	14	14		10	5	29	100	14	14	2.62
W	0																					
Y	7	14	14		29				14	14	14	14				14			14	14	100	1.57

[a] The replaceability of all Group I residues in the data base is summarized. For each parent residue (column 1, one-letter code for amino acids), the frequency with which each of the 20 alternatives is allowed (columns 3–22) is expressed as a percentage of the number of occurrences of the parent residue (column 2). Shown underlined are those replacements whose frequency is significantly ($p < 0.05$) different from that expected by chance. The significance was calculated in a 2 × 2 Chi-square test against the null hypothesis that all alternative amino acids are equally likely. Those frequencies that are significantly ($p < 0.05$) lower than expected are shown in parentheses. Zeros have been omitted except for where they are statistically significant. The last column is the average replaceability expressed as average number of allowed replacements. [Adapted from Geysen et al. (1988a). Copyright © 1988 by H. Mario Geysen, Commonwealth Serum Laboratories.]

The energy penalty associated with a void has been estimated and equates to about 1.6 kcal/mol/surface area of 40 Å^2, or more than an order of magnitude in binding affinity for a small water-sized gap (Tainer et al., 1985a). Thus, the change alanine to glycine (observed frequency 50%) is associated with an estimated energy loss of less than 1 kcal/mol, whereas valine to glycine (0%) and leucine to glycine (2%) are associated with estimated energy losses of 2.7 and 3.6 kcal/mol, respectively. The energy loss in the latter two examples is a significant proportion of the energy of binding between antigens and antibodies, which is typically in the range 8–16 kcal/mol. Thus, large side-chain to small side-chain substitutions are infrequently allowed for critical residues, which presumably provide dominant contributions to the binding energy. Alternatively, for noncritical residues and for some critical residues, induced fit of the antibody or of the protein antigen may occur to allow the packing density at the interface to remain high and dispersion forces to contribute to the interactions maintaining the complex of antigen and antibody, as discussed in detail in Section X.

X. Overview of Chemistry and Binding Energy

A complete understanding of the protein–protein interactions responsible for the binding energy of antibodies will require a synthesis of information from many different types of experiments. These experimental results should agree within their respective levels of accuracy and limitations. Whether antibody binding regions are mapped using peptides, fine structure specificity analysis, chemical modification, blocking by other antibodies, proteolytic footprinting, or X-ray crystallography, the results suggest that antibodies do not generally recognize linear sequences on proteins. Although the epitope descriptions from these different methods involve apparent discrepancies, they result, in part, from difficulties in defining protein epitopes. We have used the term sequential epitope to characterize sites identified by competitive binding of antibody to native protein antigen and peptide homologs. These sequential sites may frequently represent portions of assembled epitopes. We suggest that a reasonable level of agreement among different methods results when epitopes are considered in terms of *critical residues*, which provide dominant contributions to the binding energy, and *contact residues*, which may provide complementarity interactions that increase the overall association constant for the immune complex without being individually required.

By any definition, epitopes are the part of the antigen that interacts with the antibody. The synthesis of peptide mapping and crystallographic data allows us to approach a full description of the chemistry and mechanism of the antibody–antigen interaction. This description may also serve as a framework for the future incorporation of detailed stereochemical information on peptide interactions with Ia antigen and helper T cell receptors, which would help complete the chemical description of the immune response. In this regard, recent studies defining T cell epitopes (Francis et al., 1987; Carbone et al., 1987., Cease et al., 1987; Good et al., 1987; DeLisi and Berzofsky, 1985) and the X-ray structure determination of the HLA-A2 antigen (Bjorkman et al., 1987a,b) are major steps toward obtaining this level of stereochemical information for T cell epitopes. Here, we have considered the analysis of X-ray structural data on antibodies and antigens, the experimental determination of systematic data bases of sequential epitopes, statistically significant stereochemical biases for sequential epitopes, correlation criteria for defining dominant epitopes and collective responses, statistical evaluation of predictive methods, the application of experimental data to the design of peptide epitopes, the frequency of involvement of individual amino acids in epitopes, the degree of antigenic equivalence of amino acids, and an equation for the antibody–antigen interaction mechanism.

Based on several approaches to defining structural variability in the antibody CDRs (Section III,C), it is apparent that differential contributions to any inducible complementarity will depend on the underlying stereochemistry of the structural elements involved. For example, L1 is much more likely to show inducible complementarity than H1 (see Fig. 5A and B), and both L3 and H3 appear capable of local conformational changes. As evident from the evaluation of antibody structure, not all portions of the CDRs need to be involved in functionally important, induced complementarity for this process to occur.

For the antigen, the X-ray structures of antibody–antigen complexes suggest that side chains can show inducible complementarity (Sheriff et al., 1987b; Colman et al., 1987a) without any global changes in the antigen structure (Amit et al., 1986). The question of interest is whether mobility might bias the frequency of response, not whether ordered regions can be antigenic. For example, every residue in MHr occurs in at least one hexapeptide recognized by one or more antisera (Geysen et al., 1987c), but statistical analyses suggest a significant antigenic bias for mobile and convex areas. Highly ordered areas are generally associated with less frequent antigenic reactivity. The functional significance of mobility is supported by data suggesting inducible complementarity in the antibody inhibition of enzymes and electron transfer processes (Djavadi-Ohaniance et al., 1984; Kuo et al., 1986; Colman et al., 1987a). Functionally significant inducible changes in the antigen upon antibody binding may well be localized to portions of the interacting molecular surface, depending on the local mobility of the interacting structural elements, as suggested for the antibody CDRs.

Peptide mapping experiments have shown that the critical residues responsible for binding specificity are frequently not continuous in sequence and usually participate in three-dimensional assemblies (Geysen et al., 1987c). Furthermore, although sequential peptide homologs were used to identify antigenic sites in MHr, these sites cluster in such a way that an antibody combining site would certainly be in contact with more than one piece of linear sequence at a time (Geysen et al., 1987c). This is an expected consequence of the nature of protein surfaces, because surfaces the size of an antibody combining site are almost always formed by noncontinuous pieces of polypeptide chain (Barlow et al., 1986). Most antibodies, therefore, recognize protein conformation. Antibodies against native and denatured states can be used to determine unfolding kinetics (Chaffotte and Goldberg, 1987), and antibodies can differentiate different native structural states (Dixit et al., 1986; Wakabayashi et al., 1986).

Three levels of conformational hierarchy have been proposed in the structure of a protein antigen (Getzoff et al., 1987): (1) microassemblies of the critical amino acid side chains that form determinants for specific binding; (2) three-dimensional, usually nonsequential, clusters of microassemblies, which may correspond to assembled epitopes; and (3) the superassembly of these clusters, which defines the global surface conformation and chemistry favored for antigenic recognition (i.e., mobility, low packing density, convex surface shape, and nonneutral electrostatic environment). This hierarchy is an extension of previous observations concerning the clustering of determinants to form large antigenic surfaces (Colman et al., 1983; Wiley et al., 1981; Fanning et al., 1986). Although the complex nonbonded interaction of antibody and antigen will be unlikely to depend on a single stereochemical property, three of the four properties that correlate statistically with preferred antigenic sites in MHr are themselves highly correlated, i.e., high mobility, protruding shape, and low packing density (Geysen et al., 1987c). Although originally identified for sequential protein epitopes, the properties of mobility, shape protrusion, and charge also characterize an autoreactive epitope on alanine tRNA (Bunn and Mathews, 1987).

Superassemblies of frequently reactive sites may characterize the boundaries of topographic contact surfaces recognized and buried by interaction with antibodies. While it is unlikely that competitive binding by peptides identifies the entire interaction surface, the clusters of critical residues from sequential epitopes appear to represent, at least in part, assembled topographic epitopes, agreeing with other methods showing that most antigenic sites are assembled, rather than strictly sequential (e.g., Barlow *et al.*, 1986; Amit *et al.*, 1986; Sheriff *et al.*, 1987b; Colman *et al.*, 1987a). Thus, even when antigenic sites are probed by peptides, the results show that most sites cluster to form assembled, topographic surfaces.

Results from studies of the frequency of response to the foreign antigen MHr for several different animal species are consistent with the idea that intrinsic structural features can bias the immune response and contribute to the existence of immunodominant epitopes. Studies on proteins, such as myoglobin and cytochrome c, have clearly demonstrated that immunogenicity depends significantly on the difference between the immunogen and the responders' homologous protein and, therefore, cannot be solely an inherent property of the antigen (Benjamin *et al.*, 1984). Studies on the response from several species to the foreign protein MHr suggest that immunodominant epitopes might best be redefined as sites to which the immune response is directed most frequently. Any part of the protein surface may be antigenic, but sites reactive with the highest and lowest frequencies of response have different stereochemical properties. Despite considerable individual and cross-species variation in antigenic response, the objective definition of a valid collective response to a protein antigen appears to be possible, at least for nonhomologous, foreign proteins, such as MHr or proteins from pathogens.

Predictive methods for epitopes using amino acid sequence data are not highly successful when evaluated against peptide mapping results from MHr, myoglobin, and TMV coat protein (Section VII). In practical terms, the probability of success is affected more by peptide length than by the sequence-based predictive algorithm used. Longer peptides are more likely to produce antibodies that are cross-reactive with intact protein, simply due to statistical factors. However, the possibility that longer peptides have greater conformational similarity to the protein (Fieser *et al.*, 1987; Van Regenmortel, 1987; Moudallal *et al.*, 1985) should not be overlooked.

Sequence-based predictive methods are currently being used to find epitopes (e.g., Cierniewski and Budzynski, 1987), and they may be more useful on large and multidomain proteins in which significant portions of sequence may be buried in protein interfaces. So, it is worth considering how these methods might be improved. Both hydrophilic and hydrophobic residues are found in epitopes, so prediction methods dealing with averaged hydrophilicity may not distinguish antigenic sites (Geysen et al., 1987c). However, because loops tend to be more hydrophilic than other types of secondary structure (Getzoff et al., 1986a), hydrophilicity may help identify the exposed, antibody-accessible areas in multimeric proteins. The inclusion of His, Tyr, and Trp residues adjacent to hydrophilic neighbors may also increase the success rate for hydrophilic predictions of antigenic sites (Hopp, 1986), and other modifications, such as adding recognition factors based upon the interactions between amino acids, have been proposed (Fraga, 1982). Hydrophilic and hydrophobic regions have also been identified by a hydropathy scale (Kyte and Doolittle, 1982) and by hydrophobic moments (Eisenberg et al., 1982). Such experimental values for hydrophilicity are being frequently updated using modern techniques such as high-performance liquid chromatography in an effort to obtain better correlations with antigenicity and accessibility (Parker et al., 1986). Similarly, intron–exon boundaries, which are another sequence parameter correlated to surface exposure (Craik et al., 1982), may also be useful for selecting possible sites for antipeptide antibodies that will react with the native protein antigen (Tainer et al., 1985a). Sequence variation is another, sometimes overlooked, indication of surface exposure, local flexibility, and antigenicity (Getzoff et al., 1988a; Kieber-Emmons et al., 1987; Tainer et al., 1985a).

It has also been suggested that secondary structure may influence antigenicity. Based on their studies on TMV coat protein, Westhof et al. (1984) pointed out that the reactivity of antipeptide antibodies with the parent protein antigen is related to the mobility of sequential determinants within the protein and that this local flexibility is often associated with bends, turns, and loops. Different types of protein secondary structure may indeed have different behavior with regard to mobility. For example, loops tend to be most flexible in the middle of their sequence, but β strands tend to be more flexible at the ends. Secondary structure alone does not appear to dominate antigenicity (Geysen et al., 1987c); however, N- and C-termini, which are, like loops, frequently surface-exposed and relatively mobile (Thornton and Sibanda, 1983) are often antigenic (Tainer et al., 1985a).

Prediction of epitopes is more likely to be successful from the three-dimensional structure, if it is available. The structural parameters of shape and mobility (Geysen et al., 1987a,c; Novotný et al., 1986b; Fanning et al., 1986; Tainer et al., 1985a,b; Westhof et al., 1984; Tainer et al., 1984) appear to have a statistically significant correlation with epitopes (Geysen et al., 1987a,c). In addition, epitopes have been predicted from an assessment of local dissimilarity based on three-dimensional neighbors (Padlan, 1985). Certain sites on a molecule may not elicit an immune response, even though they possess intrinsic stereochemical features that bias the frequency of the immune response, if B cells reactive to these regions are not present as a result of tolerance or possible evolutionary effects on the germ-line repertoire, as evidenced by the variation in response to MHr in six species of animals.

Due to the sequence variability of epitopes, it is worthwhile to consider patterns of occurrence and substitution of individual amino acid side chains. Analysis of the propensity factors for the occurrence of specific amino acids as critical (Group I) residues in sequential epitopes, after normalization for overall occurrence in proteins, suggests no overwhelming distinguishing chemical parameter at the individual side-chain level, although the largest and smallest side chains appear somewhat less likely to be critical residues. Assessment of the chemistry of the antigenic equivalence for individual side chains using a replaceability matrix from a data base of 437 critical residues in sequential epitopes (Table VIII), indicates that no one chemical parameter accounts for allowed substitutions. The six least replaceable, and thus most antigenically unique, residues are Cys, Gly, His, Asn, Pro, and Tyr, which are replaceable by less than two other residues on the average. The six residues that most successfully replace other residues are Ala, Met, Gln, Arg, Ser, and Thr.

Improved methods of peptide synthesis make the construction of virtually any sequential polypeptide antigen possible. One problem for effective design of vaccines based on peptide epitopes is, therefore, the choice of appropriate sequences from proteins to synthesize and study. Results from MHr suggest that sequential epitopes on the protein (Geysen et al., 1987c) are closely related to sites that react with the corresponding antipeptide antibodies (Fieser et al., 1987; Tainer et al., 1984). Based on this observation, we suggest an experimental approach that can be used to identify peptide epitopes likely to produce antibodies reactive with the native protein (see Section VIII). Thus, for important biological problems, a reasonable plan is to map the entire protein sequence, using short overlapping peptides, and then target the synthesis of longer peptides (about 15–20 residues) to include the sequential protein epitopes that are found with high frequency. This requires the characterization of a sufficient number of antisera to identify the most reactive (most frequently responding) sequential epitopes.

For the development of peptide epitope vaccines, it may be critically important to identify the topographic relationship of sequential epitopes, which may represent portions of overlapping assembled epitopes that form superassemblies of frequently recognized sites, so (in the absence of a crystallographic structure) peptide mapping should be complemented by some footprinting technique, such as chemical modification (Burnens et al., 1987), blocking by other antibodies (Smith-Gill et al., 1982; Berzofsky et al., 1982; East et al., 1982), or proteolytic digestion (Jemmerson and Paterson, 1986a,b). Given the association of antigenic residues with sequence variability, a successful B cell peptide epitope vaccine might require a set of peptides, representing semiconserved sequence regions forming a superassembly of linear sites and consensus peptide sequences for most common critical residues in these sites (Rowlands et al., 1983). The replaceability matrix could also be used to define a consensus peptide sequence most likely to retain antigenicity. For the promotion of a memory antibody response, a T cell peptide epitope should be included to complement the B cell epitope (Cease et al., 1987; Good et al., 1987; Francis et al., 1987). The logical candidate for a peptide vaccine would then be a combination of 15- to 20- residue polypeptides, comprising a set of sequential B cell epitopes within the boundaries of one or more frequently responding topographic sites, along with one or more T cell epitopes. Although a number of difficulties remain, particularly in controlling the T cell response, it seems likely that useful synthetic vaccines based on peptides can be designed.

Besides the potential practical applications from intrinsic stereochemical biases for protein epitopes, there is the opportunity to increase our understanding of the fundamental chemical basis for this important class of protein–protein interactions. The molecular interactions of enzymes, hormones, and receptors involve specific mechanisms of action. Yet, the ability of antibodies to recognize so many different protein and nonprotein sites has commonly been attributed to a relatively rigid shape complementarity unique to each antigen–antibody complex. This is frequently expressed as a lock-and-key fit, which implies that the complete stereochemistry of the complex preexists in the individual molecules (reviewed in Mariuzza et al., 1987). There is no *a priori* reason to expect that antibodies are in a separate class from these other types of proteins that show inducible complementarity. Moreover, analysis of the basic stereochemical structure of antibodies, antibody–antigen complexes, and epitopes suggests that, although there is an essential truth in the lock-and-key description at a global level (Mariuzza et al., 1987), it is too simple a paradigm for understanding the energetics of the antibody–antigen interaction, which can apparently include precollision orientation by electrostatic forces and local induced complementarity. The most apt metaphors for this interaction are probably biological ones, such as the handshake suggested by Colman (1988), which can accommodate a multistep interaction mechanism.

Antibody binding to a protein antigen can proceed by a mechanism involving inducible complementarity without requiring all interacting regions of antibody and epitope to contribute to mobility. In global terms, the usual first stage of antigen–antibody binding is an association of rigid bodies, as previously proposed (Amit et al., 1986). In subsequent steps, side-chain rearrangements at the contact surface can usually be expected, and any antibody CDRs that clasp (wrap around) the epitope can fold into their final conformation only after antibody binding. Of the antibody variable domain CDRs, heavy chain H3 and light chain L1 and L3 are most likely to have conformational adjustments after binding the antigen.

Overall, the available data suggest that antibodies are uniquely successful in recognizing and forming stable complexes with multiple sites on proteins, because antibody CDRs can drive toward a final state in which the interacting molecular surfaces of antibody and antigen resemble the interfaces buried in other stable protein–protein complexes. Some immunodominant regions may already have stereochemical features that are similar to protein interfaces, otherwise the antibody must supply sufficient binding energy to pay the cost of changing the nature of the exposed surface. Given the apparent local flexibility of interacting surfaces and the relatively poorly understood entropic contributions from the release of solvent, an accurate chemical reaction equation for antibody–antigen interaction is undoubtedly complex and only approximated by gas-phase interaction equations that neglect solvation. Still, it is worth posing a consistent and testable equation for the interaction mechanism for antibody (Ab) binding to protein antigens (Ag).

Based on the synthesis of biochemical and biophysical data presented here, the antigen–antibody interaction can be divided into recognition and binding components as follows:

$$Ab + Ag \rightleftharpoons Ab \cdots Ag \rightleftharpoons Ab' \cdot Ag'$$

The recognition process depends on the antibody's ability to form a preassociation complex (Ab $\cdots$ Ag) through contact with critical exposed side chains in the antigen. The observed association of the most highly reactive sites with nonneutral electrostatic regions and the least reactive sites with surface grooves containing bound waters (Geysen et al., 1987c) suggests that relatively long-range electrostatic and polar interactions between exposed side chains can allow recognition and initial complex formation with concomitant release of water molecules. Results from X-ray crystallography (Sheriff et al.,

1987b; Colman et al., 1987a; Amit et al., 1986) and critical residue studies (Getzoff et al., 1987) suggest that hydrogen bonds play a critical role in the interaction chemistry of the final complex.

The mobility of the interacting structural elements has implications for the stability of the initial complex and for functionally important inducible complementarity. Mobile structural elements of the antibody and antigen exhibit more solvent interactions and fewer intramolecular interactions (lower packing pressure) than highly ordered regions and are, therefore, able to contribute a greater percentage of atomic interactions to the formation of the initial complex than highly ordered regions. Thus, mobile elements might both stabilize the initial complex relative to the free antigen and antibody, and also more readily allow induced complementarity in the final complex.

In MHr, the most frequently recognized antigenic sites tend to have significant negative electrostatic potentials. This may be biased by the net negative charge on MHr, as many antibodies are charge complementary to their antigens (Benjamin et al., 1984; Berzofsky, 1985; Crumpton, 1974; Karush, 1962). Many haptens, such as nitrates, carboxylates, sulfonates, arsonates, amides, and phosphate compounds, are charged, and mutants that escape monoclonal antibody binding often appear to have charge changes. In the crystallographic structure of the HyHEL5–lysozyme complex, two arginine residues of lysozyme form electrostatic interactions with two glutamic acid residues of the antibody, suggesting that such electrostatic interactions can provide significant contributions to the binding energy (Sheriff et al., 1987b). Electrostatic attraction may increase the rate of antibody–antigen complex formation by facilitated diffusion and may also stabilize an initial complex, allowing subsequent slower steps to occur before the antigen and antibody diffuse apart. Electrostatic forces are very strong in the absence of water and may stabilize the final antigen–antibody interface (Ab' · Ag') significantly, as in other known protein–protein interfaces (Getzoff et al., 1986b; Chothia and Janin, 1975; Crumpton, 1974). Thus, the immune response is apparently sensitive to electrostatic forces, and unpaired charge donors or acceptors can provide specificity; however, antibodies to protein sites that do not include critical charged residues will not emphasize electrostatic forces for their interaction chemistry.

The observed critical role of hydrophobic side chains for antibody binding finds support in analyses of known protein complexes (Miller et al., 1987; Getzoff et al., 1986b; Chothia and Janin, 1975), indicating that binding affinity should depend in part on the antibody's ability to maximize the buried hydrophobic surface area in the final complex (Ab' · Ag'). Based on the architecture of the individual microassemblies and the global stereochemistry of the superassemblies of sequential epitopes in MHr, antibody binding may drive the local destabilization of the free antigen side-chain conformations and develop optimal intermolecular complementarity by two processes: (1) the disruption of surface-exposed side-chain to side-chain interactions, such as intraprotein salt-bridges and hydrogen bonds; and (2) the entropically favored removal of the bound water molecules that act to stabilize the polar side-chain positions and surface topography. This proposed separation of antibody recognition and binding is consistent with observed antigenic biases for high mobility and convex shape (Hogrefe et al., 1987; Geysen et al., 1987a,c; Novotný et al., 1986b; Tainer et al., 1985a, 1984; Westhof et al., 1984; Williams and Moore, 1985; Thornton et al., 1986).

Antigen and antibody proteins are conformationally stable in solution, so large (and presumably energetically costly) conformational changes are unlikely. Yet, protein surfaces are inherently flexible and, by analogy to enzymes, one could expect antibody binding to a protein molecule to elicit atomic shifts, ranging from less than 1 Å to more than 10 Å, that would be important to the stabilization of the complex (Koshland, 1970, 1976; Creighton, 1983). Proteins can bind simple substrates and small metal compounds with relatively little induced conformational change (Tainer et al., 1983; Getzoff and Tainer, 1985), as has been observed for antibody–hapten interactions (Segal et al., 1974; Amzel et al., 1974). Yet, the transition from the 4Zn to the 2Zn form of insulin, in which the conformation of the first eight residues changes from α-helix to extended conformation (Chothia et al., 1983; Smith et al., 1984), requires atomic shifts of up to 21 Å. Protein–protein interactions are more likely to show functionally important induced conformational changes, because their large interacting surfaces are driven toward maximizing stereochemical complementarity. Such changes in the molecular surface that improve stereochemical complementarity between interacting molecules may reduce unpaired hydrogen bonding residues [costing from about 0.5–1.0 kcal, uncharged, to 3.5–4.5 kcal, charged, per unmade hydrogen bond (Bartlett and Marlowe, 1987; Creighton, 1983; Fersht, 1972)] and gaps in the interface [costing about 1.6 kcal for a gap the size of one water molecule (Tainer et al., 1985a)]. Any unfavorable steric collisions not corrected by inducible local rearrangements would be very costly energetically and, hence, provide high specificity by restricting binding.

Although antibodies can be designed to have enzymatic activity and undergo catalytic turnover (Lerner and Tramontano, 1988, 1987), the biologically important activity of natural antibodies involves formation of a stable complex with antigen. Unlike an enzyme, which is purely a catalyst, an antibody may act to increase the energy of the antigen and/or antibody in solution, so long as the free energy of the complex (Ab' · Ag') is lower than that of free proteins in solution. The protein avidin, binds the smaller molecule, biotin, with an association constant of 10^{15}, representing about 22 kcal/mol (Green, 1975); so the considerably larger interface formed by the antibody–antigen complex could provide sufficient binding energy to pay for local rearrangements and maintain high affinity. Although it is not possible to isolate the individual energetic components for antibody binding to sites in MHr, the antibody affinities (10^7–10^8) associated with the energy remaining after any rearrangements (over 10 kcal/mol) correspond to a significant portion of the observed net stability for a protein [about 16 kcal/mol (Creighton, 1983)].

Antibodies share with other proteins the ability to bind with high affinities to small molecules (Karush, 1978), presumably without requiring significant conformational change. For protein antigens, the necessary specificity could arise from the contributions of many specific groups over a large interaction area, about 600–1000 Å^2 from the rough dimensions of an antibody binding region and buried areas in D1.3 and HyHEL5 (Sheriff et al., 1987b; Amit et al., 1986; Davies and Metzger, 1983). Thus, antibody–antigen complexes have about the same buried surface as that found in the subunit interfaces of stable oligomeric proteins (Getzoff et al., 1986b; Miller et al., 1987). Recognition of each of the many stereochemical variants possible for a surface of this size would require an impossibly large number of antibodies, if antigen–antibody interactions proceed by a hard body, lock-and-key mechanism. However, if these interactions can proceed by recognition of a smaller number of critical residues and subsequent adaptation of the molecular surface resulting from side-chain movements in the epitope, each antibody can recognize more protein sites without losing the required immunological specificity, which is inherent in the stereochemistry of each microassembly of critical residues in the antigen.

The biological significance of the observed three-dimensional continuity of the immunologically reactive surface, which includes critical interactions between the protein's secondary structural elements, is that the interaction of polyclonal antibodies with protein antigens can be envisioned as a cascade, which can function to uncouple side-chain interactions and, perhaps, destabilize the original protein conformation. Thus, the binding of an antibody to one epitope would help induce cooperative conformational changes in adjoining epitopes, making the binding of antibodies to the second site more favorable, while disrupting the stability and/or activity of the protein antigen. This effect has, in fact, been observed (Paterson, 1985; Parham, 1984; Lubeck and Gerhard, 1982; Celada and Strom, 1972).

The synthesis of results from the combination of X-ray data with those from other approaches is useful for probing general questions in protein chemistry. The implied ability of hydrogen bonds and electrostatic forces to influence antigenic recognition supports their general importance in the specificity of biological interactions. Similarly, the importance of local mobility and surface shape in many types of molecular interaction has been recognized; however, analysis of the results from antigenicity experiments on MHr suggests that these two correlated parameters may influence the process of induced fit in different ways, with shape potentially dominating the kinetics of water release and mobility the energetic cost of concerted conformational changes (Geysen et al., 1987c).

One intriguing implication of inducible complementarity concerns protein folding. The pathway for complex formation converts a folded conformation in one three-dimensional environment to an altered conformation in a different environment. Whether transition states fall along the pathway for a concerted mechanism or form barriers between intermediate states, they may closely resemble local protein folding intermediates. If so, they may be used to extend previous work (Anfinsen and Scheraga, 1975) on the use of antibodies as molecular tools to address questions of protein folding.

Every antibody–antigen complex may have a different specific mechanism, which is impossible to determine for each of the potential 10 million antibodies generated by the immune system (Max, 1984; Steward and Steensgaard, 1983). The nature of the immune response also changes over time (Berek and Milstein, 1987). Therefore, the repertoire of physicochemical components that underlie the mechanisms of antibody binding may rival the diversity of antibodies themselves. Nevertheless, the issue in defining interaction chemistry is to search out common, basic features that characterize antibody–antigen interaction, knowing that each case is somewhat different in its particulars. For instance, because of enthalpic and entropic considerations, one might expect the binding of antibodies with the highest affinities to involve the least rearrangements at the site, but this need not be the case, if such changes decrease the dissociation rate more than the association rate for the complex. For example, data from following antifluorescein antibodies during early, intermediate, and late stages of the immune response suggest that the association rate increases by a factor of only 3, whereas the dissociation rate decreases by a factor of over 1000, and the interaction kinetics are consistent with conformational rearrangements and loss of water in the formation of the complex (Levinson et al., 1975).

The surprise in the integration of results from different biochemical and biophysical approaches is not in their discrepancies and variations, but in their commonalities, which suggest that understanding and ultimately control of antibody–antigen interactions will result from basic knowledge about the chemistry of these interacting molecules. The basic chemical principles identified by intensive multidisciplinary studies on antibody–antigen interaction should be applicable to ongoing studies on T cell epitopes and to protein interactions in general.

Acknowledgments

We thank Hannah Alexander, Talapady Bhat, Nadine Chien, Gerson Cohen, Miroslaw Cygler, David Davies, Allen Edmundson, Arnold Feinstein, Terry Fieser, Cindy Fisher, Tom Mason, Arthur Olson, Eduardo Padlan, Michael Pique, Roberto Poljak, Victoria Roberts, Matthew Scharff, Steven Sheriff, Enid Silverton, Sandra Smith-Gill, and Peter Wright for helping to make this review current by contributing criticisms, data, and often unpublished results. This work was supported in part by NIH (RO1 GM34338 to E. D. G; RO1 AI22160 to J. A. T.) and DOD (DAM17 to J. A. T.).

References

Alexander, H., Johnson, D. A., Rosen, J., Jerabek, L., Green, N., Weissman, I. L., and Lerner, R. A. (1983). Mimicking the alloantigenicity of proteins with chemically synthesized peptides differing in single amino acids. *Nature (London)* **306,** 697–699.

Alexander, H., Heffron, F. L., Fieser, T. M., Hay, B. N., Getzoff, E. D., Tainer, J. A., and Lerner, R. A. (1988). Probing the folding and antigenic profile of myohemerythrin by site directed mutagenesis. *Proc. Natl. Acad. Sci. U.S.A.* In preparation.

Amit, A. G., Mariuzza, R. A., Phillips, S. E. V., and Poljak, R. J. (1985). Three-dimensional structure of an antigen–antibody complex at 6 Å resolution. *Nature (London)* **313,** 156–158.

Amit, A. G., Mariuzza, R. A., Phillips, S. E. V., and Poljak, R. J. (1986). Three-dimensional structure of an antigen–antibody complex at 2.8 Å resolution. *Science* **233,** 747–753.

Amzel, L. M., and Poljak, R. J. (1979). Three-dimensional structure of immunoglobulins. *Annu. Rev. Biochem.* **48,** 961–997.

Amzel, L. M., Poljak, R. J., Saul, F., Varga, J. M., and Richards, F. F. (1974). The three-dimensional structure of a combining region–ligand complex of immunoglobulin NEW at 3.5-Å resolution. *Proc. Natl. Acad. Sci. U.S.A.* **71,** 1427–1430.

Anfinsen, C. B., and Scheraga, H. A. (1975). Experimental and theoretical aspects of protein folding. *Adv. Protein Chem.* **29,** 205–300.

Artymiuk, P. J., Blake, C. C. F., Grace, D. E. P., Oatley, S. J., Philips, D. C., and Sternberg, M. J. E. (1979). Crystallographic studies of the dynamic properties of lysozyme. *Nature (London)* **280,** 563–568.

Audibert, F., Jolivert, M., Chedid, L., Alouf, J. E., Boquet, P., Rivaille, P., and Siffert, O. (1981). Active antitoxic immunization by a diptheria toxin synthetic oligopeptide. *Nature (London)* **289,** 593–594.

Bahraoui, E. M., Granier, C., Van Reitschoten, J., Rochat, H., and El Ayeb. M. (1986). Specificity and neutralizing capacity of antibodies elicited by a synthetic peptide of scorpion toxin. *J. Immunol.* **136,** 3371–3377.

Barlow, D. J., Edwards, M. S., and Thornton, J. M. (1986). Continuous and discontinuous antigenic determinants. *Nature (London)* **322,** 747–748.

Bartlett, P. A., and Marlowe, C. K. (1987). Evaluation of intrinsic binding energy from a hydrogen bonding group in an enzyme inhibitor. *Science* **235**, 569–571.

Beachey, E. H., Seyer, J. M., Dale, J. B., Simpson, W. A., and Kang, A. H. (1981). Type-specific protective immunity evoked by synthetic peptide of *Streptococcus pyogenes* M protein. *Nature (London)* **292**, 457–459.

Berek, C., and Milstein, C. (1987). Mutation drift and repertoire shift in the maturation of the immune response. *Immunol. Rev.* **96**, 23–41.

Benjamin, D. C., Berzofsky, J. A., East, I. J., Gurd, F. R. N., Hannum, C., Leach, S. J., Margoliash, E., Michael, J. G., Miller A., Prager, E. M., Reichlen, M., Sercarz, E. E., Smith-Gill, S. J., Todd, P. E., and Wilson, A. C. (1984). The antigenic structure of proteins: A reappraisal. *Annu. Rev. Immunol.* **2**, 67–101.

Bernstein, F. C., Koetzle, T. F., Williams, G. J. B., Meyer, E. F., Brice, M. D., Rodgers, J. R., Kennard, O., Shimanouchi, T., and Tasumi, M. (1977). The protein data bank: A computer-based archival file for macromolecular structures. *J. Mol. Biol.* **112**, 535–542.

Berzofsky, J. A. (1985). Intrinsic and extrinsic factors in protein antigenic structure. *Science* **229**, 932–940.

Berzofsky, J. A., Buckenmeyer, G. K., Hicks, G., Gurd, F. R. N., Feldman, R. J., and Minna, J. (1982). Topographic antigenic determinants recognized by monoclonal to sperm whale myoglobin. *J. Biol. Chem.* **257**, 3189.

Berzofsky, J. A., Cease, K. B., Cornette, J. L., Spouge, J. L., Margalit, H., Berkower, I. J., Good, M. F., Miller, L. H., and DeLisi, C. (1987). Protein antigenic structures recognized by T cells: Potential applications to vaccine design. *Immunol. Rev.* **98**, 9–52.

Bittle, J. L., Houghten, R. A., Alexander, H., Shinnick, T. M., Sutcliffe, J. G., Lerner, R. A., Rowlands, D. J., and Brown, F. (1982). Protection against foot-and-mouth disease by immunization with a chemically synthesized peptide predicted from the viral nucleotide sequence. *Nature (London)* **298**, 30–33.

Bjorkman, P. J., Sapir, M. A., Samraoui, B., Bennett, W. S., Strominger, J. L., and Wiley, D. C. (1987a). Structure of the human class I histocompatibility antigen, HLA-A2 *Nature (London)* **329**, 506–512.

Bjorkman, P. J., Sapir, M. A., Samraoui, B., Bennett, W. S., Strominger, J. L., and Wiley, D. C. (1987b). The foreign antigen binding site and T-cell recognition regions of Class I histocompatibility antigens. *Nature (London)* **329**, 512–518.

Blaney, J. M., Weiner, P. K., Dearing, A., Kollman, P. A., Jorgensen, E. C., Oatley, S. J., Burridge, J. M., and Blake, C. C. F. (1982). Molecular mechanics simulation of protein–ligand interactions: Binding of thyroid hormone analogues to prealbumin. *J. Am. Chem. Soc.* **104**, 6424–6434.

Bunn, C. C., and Mathews, M. B. (1987). Autoreactive epitope defined as the anticodon region of alanine transfer RNA. *Science* **238**, 1116–1119.

Burnens, A., Demotz, S., Corradin, G., Binz, H., and Bosshard, H. R. (1987). Epitope mapping by chemical modification of free and antibody-bound protein antigen. *Science* **235**, 780–783.

Carbone, F. R., Fox, B. S., Schwatrz, R. H., and Paterson, Y. (1987). The use of hydrophobic α helix defined peptides in delineating the T cell determinant for pigeon cytochrome c. *J. Immunol.* **138**, 1838–1844.

Cease, K. B., Margalit, H., Cornette, J. L., Putney, S. D., Robey, W. G., Ouyang, C., Streicher, H. Z., Fischinger, P. J., Gallo, R. C., DeLisi, C., and Berzofsky, J. A. (1987). Helper T-cell antigenic site identification in the acquired immunodeficiency syndrome virus gp120 envelope protein and induction of immunity in mice to the native protein using a 16-residue synthetic peptide. *Proc. Natl. Acad. Sci. U.S.A.* **84,** 4249–4253.

Celada, F., and Strom, R. (1972). Antibody-induced conformational changes in proteins. *Q. Rev. Biophys.* **5,** 395–425.

Chaffotte, A-F., and Goldberg, M. E. (1987). Kinetics of the spontaneous transient unfolding of a native protein studies with monoclonal antibodies. *J. Mol. Biol.* **197,** 131–140.

Chang, C-H, Short, M. T., Westholm, F. A., Stevens, F. J., Wang, B-C., Furey, W., Solomon, A, and Schiffer, M. (1985). Novel arrangement of immunoglobulin variable domains: X-ray crystallographic analysis of the λ-chain dimer Bence–Jones protein loc. *Biochemistry* **24,** 4890–4897.

Chang, C.-H., Carperos, W. E., Ainsworth, C. F., Olsent, K. W., and Schiffer, M. (1987). Pronounced effect of the solvent of crystallization on the structure of a multi-domain protein. *Fourteenth International Congress of Crystallography* (collected abstracts, C-19).

Chien, N. C., Roberts, V. A., Guisti, A., Scharff, M. D., and Getzoff, E. D. (1988). Disruption of an antigen binding site by a distant amino acid substitution: Proposal of a structural mechanism. *Cell*. Submitted.

Chothia, C., and Janin, J. (1975). Principles of protein–protein recognition. *Nature (London)* **256,** 705–708.

Chothia, C., and Lesk, A. M. (1987). Canonical structures for the hypervariable regions of immunoglobulins. *J. Mol. Biol.* **196,** 901–917.

Chothia, C., Lesk, A. M., Dodson, G. G., and Hodgkin, D. C. (1983). Transmission of conformational change in insulin. *Nature (London)* **302,** 500–505.

Chothia, C., Novotny, J., Bruccoleri, R., and Karplus, M. (1985). Domain association in immunoglobulin molecules: The packing of variable domains. *J. Mol. Biol.* **186,** 651–663.

Chothia, C., Lesk, A. M., Levitt, M., Amit, A. G., Mariuzza, R. A., Phillips, S. E. V., and Poljak, R. J. (1986). The predicted structure of immunoglobulin D1.3 and its comparison with the crystal structure. *Science* **233,** 755–758.

Cierniewski, C. S., and Budzynski, A. Z. (1987). Conformational equilibria in the γ chain COOH terminus of human fibrinogen. *J. Biol. Chem.* **262,** 13896–13901.

Colman, P. M. (1988). Structure of antibody–antigen complexes: Implications for immune recognition. *Adv. Immunol.* **43,** 99–132.

Colman, P. M., Deisenhofer, J., Huber, R., and Palm, W. (1976). Structure of the human antibody molecule Kol (immunoglobulin G1): An electron density map at 5 Å resolution. *J. Mol. Biol.* **100,** 257–282.

Colman, P. M., Vargese, J. N, and Laver, W. G. (1983). Structure of the catalytic and antigenic sites in influenza virus neuraminidase. *Nature (London)* **303,** 41–44.

Colman, P. M., Laver, W. G., Varghese, J. N., Baker, A. T., Tulloch, P. A., Air, G. M., and Webster, R. G. (1987a). Three-dimensional structure of a complex of antibody with influenza virus neuraminidase. *Nature (London)* **326,** 358–363.

Colman, P. M., Air, G. M., Webster, R. G., Varghese, J. N., Baker, A. T., Lentz, M. R., Tulloch, P. A., and Laver, W. G. (1987b). How antibodies recognize virus proteins. *Immunol. Today* **8,** 323–326.

Connolly, M. L. (1983a). Solvent-accessible surfaces of proteins and nucleic acids. *Science* **221,** 709–713.

Connolly, M. L. (1983b). Analytical molecular surface calculation. *J. Appl. Crystallogr.* **16,** 548–558.

Connolly, M. L. (1985). Depth-buffer algorithms for molecular modelling. *J. Mol. Graphics* **3,** 19–24.

Connolly, M. L., and Olson, A. J. (1984). GRANNY, a companion to GRAMPS for the real-time manipulation of macromolecular molecules. *Comput. Chem.* **9,** 1–6.

Cooper, H. M., Jemmerson, R., Hunt, D. F., Griffin, P. R., Yates, J. R., Shabanowitz, J., Zhu, N., and Paterson, Y. (1987). Site-directed chemical modification of horse cytochrome c results in changes in antigenicity due to local and long range conformational perturbations. *J. Biol. Chem.* **262,** 11591–11597.

Craik, R. J., Sprang, S., Fletterick, R., and Rutter, W. (1982). Intron–exon splice junctions map at protein surfaces. *Nature (London)* **299,** 180–182.

Creighton, T. E. (1983). An empirical approach to protein conformation stability and flexibility. *Biopolymers* **22,** 49–58.

Crumpton, M. J. (1974). Antigenicity and immunogenicity. *Antigens* **2,** 133–158.

Cygler, M., Boodhoo, A., Lee, J. S., and Anderson, W. F. (1987). Crystallization and structure determination of an autoimmune antipoly (dT) immunoglobulin Fab fragment at 3.0 Å resolution. *J. Biol. Chem.* **262,** 643–648.

Darsley, M. J., and Rees, A. R. (1985). Three distinct epitopes within the loop region of hen egg lysozyme defined with monoclonal antibodies *EMBO J.* **4,** 383–392.

Davies, D. R., and Metzger, H. (1983). Structural basis of antibody function. *Annu. Rev. Immunol.* **1,** 87–117.

Davies, D. R., Padlan, E. A., and Segal, D. M. (1975). Three-dimensional structure of immunoglobulins. *Annu. Rev. Biochem.* **44,** 639–667.

Davies, D. R., Sheriff, S., and Padlan, E. A. (1988). Antibody–antigen complexes. *J. Biol. Chem.* **263,** in press.

Dean, P. M. (1981). Drug–receptor recognition: Electrostatic field lines at the receptor and dielectric effects. *Br. J. Pharmacol.* **74,** 39–46.

de la Paz, P., Sutton, B. J., Darsley, J., and Rees, A. R. (1986). Modeling of the combining sites of three anti-lysozyme monoclonal antibodies and of the complex between one of the antibodies and its epitope. *EMBO J.* **5,** 415–425.

DeLisi, C., and Berzofsky, J. A. (1985). T-cell antigenic sites tend to be amphipathic structures. *Proc. Natl. Acad. Sci. U.S.A.* **82,** 7048–7052.

Dixit, V. M., Galvin, N. J., O'Rourke, K. M., and Frazier, W. A. (1986). Monoclonal antibodies that recognize calcium-dependent structures of human thrombospodin: Characterization and mapping of their epitopes. *J. Biol. Chem.* **261,** 1962–1968.

Djavadi-Ohaniance, L., Friguet, B., and Goldberg, M. E. (1984). Structural and functional influence of enzyme–antibody interactions: Effects of eight different monoclonal antibodies on the enzymatic activity of *Escherichia coli* tryptophan synthase. *Biochemistry* **23,** 97–104.

Dorow, D. S., Shi, P-T., Carbone, F. R., Minasian, E., Tadd, P. E. E., and Leach, S. J. (1985). Two large immunogenic and antigenic myoglobin peptides and the effects of cyclisation. *Mol. Immunol.* **22,** 1255–1264.

Dower, S. K., Wain-Hobson, S., Gettins, P., Givol, D., Jackson, W. R. C., Perkins, S. J., Sunderland, C. A., Sutton, B. J., Wright, C. E., and Dwek, R. A. (1977). The combining site of the dinitrophenyl-binding immunoglobulin A myeloma protein MOPC 315. *Biochem. J.* **165,** 207–225.

Dyrberg, T., and Oldstone, M. B. A. (1986). Peptides as antigens: Importance of orientation. *J. Exp. Med.* **164,** 1344–1349.

Dyson, H. J., Cross, K. J., Houghten, R. A., Wilson, I. A., Wright, P. E., and Lerner, R. A. (1985). The immunodominant site of a synthetic immunogen has a conformational preference in water for a type II reverse turn. *Nature (London)* **318,** 480–483.

Dyson, H. J., Rance, M., Houghten, R. A., Wright, P. E., and Lerner, R. A. (1988). Folding of immunogenic peptide fragments of proteins in water solution: II. The nascent helix. *J. Mol. Biol.* **201,** 201–217.

East, I. J., Hurrell, J. G. R., Todd, P. E. E., and Leach, S. J. (1982). Antigenic specificity of monoclonal antibodies to human myoglobin. *J. Biol. Chem.* **257,** 3199–3202.

Edmundson, A. B., and Ely, K. R. (1985). Binding of N-formylated chemotactic peptides in crystals of the Mcg light chain dimer: Similarities with neutrophil receptors. *Mol. Immunol.* **22,** 463–475.

Edmundson, A. B., and Ely, K. R. (1986). Three-dimensional analyses of the binding of synthetic chemotactic and opioid peptides in the Mcg light chain dimer. *Ciba Found. Symp.* **119,** 107–129.

Edmundson, A. B., Ely, K. R., Girling, R. L., Abola, E. E., Schiffer, M., Westholm, F. A., Fausch, M. D., and Deutsch, H. F. (1974). Binding of 2,4-dinitrophenyl compounds and other small molecules to a crystalline λ-type Bence–Jones dimer. *Biochemistry* **13,** 3816–3827.

Edmundson, A. B., Ely, K. R., and Herron, J. N. (1984). A search for site-filling ligands in the Mcg Bence–Jones dimer: Crystal binding studies of fluorescent compounds. *Mol. Immunol.* **21,** 561–576.

Ehrlich, P. (1900). The Croonian lecture: On immunity. *Proc. R. Soc. London* **66,** 424–439.

Eisenberg, D., Weiss, R. M., Terwilliger, T. C., and Wilcox, W. (1982). Hydrophobic moments and protein structure. *Faraday Symp. Chem. Soc.* **17,** 109–120.

Ely, K. R., Firca, J. R., Williams, K. J., Abola, E. E., Fenton, J. M., Schiffer, M., Panagiotopoulos, N. C., and Edmundson, A. B. (1978). Crystal properties as indicators of conformational changes during ligand binding or interconversion of Mcg light chain isomers. *Biochemistry* **17,** 158–167.

Ely, K. R., Peabody, D. C., Holm, T. R., Cheson, B. D., and Edmundson, A. B. (1985). Accessible intrachain disulfide bonds in hybrids of light chain. *Mol. Immunol.* **22,** 85–92.

Ephrussi, A., Church, G. M., Tonegawa, S., and Gilbert, W. (1985). B lineage-specific interations of an immunoglobulin enhancer with cellular factors in vivo. *Science* **227,** 134–140.

Epp, O., Colman, P., Fehlhammer, H., Bode, W., Schiffer, M., Huber, R., and Palm, W. (1974). Crystal and molecular structure of a dimer composed of the variable portions of the Bence–Jones REI. *Eur. J. Biochem.* **45,** 513–524.

Fanning, D. W., Smith, J. A., and Rose, G. D. (1986). Molecular cartography of globular proteins with application to antigenic sites. *Biopolymers* **25,** 863–883.

Feldmann, R. J., Potter, M., and Glaudemans, C. P. J. (1981). A hypothetical space-filling model of the v-regions of the galactan-binding myeloma immunoglobulin J539. *Mol. Immunol.* **18,** 683–698.

Fersht, A. R. (1972). Conformational equilibria in α- and γ-chymotrypsin: The energetics and importance of the salt bridge. *J. Mol. Biol.* **64,** 497–509.

Fieser, T. M., Tainer, J. A., Geysen, H. M., Houghten, R. A., and Lerner, R. A. (1987). Influence of protein flexibility and peptide conformation on reactivity of monoclonal anti-peptide antibodies with a protein α-helix. *Proc. Natl. Acad. Sci. U.S.A.* **84,** 8568–8572.

Fine, R. M., Wang, H., Shenkin, P. S., Yarmush, D. L., and Levinthal, C. (1986). Prediction of complementarity determining region (CDR) structures from services. *Proteins: Struct. Funct. Genet.* **1,** 342–363.

Fraga, S. (1982). Theoretical prediction of protein antigenic determinants from amino acid sequences. *Can. J. Chem.* **60,** 2606–2610.

Francis, M. J., Hastings, G. Z., Syred, A. D., McGinn, B., Brown, F., and Rowlands, D. J. (1987). Non-responsiveness to a foot-and-mouth disease virus peptide overcome by addition of foreign helper T-cell determinants. *Nature (London)* **330,** 168–170.

Furey, W., Jr., Wang, B. C., Yoo, C. S., and Sax, M. (1983). Structure of a novel Bence–Jones protein (Rhe) fragment at 1.6 Å resolution. *J. Mol. Biol.* **167,** 661–692.

Gerin, J. L., Alexander, H., Shih, J. W-K., Purcell, R. H., Dapolito, G., Engle, R., Green, N., Sutcliffe, J. G., Shinnick, T. M., and Lerner, R. A., (1983). Chemically synthesized peptides of hepatitus B surface antigen duplicate the d/y specificites and induce subtype-specific antibodies in chimpanzees. *Proc. Natl. Acad. Sci. U.S.A.* **80,** 2365–2369.

Gettins, P., Potter, M., Leatherbarrow, R. J., and Dwek, R. A. (1982). A combined proton and phosphorus-31 nuclear magnetic resonance investigation of the combining site of M603, a phosphocholine-binding myeloma protein. *Biochemistry* **21,** 4927–4931.

Getzoff, E. D. (1985). Electrostatic recognition between antigen and antibody. In "Molecular Dynamics and Protein Structure" (J. Hermans, ed.), pp. 115–118. UNC Press, Chapel Hill, North Carolina.

Getzoff, E. D., and Tainer, J. A. (1985). Superoxide dismutase as a model ion channel. In "Ion Channel Reconstitution" (C. Miller, ed.), pp. 57–74. Plenum, New York.

Getzoff, E. D., Tainer, J. A., Weiner, P. K., Kollman, P. A., Richardson, J. S., and Richardson, D. C. (1983). Electrostatic recognition between superoxide and copper, zinc superoxide dismutase. *Nature (London)* **306,** 287–290.

Getzoff, E. D., Tainer, J. A., and Lerner, R. A. (1985). The chemistry of antigen–antibody union. In "Immune Regulation" (M. Feldmann and N. A. Mitchison, eds.), pp. 243–258. Humana, Clifton, New Jersey.

Getzoff, E. D., Hallewell, R. A., and Tainer, J. A. (1986a). Structural implications for macromolecular recognition and redesign. In "Protein Engineering: Applications in Science, Industry, and Medicine" (M. Inouye, ed.), pp. 41–69. Academic Press, New York.

Getzoff, E. D., Tainer, J. A., and Olson, A. J. (1986b). Recognition and interactions controlling the assemblies of β barrel domains. *Biophys. J.* **49,** 191–206.

Getzoff, E. D., Geysen, H. M., Rodda, S. J., Alexander, H., Tainer, J. A., and Lerner, R. A. (1987). Mechanisms of antibody binding to a protein. *Science* **235,** 1191–1196.

Getzoff, E. D. McRee, D. E., Parge, H. E., Capozza, M. A., Bernstein, S. L., and Tainer, J. A. (1988a). Structural implications for antigenic recognition. In "Vaccines: New Concepts and Developments" (H. Kohler and P. T. La Verde, eds.), pp. 29–41. Longman, London.

Getzoff, E. D., Parge, H. E., McRee, D. E., and Tainer, J. T. (1988b). Understanding the structure and antigenicity of gonococcal pili. *Rev. Infect. Dis.* **10,** in press.

Geysen, H. M., Meloen, R. H., and Barteling, S. J. (1984). Use of peptide synthesis to probe viral antigens for epitopes to a resolution of a single amino acid. *Proc. Natl. Acad. Sci. U.S.A.* **81,** 3998–4002.

Geysen, H. M., Barteling, S. J., and Meloen, R. H. (1985a). Small peptides induce antibodies with a sequence and structural requirement for binding antigen comparable to antibodies raised against the native protein. *Proc. Natl. Acad. Sci. U.S.A.* **82,** 178–182.

Geysen, H. M., Mason, T. J., Rodda, S. J., Meloen, R. H., and Barteling, S. J. (1985b). Amino acid composition of antigenic determinants: Implication for antigen processing by the immune system of animals. In "Vaccines 85" (R. A. Lerner, R. M. Chanock, and F. Brown, eds.), pp. 133–137. Cold Spring Harbor Laboratory, New York.

Geysen, H. M., Rodda, S. J., Mason, T. J., Tainer, J. A., Alexander, H., Getzoff, E.D., and Lerner, R. A. (1987a). Antigenicity of myohemerythrin. *Science* **238,** 1584–1586.

Geysen, H. M., Rodda, S. J., Mason, T. J., Tribbick, G., and Schoofs, P. G. (1987b). Strategies for epitope analysis using peptide synthesis. *J. Immunol. Methods* **102**, 259–274.

Geysen, H. M., Tainer, J. A., Rodda, S. J., Mason, T. J., Alexander, H., Getzoff, E. D., and Lerner, R. A. (1987c). Chemistry of antibody binding to a protein. *Science* **235**, 1184–1190.

Geysen, H. M., Mason, T. J., and Rodda, S. J. (1988a). Cognitive features of continuous antigenic determinants. *J. Mol. Recognition* **1**, 32–41.

Geysen, H. M., Rodda, S. J., and Mason, T. J. (1988b). A synthetic strategy for epitope mapping. *Proc. Am. Pept. Symp. 10th*, St. Louis, Missouri, pp. 519–523.

Gnann, J. W., McCormick, J. B., Mitchell, S., Nelson, J. A., and Oldstone, Michael B. A. (1987). Synthetic peptide immunoassay distinguishes HIV type I and HIV type 2 infections. *Science* **237**, 1346–1349.

Good, M. F., Maloy, W. L., Lunde, M. N., Margalit, H., Cornette, J. L., Smith, G. L., Moss, B., Miller, L. H., and Berzofsky, J. A. (1987). Construction of synthetic immunogen: Use of new T-helper epitope on malaria circumsporozoite protein. *Science* **235**, 1059–1062.

Green, N. M. (1975). Avidin. *Adv. Protein Chem.* **29**, 85–133.

Greenspan, N. S., and Monafo, W. J. (1987). Topographic analysis with monoclonal antiidiotypes: Probing the functional anatomy of immunoglobulin variable domains. *Int. Rev. Immunol.* **2**, 391–417.

Hagblom, P., Segal, E., Billard, E., and So, M. (1985). Intragenic recombination leads to pilus antigenic variation in *Neisseria gonorrhoeae*. *Nature (London)* **315**, 156–158.

Hamel, P. A., Isenman, D. E., Klein, M. H., Luedtke, R., and Dorrington, K. (1984). Structural basis for the preferential association of autologous immunoglobulin subunits: Role of the J region of the light chain. *Mol. Immunol.* **21**, 227–283.

Hogrefe, H. H., Griffith, J. P., Rossmann, M. G., and Goldberg, E. (1987). Characterization of the antigenic sites on the refined 3-Å resolution structure of mouse testicular lactate dehydrogenase. *Can. J. of Biol. Chem.* **262**, 13155–13162.

Holowka, D. A., Strosberg, A. D., Kimball, J. W., Harber, E., and Cathou, R. E. (1972). Changes in intrinsic circular dichroism of several homogeneous anti-type III pneumococcal antibodies on binding of a small hapten. *Proc. Natl. Acad. Sci. U.S.A.* **69**, 3399–3403.

Honjo, T. (1983). Immunoglobulin genes. *Annu. Rev. Immunol.* **1**, 499–528.

Hopp, T. P. (1986). Protein surface analysis: Methods for identifying antigenic determinants and other interaction sites. *J. Immunol. Methods* **88**, 1–18.

Hopp, T. P., and Woods, K. R. (1981). Prediction of protein antigenic determinants from amino acid sequences. *Proc. Natl. Acad. Sci. U.S.A.* **78**, 3824–3828.

Hopp, T. P. and Woods, K. R. (1983). A computer program for predicting antigenic determinants. *Mol. Immunol.* **20**, 483–489.

Houghten, R. A. (1985). General method for the rapid solid-phase synthesis of large numbers of peptides: Specificity of antigen–antibody interaction at the level of individual amino acids. *Proc. Natl. Acad. Sci. U.S.A.* **82**, 5131–5135.

Houghten, R. A., DeGraw, S. T., Bray, M. K., Hoffmann, S. R., and Frizzell, N. D. (1986). Simultaneous multiple peptide synthesis: The rapid preparation of large numbers of discrete peptides for biological, immunological, and methodological studies. *BioTechniques* **4**, 522–528.

Howell, E. E., Villafranca, J. E., Warren, M. S., Oatley, S. J., and Kraut. J. (1986). Functional role of aspartic acid-27 in dihydrofolate reductase revealed by mutagenesis. *Science* **231**, 1123–1128.

Huber, R. (1986). Structural basis for antigen-antibody recognition. *Science* **233**, 702-703.
Huber, R., and Bennett, W. S., Jr. (1983). Functional significance of flexibility in proteins. *Biopolymers* **22**, 261-279.
Ichiye, T., Olafson, B. D., Swaminathan, S., and Karplus, M. (1986). Structure and internal mobility of proteins: A molecular dynamics study of hen egg white lysozyme. *Biopolymers* **25**, 1909-1937.
Jacob, C. O., Sela, M., and Arnon, R. (1983). Antibodies against synthetic peptides of the B subunit of cholera toxin: Crossreaction and neutralization of the toxin. *Proc. Natl. Acad. Sci. U.S.A.* **80**, 7611-7615.
Jacobs, J., Schultz, P. G., Sugasawara, R., and Powell, M. (1987). Catalytic antibodies. *J. Am. Chem. Soc.* **109**, 2174-2176.
Jemmerson, R., and Paterson, Y. (1986a). Mapping epitopes on a protein antigen by the proteolysis of antigen-antibody complexes. *Science* **232**, 1001-1004.
Jemmerson, R., and Paterson, Y. (1986b). Mapping antigenic sites on proteins: Implications for the design of synthetic vaccines. *BioTechniques* **4**, 18-31.
Jones, P. T., Dear, P. H., Foote, J., Neuberger, M. S., and Winter, G. (1986). Replacing the complementarity-determining regions in a human antibody with those from a mouse. *Nature (London)* **321**, 522-525.
Kabat, E. A. (1978). The structural basis of antibody complementarity. *Adv. Protein Chem.* **32**, 1-75.
Kabat, E. A. (1983). The antibody combining site. *Prog. Immunol. Int. Congr. Immunol. 5th,* pp. 67-85.
Kabat, E. A., Wu, T. T., and Bilofsky, H. (1976). Attempts to locate residues in complementarity-determining regions of antibody combining sites that make contact with antigen. *Proc. Natl. Acad. Sci. U.S.A.* **73**, 617-619.
Kabat, E. A., Wu, T. T., Reid-Miller, M., Perry, H. M., and Gottesman, K. S. (1987). "Sequences of Proteins of Immunological Interest," 4th Ed. United States Department of Health and Human Services, National Institutes of Health, Bethesda, Maryland.
Karplus, P. A., and Schulz, G. E. (1985). Prediction of chain flexibility in proteins. (1985). *Naturwissenschaften* **72**, 212-213.
Karush, F. (1962). Immunologic specificity and molecular structure. *Adv. Immunol.* **2**, 1-40.
Karush, F. (1978). The affinity of antibody: Range, variability, and the role of multivalence. *In* "Comprehensive Immunology" (G. W. Litman and R. A. Good, eds.), pp. 85-116. Plenum, New York.
Kieber-Emmons, T., and Köhler, H. (1986a). Evolutionary origin of autoreactive determinants (autogens). *Proc. Natl. Acad. Sci. U.S.A.* **83**, 2521-2525.
Kieber-Emmons, T., and Köhler, H. (1986b). Towards a unified theory of immunoglobulin structure-function relations. *Immunol. Rev.* **90**, 29-48.
Kieber-Emmons, T., Getzoff, E., and Köhler, H. (1987). Perspectives on antigenicity and idiotypy. *Int. Rev. Immunol.* **2**, 339-356.
Kim, P. S., and Baldwin, R. L. (1984). A helix stop signal in the isolated s-peptide of ribonuclease A. *Nature (London)* **307**, 329-334.
Klippenstein, G. L., Van Riper, D. A., and Oosterom, E. A. (1972). A comparative study of the oxygen transport proteins of *Dendrostomum pyroides*. *J. Biol. Chem.* **247**, 5959-5963.
Kohler, G., and Milstein, C. (1975). Continuous cultures of fused cells secreting antibody of predetermined specificity. *Nature (London)* **256**, 495-497.

Koshland, D. E., Jr., (1970). The molecular basis for enzyme regulation. In "The Enzymes," Vol. I, 3rd Ed. (P. D. Boyer, ed.) pp. 342–397. Academic Press, New York.

Koshland, D. E., Jr. (1976). Role of flexibility in the specificity, control, and evolution of enzymes. FEBS Lett. 62, E47–E52.

Kuo, L., Davies, H. C., and Smith, L. (1986). Monoclonal antibody to human cytochrome c: Effect on electron-transfer reactions. Biochim. Biophys. Acta 848, 247–255.

Kyte J., and Doolittle, R. F. (1982). A simple method for displaying the hydropathic character of a protein. J. Mol. Biol. 157, 105–132.

Lancet, D., and Pecht, I. (1976). Kinetic evidence for hapten-induced conformational transition in immunoglobulin MOPC 460. Proc. Natl. Acad. Sci. U.S.A. 73, 3549–3553.

Lando, G., and Reichlin, M. (1982). Antigenic structure of sperm whale myoglobin: II. Characterization of antibodies preferentially reactive with peptides arising in response to immunization with the native protein. J. Immunol. 129, 212–216.

Lando, G., Berzofsky, J. A., and Reichlin, M. (1982). Antigenic structure of sperm whale myoglobin: I. Partition of specificities between antibodies reactive with peptides and native protein. J. Immunol. 129, 206–211.

Lee, B., and Richards, F. M. (1971). The interpretation of protein structures: Estimation of static accessibility. J. Mol. Biol. 55, 379–400.

Lee, J. S., Dombroski, D. F., and Mosmann, T. R. (1982). Specificity of autoimmune monoclonal Fab fragments binding to single-stranded deoxyribonucleic acid. Biochemistry 21, 4940–4945.

Lerner, R. A. (1982). Tapping the immunological repertoire to produce antibodies of predetermined specificity. Nature (London) 299, 592–596.

Lerner, R. A. (1983). Synthetic vaccines. Sci. Am. 248, 66–74.

Lerner, R. A. (1984). Antibodies of predetermined specificity in biology and medicine. Adv. Immunol. 36, 1–44.

Lerner, R. A., and Tramontano, A. (1987). Antibodies as enzymes. Trends Biochem. Sci. Pers. Ed. 12, 427–430.

Lerner, R. A., and Tramontano, A. (1988). Catalytic antibodies. Sci. Am. 258, 58–70.

Lerner, R. A., Green, N., Alexander, H., Liu, F-T., Sutcliffe, J. G., and Shinnick, T. M. (1981a). Chemically synthesized peptides predicted from the nucleotide sequence of the hepatitus B virus genome elicit antibodies reactive with the native envelope protein of Dane particles. Proc. Natl. Acad. Sci. U.S.A. 78, 3403–3407.

Lerner, R. A., Sutcliffe, J. G., and Shinnick, T. M. (1981b). Antibodies to chemically synthesized peptides predicted from DNA sequence as probes of gene expression. Cell 23, 309–310.

Levinson, S. A., Hicks, A. N., Portmann, A. J., and Dandliker, W. B. (1975). Fluorescence polarization and intensity kinetic studies of antifluorescein antibody obtained at different stages of the immune response. Biochemistry 14, 3778–3786.

Lubeck, M., and Gerhard, W. (1982). Conformational change at topologically distinct antigenic sites on the influenza A/PR/8/34 virus HA molecule are induced by binding of monoclonal antibodies. Virology 118, 1–7.

Lugwig, D. S., Finkelstein, R. A., Karu, A. E., Dallas, W. S., Ashby, E. R., and Schoolnik, G. K. (1987). Anti-idiotypic antibodies as probes of protein active sites: Application to cholera toxin subunit B. Proc. Natl. Acad. Sci. U.S.A. 84, 3673–3677.

Mainhart, C. R., Potter, M., and Feldmann, R. J. (1984). A refined model for the variable domains (Fv) of the J539 $\beta(1,6)$-D-galactan-binding immunoglobulin. Mol. Immunol. 21, 469–478.

Mariuzza, R. A., Phillips, S. E., and Poljak, R. J. (1987). The strutural basis of antigen–antibody recognition. Annu. Rev. Biophys. Chem. 16, 139–159.

Marquart, J., Deisenhofer, J., Huber, R., and Palm, W. (1980). Crystallographic refinement and atomic models of the interact immunoglobulin molecule kol and its antigen-binding fragment at 3.0 Å and 1.9 Å resolution. *J. Mol. Biol.* **141,** 369–391.

Marquart, M., and Deisenhofer, J. (1982). The three-dimensional structure of antibodies. *Immunol. Today* **3,** 160–167.

Mason, T. J., and Geysen, H. M. (1988). Antibody response to tobacco mosaic virus coat protein. *J. Immunol. Methods.* In preparation.

Matsushima, M., Marquart, M., Jones, T. A., Colman, P. M., Bartels, K., and Huber, R. (1978). Crystal structure of the human Fab fragment Kol and its comparison with the intact Kol molecule. *J. Mol. Biol.* **121,** 441–459.

Matthew, J. B. (1985). Electrostatic effects in proteins. *Annu. Rev. Biophys. Biophys. Chem.* **14,** 387–417.

Matthew, J. B., and Gurd, F. R. N. (1986). Stabilization and destabilization of protein structure by charge interactions. *In* "Methods in Enzymology," Vol. 130 (C. H. W. Hirs and S. N. Timasheff, eds.), pp. 437–453. Academic Press, Orlando.

Max, E. D. (1984). Immunoglobulins: Molecular genetics. *In* "Fundamental Immunology" (W. E. Paul, ed.), pp. 167–204. Raven, New York.

Max, N. L., and Getzoff, E. D. (1988). Spherical harmonic molecular surfaces. *IEEE Comput. Graphics Appl.* **8,** 42–50.

McCray, J., and Werner, G. (1987). Different rhinovirus serotypes neutralized by antipeptide antibodies. *Nature (London)* **329,** 736–738.

Mehra, V., Sweetser, D., and Young, R. A. (1986). Efficient mapping of protein antigenic determinants. *Proc. Natl. Acad. Sci. U.S.A.* **83,** 7013–7017.

Metzger, D. W., Ch'ng, L-K., Miller, A., and Sercarz, E. E. (1984). The expressed lysozyme-specific B cell repertoire I. Heterogeneity in the monoclonal anti-hen egg white lysozyme specificity repertoire, and its difference from the *in situ* repertoire. *Eur. J. Immunol.* **14,** 87–93.

Metzger, H. (1978). The effect of antigen on antibodies: Recent studies. *Contemp. Top. Mol. Immunol.* **7** 119–152.

Miller, S., Lesk, A. M., Janin, J., and Chothia, C. (1987). The accessible surface area and stability of oligomeric proteins. *Nature (London)* **328,** 834–836.

Moews, P. C., Know, J. R., Waxman, D. J., and Strominger, J. L. (1981). Secondary structure relations between beta-lactamases and penicillin-sensitive D-alanine carboxypeptidases. *Int. J. Pept. Protein Res.* **17,** 211–218.

Moore, G. R., and Williams, R. J. P. (1980). Comparison of the structures of various eukaryotic ferricytochromes c and ferrocytochromes and their antigenic differences. *Eur. J. Biochem.* **103,** 543–550.

Moudallal, Z. Al., Briand, J. P., and Van Regenmortel, M. H. V. (1985) A major part of the polypeptide chain of tobacco mosaic virus protein is antigenic. *EMBO J.* **4,** 1231–1235.

Müller, G. M., Shapira, M., and Arnon, R. (1982). Anti-influenza response achieved by immunization with a synthetic conjugate. *Proc. Natl. Acad. Sci. U.S.A.* **79,** 569–573.

Napper, A., Benkovic, S. J., Tramontano, A., and Lerner, R. A. (1987). A stereospecific cyclization catalyzed by an antibody. *Science* **237,** 1041–1043.

Norrby, E., Biberfeld, G., Chiodi, F., Gegerfeldt, A. V., Naucler, A., Parks, E., and Lerner, R. A. (1987). Discrimination between antibodies to HIV and to related retroviruses using site-directed serology. *Nature (London)* **329,** 248–250.

Nossal, G. J. V. (1983). Cellular mechanisms of immunological tolerance. *Annu. Rev. Immunol.* **1,** 33–62.

Novotný, J., Bruccoleri, R., Newell, J., Murphy, D., Haber, E., and Karplus, M. (1983). Molecular anatomy of the antibody binding site. *J. Biol. Chem.* **258,** 14433–14437.

Novotný, J., Handschumacher, M., and Haber, E. (1986a). Location of antigenic epitopes on antibody molecules. *J. Mol. Biol.* **189,** 715–721.

Novotný, J., Handschumacher, M., Haber, E., Bruccoleri, R. E., Carlson, W. B., Fanning, D. W., Smith, J. A., and Rose, G. D. (1986b). Antigenic determinants in proteins coincide with surface regions accessible to large probes (antibody domains). *Proc. Natl. Acad. Sci. U.S.A.* **83,** 226–230.

O'Donnell, T. J., and Olson, A. J. (1981). GRAMPS—a graphics language interpreter for real-time interactive three-dimensional picture editing and animation. *Computer Graphics* **15,** 133–142.

Ohlendorf, D. H., and Matthew, J. B. (1985). Electrostatics and flexibility in protein–DNA interactions. *Adv. Biophys.* **20,** 137–151.

Okada, A., Nakanishi, M., Tsurui, H., Wada, A., Terashima, M., and Osawa, T. (1985). A hapten-induced conformational change accompanies the cryoprecipitation of an immunoglobulin. *Mol. Immunol.* **22,** 715–718.

Oldstone, Michael B. A. (1987). Molecular mimicry and autoimmune disease. *Cell* **50,** 819–820.

Padlan, E. A. (1977). Structural basis for the specificity of antibody–antigen reactions and structural mechanisms for the diversification of antigen-binding specificities. *Q. Rev. Biophys.* **10,** 35–65.

Padlan, E. A. (1985). Quantitation of the immunogenic potential of protein antigens. *Mol. Immunol.* **22,** 1243–1254.

Padlan, E. A., Davies, D. R., Rudikoff, S., and Potter, M. (1976a). Structural basis for the specificity of phosphorylcholine-binding immunoglobulins. *Immunochemistry* **13,** 945–949.

Padlan, E. A., Davies, D. R., Pecht, I., Givol, D., and Wright, C. (1976b). Model-building studies of antigen-binding sites: The hapten-binding site of MOPC-315. *Cold Spring Harbor Symp. Quant. Biol.* **41,** 627–637.

Parge, H. E., McRee, D. E., Capozza, M. A., Bernstein, S. L., Getzoff, E. D., and Tainer, J. A. (1987). Three dimensional structure of bacterial pili. *Antonie van Leeuwenhoek J. Microbiol.* **53,** 447–453.

Parham, P. (1984). Changes in conformation with loss of alloantigenic determinants of a histocompatibility antigen (HLA-B7) induced by monoclonal antibodies. *J. of Immunol.* **132,** 2975–2983.

Parker, J. M. R., Guo, D., and Hodges, R. S. (1986). New hydrophilicity scale derived from high-performance liquid chromatography peptide retention data: Correlation of predicted surface residues with antigenicity and X-ray derived accessible sites. *Biochemistry* **25,** 5425–5432.

Paterson, Y. (1985). Delination and conformational analysis of two synthetic peptide models of antigenic sites on rodent cytochrome c. *Biochemistry* **24,** 1048–1055.

Paterson, Y. (1988). Structural differences in B- and T-cell determinants on protein antigens. *Int. Rev. Immunol.* In press.

Perutz, M. F. (1978). Electrostatic effects in proteins. *Science* **201,** 1187–1191.

Phillips, D. C. (1974). Crystallographic studies of lysozyme and its interactions with inhibitors and substrates. *In* "Lysozyme" (E. F. Osseman, R. E. Canfield, and S. Beychok, eds.), pp. 9–30. Academic Press, New York.

Poljak, R. J., Amzel, L. M., Chen, B. L., Phizackerley, R. P., and Saul, F. (1974). The 3-dimensional structure of the Fab' fragment of a human myeloma immunoglobulin at 2.0-Å resolution. *Proc. Natl. Acad. Sci. U.S.A.* **71**(9), 3440–3444.

Pollack, S. J., Jacobs, J. W., and Schultz, P. G. (1986). Selective chemical catalysis by an antibody. *Science* **234,** 1570–1573.

Prince, A. M., Ikram, H., and Hopp, T. P. (1982). Hepatitus B virus vaccine: Identification of HBsAg/a and HBsAg/d but not HBsAg/y subtype antigenic determinants on a synthetic immunogenic peptide. *Proc. Natl. Acad. Sci. U.S.A.* **79**, 579–582.

Pullman, B., Lavery, R., and Pullman, A. (1982). Two aspects of DNA polymorphism and microheterogeneity: Molecular electrostatic potential and steric accessibility. *Eur. J. Biochem.* **124**, 229–238.

Rao, S. T., Hogle, J., and Sundaralingam, M. (1983). Studies of monoclinic hen egg white lysozyme. The refinement at 2.5 Å resolution–conformational variability between the two independent molecules. *ACTA Crystallogr. Sect. C: Cryst. Struct. Commun.* **39**, 237–240.

Rees, A. R., and de la Paz, P. (1986). Investigating antibody specificity using computer graphics and protein engineering. *Trends Biochem. Sci.* **11**, 144–148.

Richards, F. M. (1977). Areas, volumes, packing, and protein structure. *Annu. Rev. Biophys. Bioeng.* **6**, 151–176.

Richards, F. M. (1985). Calculation of molecular volumes and areas for structures of known geometry. *In* "Methods in Enzymology," Vol. 115 (H. W. Wyckoff, C. H. W. Hirs, and S. N. Timasheff, eds.), pp. 440–464. Academic Press, New York.

Richardson, J. S. (1981). The anatomy and taxonomy of protein structure. *In* "Advances in Protein Chemistry" (C. B. Anfinsen, J. T. Edsall, and F. M. Richards, eds.), pp. 168–339. Academic Press, New York.

Roberts, S., Cheetham, J. C., and Rees, A. R. (1987). Generation of an antibody with enhanced affinity and specificity for its antigen by protein engineering. *Nature (London)* **328**, 731–734.

Rodda, S. J., Geysen, H. M., Mason, T. J., and Schoofs, P. G. (1986a). The antibody response to myoglobin. I. Systematic synthesis of myoglobin peptides reveals location and substructure of species-dependent continuous antigenic determinants. *Mol. Immunol.* **23**, 603–610.

Rodda, S. J., Geysen, H. M., Mason, T. J., and Tribbick, G. (1986b). Probing the antigen combining sites of antibodies using synthetic peptides. *Protides Biol. Fluids* **34**, 91–93.

Rowlands, D. J., Clarke, B. E., Carroll, A. R., Brown, F., Nicholson, B. H., Bittle, J. L., Houghten, R. A., and Lerner, R. (1983). Chemical basis of antigenic variation in foot-and-mouth disease virus. *Nature (London)* **306**, 694–697.

Satow, Y., Cohen, H. G., Padlan, E. A., and Davies, D. R. (1986). Phosphocholine binding immunoglobulin Fab McPC603: An x-ray diffraction study at 2.7 Å. *J. Mol. Biol.* **190**, 593–604.

Saul, F. A., and Poljak, R. J. (1985). Three-dimensional structure and function of immunoglobulins. *Ann. Inst. Pasteur Paris* **136**, 259–294.

Saul, F. A., Amzel, L. M., and Poljak, R. J. (1978). Preliminary refinement and structural analysis of the Fab fragment from human immunoglobulin new at 2.0 Å resolution. *J. Biol. Chem.* **253**, 585–597.

Scharff, M. D., Chien, N., Giusti, A. M., Zack, D., and Diamond, B. (1985). Single amino acid substitutions cause changes in antibody affinity and specificity. *In* "Immune Recognition of Protein Antigens" (W. G. Laver, and G. M. Air, eds.), pp. 44–47.

Schlessinger, J., Steinberg, I. Z., Givol, D., Hockman, J., and Pecht, I. (1975). Antigen-induced conformational changes in antibodies and their Fab fragments studied by circular polarization of fluorescence. *Proc. Natl. Acad. Sci. U.S.A.* **72**, 2775–2779.

Schmidt, M. A., O'Hanley, P., Lark, D., and Schoolnik, G. K. (1988). Synthetic peptides corresponding to protective epitopes of *Escherichia coli* digalactoside-binding pilin prevent infection in a murine pyelonephritis model. *Proc. Natl. Acad. Sci. U.S.A.* **85**, 1247–1251.

Schoofs, P. G., Geysen, H. M., Jackson, D. C., Brown, L. E., Tang, X-L., and White, D. O.

(1988). Epitopes of an influenza viral peptide recognized by antibody at single amino acid resolution. *J. Immunol.* **140,** 611–616.

Schulze-Gahmen, U., Prinz, H., Glatter, U., and Beyreuther, K. (1985). Towards assignment of secondary structures by anti-peptide antibodies. Specificity of the immune response to a β-turn. *EMBO J.* **4,** 1731–1737.

Segal, D. M., Padlan, E. A., Cohen, G. H., Rudikoff, S., Potter, M., and Davies, D. R. (1974). The three-dimensional structure of a phosphorylcholine-binding mouse immunoglobulin Fab and the nature of the antigen binding site. *Proc. Natl. Acad. Sci. U.S.A.* **71,** 4298–4302.

Sheridan, R. P., and Allen, L. C. (1981). The active site electrostatic potential of human carbonic anhydrase. *J. Am. Chem. Soc.* **103,** 1544–1550.

Sheriff, S., Hendrickson, W. A., Stenkamp, R. E., Sieker, L. C., and Jensen, L. H., (1985). Influence of solvent accessibility and intermolecular contacts on atomic mobilities in hemerythrins. *Proc. Natl. Acad. Sci. U.S.A.* **82,** 1104–1107.

Sheriff, S., Hendrickson, W. A., and Smith, J. L. (1987a). Structure of myohemerythrin in the azidomet state at 1.7/1.3 Å resolution. *J. Mol. Biol.* **197,** 273–296.

Sheriff, S., Silverton, E. W., Padlan, E. A., Cohen, G. H., Smith-Gill, S. J., Finzel, B. C., and Davies, D. R. (1987b). Three-dimensional structure of an antibody–antigen complex. *Proc. Natl. Acad. Sci. U.S.A.* **84,** 8075–8079.

Shinnick, T. M., Sutcliffe, J. G., Green, N., and Lerner, R. A. (1983). Synthetic peptide immunogens as vaccines. *Annu. Rev. Microbiol.* **37,** 425–446.

Shrake, A., and Rupley, J. A. (1973). Environment and exposure to solvent of protein atoms. Lysozyme and insulin. *J. Mol. Biol.* **79,** 351–371.

Silverton, E. W., Navia, M. A., and Davies, D. R. (1977). Three-dimensional structure of an intact human immunoglobulin. *Proc. Natl. Acad. Sci. U.S.A.* **74,** 5140–5144.

Smith, G. D., Swenson, D. C., Dodson, E. J., Dodson, G. G., and Reynolds, C. D. (1984). Structural stability in the 4-zinc human insulin hexamer. *Proc. Natl. Acad. Sci. U.S.A.* **81,** 7093–7097.

Smith-Gill, S. J., Wilson, A. C., Potter, M., Prager, E. M., Feldmann, R. J., and Mainhart, C. R. (1982). Mapping the antigenic epitope for a monoclonal antibody against lysozyme. *J. Immunol.* **128,** 314–322.

Smith-Gill, S. J., Lavoie, T. B., and Mainhart, C. R. (1984). Antigenic regions defined by monoclonal antibodies corresponding to structural domains of avian lysozyme. *J. Immunol.* **133,** 384–393.

Smith-Gill, S. J., Hamel, P. A., Klein, M. H., Rudikoff, S., and Dorrington, K. J. (1986). Contribution of the V_K4 light chain to antibody specificity for lysozyme and $\beta(1,6)$D-galctan. *Mol. Immunol.* **23,** 919–926.

Smith-Gill, S. J., Hamel, P. A., Lavoie, T. B., and Dorrington, K. J. (1987a). Contributions of immunoglobulin heavy and light chains to antibody specificity for lysozyme and two haptens. *J. Immunol.* **139,** 4135–4144.

Smith-Gill, S. J., Mainhart, C., Lavoie, T. B., Feldmann, R. J., Drohan, W., and Brooks, B. R. (1987b). Three-dimensional model of an anti-lysozyme antibody. *J. Mol. Biol.* **194,** 713–724.

Stanford, J. M., and Wu, T. T. (1981). A predictive method for determining possible three-dimensional foldings of immunoglobulin backbones around antibody combining sites. *J. Theor. Biol.* **88,** 421–439.

Stevens, F. J., Westholm, F. A., Solomon, A., and Schiffer, M. (1980). Self-association of human immunoglobulin κI light chains: Role of the third hypervarible region. *Proc. Natl. Acad. Sci. U.S.A.* **77,** 1144–1148.

Steward, M. W., and Steensgaard, J. (1983). "Antibody Affinity." CRC Press, Inc., Boca Raton, Florida.

Suh, W. W., Bhat, T. N., Navia, M. A., Cohen, G. H., Rao, D. H., Rudikoff, S., and Davies, D. R. (1986). The galactan-binding immunoglobulin Fab J539: An x-ray diffraction study at 2.6 Å resolution. *Proteins: Struct. Funct. Genet.* **1,** 74–80.

Sutcliffe, J. G., Shinnick, T. M., Green, N., Liu, F. T., Niman, H. L., and Lerner, R. A. (1980). Chemical synthesis of a polypeptide predicted from nucleotide sequence allows detection of a new retroviral gene product. *Nature (London)* **287,** 801–805.

Sutcliffe, J. G., Shinnick, T. M., Green, N., and Lerner, R. A. (1983). Antibodies that react with predetermined sites on proteins. *Science* **219,** 660–666.

Tainer, J. A., Getzoff, E. D., Richardson, J. S., and Richardson, D. C. (1983). Structure and mechanism of Cu, Zn superoxide dismutase. *Nature (London)* **306,** 284–287.

Tainer, J. A., Getzoff, E. D., Alexander, H., Houghten, R. A., Olson, A. J., Lerner, R. A., and Hendrickson, W. A. (1984). The reactivity of anti-peptide antibodies is a function of the atomic mobility of sites in a protein. *Nature (London)* **312,** 127–133.

Tainer, J. A., Getzoff, E. D., Paterson, Y., Olson, A. J., and Lerner, R. A. (1985a). The atomic mobility component of protein antigenicity. *Annu. Rev. Immunol.* **3,** 501–535.

Tainer, J. A., Getzoff, E. D., Sayre, J., and Olson, A. J. (1985b). Modeling intermolecular interactions: Topography, mobility, and electrostatic recognition. *J. Mol. Graphics* **3,** 103–105.

Teillaud, J-L., Desaymard, C., Giusti, A. M., Haseltine, B., Pollock, R. R., Yelton, D. E., Zack, D. J., and Scharff, M. D. (1983). Monoclonal antibodies reveal the structural basis of antibody diversity. *Science* **222,** 721–726.

Thornton, J. M., and Sibanda, B. L. (1983). Amino and carboxy-terminal regions in globular proteins. *J. Mol. Biol.* **167,** 443–460.

Thornton, J. M., Edwards, M. S., Taylor, W. R., and Barlow, D. J. (1986). Location of 'continuous' antigenic determinants in the protruding regions of proteins. *EMBO J.* **5,** 409–413.

Tonegawa, S. (1983). Somatic generation of antibody diversity. *Nature (London)* **302,** 575–581.

Tramontano, A., Janda, K. D., and Lerner, R. A. (1986). Catalytic antibodies. *Science* **234,** 1566–1570.

Tramontano, A., Ammann, A. A., and Lerner, R. A. (1988). Antibody catalysis approaching the activity of enzymes. *J. Am. Chem. Soc.* **110,** 2282–2286.

Tribbick, G., Getzoff, E. D., and Geysen, H. M. (1988). Comparison of the immune response among different species. In preparation.

Van Eldik, L. J., and Lukas, T. J. (1987). Site-directed antibodies to vertebrate and plant calmodulins. *In* "Methods in Enzymology" Vol. 139 (A. R. Means and P. M. Conn, eds.), pp. 393–405. Academic Press, Orlando.

Van Regenmortel, M. H. V. (1987). Antigenic cross-reactivity between proteins and peptides: New insights and applications. *Trends Biochem. Sci.* **12,** 237–240.

Van Regenmortel, M. H. V., Muller, S., Quesniaux, V. F., Altschuh, D., and Briand, J. P. (1988). Operational aspects of epitope identification: Structural features of proteins recognized by antibodies. *In* "Vaccines: New Concepts and Developments" (H. Kohler, and P. T. LaVerde, eds.), pp. 113–122. Longman, London.

Victor-Korbin, C., Manser, T., Moran, T., Imanishi-Kari, T., Gefter, M., and Bona, C. (1985). Shared idiotopes among antibodies encoded by heavy-chain variable region (V_H) gene members of the J558 V_H family as basis for cross-reactive regulation of clones with different antigen specificity. *Proc. Natl. Acad. Sci. U.S.A.* **82,** 7696–7700.

Wakabayashi, K, Sakata, Y., and Aoki, N. (1986). Conformation-specific monoclonal antibodies to the calcium-induced structure of protein C. *J. Biol. Chem.* **261,** 11097–11105.

Wall, R., and Kuehl, M. (1983). Biosynthesis and regulation of immunoglobulins. *Annu. Rev. Immunol.* **1**, 393–422.

Warshel, A., and Levitt, M. J. (1976). Theoretical studies of enzyme reactions: Dielectric, electrostatic, and steric stabilization of the carbonium ion in the reaction of lysozyme. *J. Mol. Biol.* **103**, 227–249.

Weiner, P. K., Langridge, R., Blaney, J. M., Schaefer, R., and Kollman, P. A. (1982). Electrostatic potential molecular surfaces. *Proc. Natl. Acad. Sci. U.S.A.* **79**, 3754–3758.

Welling, G. W., Weijer, W. J., van der Zee, R., and Welling-Wester, S. (1985). Prediction of sequential antigenic regions in proteins. *FEBS Lett.* **188**, 215–218.

Wells, T. N. C., and Fersht, A. R. (1985). Hydrogen bonding in enzymatic catalysis analysed by protein engineering. *Nature (London)* **316**, 656–657.

Westhof, E., Altschuh, D., Moras, D., Bloomer, A. C., Mondragon, A., Klug, A., and Van Regenmortel, M. H. V. (1984). Correlation between segmental mobility and the location of antigenic determinants in proteins. *Nature (London)* **311**, 123–126.

Wetlaufer, D. B. (1981). Folding of protein fragments. *Adv. Protein Chem.* **34**, 61–92.

Wiley, D. C., Wilson, I. A., and Skehel, J. (1981). Structural identification of the antibody-binding sites of Hong Kong influenza haemagglutinin and their involvement in antigenic variation. *Nature (London)* **289**, 373–378.

Williams, R. J. P., and Moore, G. R. (1985). Protein antigenicity, organization and mobility. *Trends Biochem. Sci.* **10**, 96–97.

Wilson, I. A., Haft, D. H., Tainer, J. A., Getzoff, E. D., Lerner, R. A., and Brenner, S. (1984). Identical short peptide sequences in unrelated proteins can have different conformations: A testing ground for theories of immune recognition. *Proc. Natl. Acad. Sci. U.S.A.* **82**, 5255–5259.

Wolfenden, R. V., Cullis, P. M., and Southgate, C. C. F. (1979). Water, protein folding, and the genetic code. *Science* **206**, 575–577.

Wu, T. T., and Kabat, E. A. (1970). An analysis of the sequences of the variable regions of Bence–Jones proteins and myeloma light chains and their implications for antibody complementarity *J. Exp. Med.* **132**, 211–250.

Zidovetzki, R., Blatt, Y., and Pecht, I. (1981). A heterologous immunoglobulin chain recombinant carries a distinct site for dinitrophenyl and obeys the common hapten binding mechanism. *Biochemistry* **20**, 5011–5018.

Structure of Antibody–Antigen Complexes: Implications for Immune Recognition

P. M. COLMAN

Commonwealth Scientific and Industrial Research Organization (CSIRO), Division of Protein Chemistry, Parkville 3052, Victoria, Australia

I. Introduction

The basis of antibody–antigen interactions has long been believed to reside in complementary structures on the two molecules (Ehrlich, 1900). Similarly, complementarity is at the heart of enzyme–substrate interactions (Fischer, 1894), and indeed, of many chemical and biological phenomena (Pauling, 1948). This basic hypothesis has required little modification over the years and is now further substantiated by the description of three antibody–antigen complexes at atomic resolution (Amit *et al.*, 1986; Colman *et al.*, 1987a; Sheriff *et al.*, 1987). In each of these structures, the antigen is protein and only protein antigens will be considered here. No direct structural data are available for macromolecular carbohydrate or glycolipid antigens bound to antibody.

For protein antigens, the antibody–antigen interaction is but another example of a protein–protein interaction. A substantial data base, drawn from high-resolution X-ray diffraction studies, has enabled many of the fundamental general principles of these interactions to be enumerated. One of the central questions that this article seeks to address is, do antibody–antigen interactions demonstrate or extend these same principles?

In developing this theme, it will be necessary to review aspects of protein–protein interactions together with the structures of both antibody and antigen. A major focus will be on the structure of antibodies and the design principles of the antigen combining site, leading to commentary on the issue of antigen-induced conformational change in antibody, and vice versa.

II. Principles of Protein–Protein Interactions

A. THERMODYNAMICS

Protein folding is itself an example of a protein–protein interaction. The free energy of the folded state is a result of the balance between competing thermodynamic effects, conformational entropy on the one hand, which favors the unfolded state, and the free energy of hydrophobic interactions, of hydrogen bonds, and of van der Waals' contacts on the other, favoring the folded state (Kauzmann, 1959). Similarly, when two proteins bind to each other, the resulting difference in free energy between the bound and unbound states is a sum of entropy terms (e.g., changes in the entropy of water, the loss of independent motion of the interactants, and possibly, a reduction in flexibility of interacting segments) and enthalpy terms.

B. BURIED SURFACES

Since the hydrogen-bonding and electrostatic contributions to the free energy of surface regions on the protein are similar, either in interactions with water or with another protein molecule, it can be argued that the free energy of the interaction between two proteins is mainly a function of the buried surface area in the interface (Chothia, 1974, 1975). One rider to this simplification is that electrostatic contributions to the enthalpy will depend on the local value of the dielectric constant, near 80 if the environment is aqueous, or some lower value, if the charges are buried in a protein–protein interface. The reduction in hydrophobic free energy, resulting from the removal of 1 $Å^2$ of protein surface area from aqueous solvent, is approximately 25 cal/mol (Chothia, 1974), and in typical protein–protein complexes, each molecule or subunit has 600 $Å^2$ or more of its surface area abstracted from solvent (Chothia and Janin, 1975). In this simplified model of protein–protein interaction, the specificity is determined by enthalpy, i.e., complementary atom–atom interactions in the interface, and the binding energy is determined by hydrophobicity, i.e., the total area of buried surface in the complex.

One demonstration of the approximation embodied in this model comes from studies of antibody complexes with wild-type and with variant antigens. Certain single amino acid mutants of influenza virus neuraminidase bind monoclonal antibodies with reduced affinities, and yet in one case, it has now been shown that the binding of the antibody to wild-type and to a variant neuraminidase is isosteric (Laver *et al.*, 1987; Colman *et al.*, 1987a). In that case, and probably in others in which single amino acid changes in the interface influence binding energy [for example, heteroclitic antibodies, as discussed in Amit *et al.* (1986), and a reported change of three orders of magnitude in the binding affinity of an antibody as a result of an Arg-to-Lys substitution in the antigen (Sheriff *et al.*, 1987)], the alteration to the buried surface area is negligible, and the altered binding energy derives from changed atom–atom interactions (see further discussion in Section V,C). One striking high-resolution structure demonstration of the contribution of specific interactions to binding energy is the study of inhibitors of thermolysin (Bartlett and Marlowe, 1987; Tronrud *et al.*, 1987) in which the incorporation of one additional hydrogen bond into the enzyme–ligand interface lowers the K_i by a factor of nearly 1000.

A more recent study of the size of buried surfaces in oligomeric proteins (Miller *et al.*, 1987) confirms that, in stable protein–protein complexes, at least 600 Å^2 of surface area from each molecule or subunit is buried by the interaction. We might then expect that, when an antigen–antibody complex is formed, at least 600 Å^2 of surface area from both antigen and antibody will be removed from contact with solvent and buried in the interface (see Sections III,B,1 and V,A), presuming that the antigen can present a suitably large area of interface to complement the antibody.

A better method of estimating the solvation energy, either in protein folding or in protein–protein complexes, takes into consideration the atomic position and the atomic solvation parameter for individual atoms in a structure (Eisenberg and McLachlan, 1986). Atomic solvation parameters for nonpolar atoms, evaluated in that work, are in the range of 16–21 cal/Å^2/mol. Such analyses offer the possibility of accounting, in detail, for altered binding energy deriving from single amino acid sequence changes in the interface, as discussed above.

The equilibrium dissociation constant, K_d, for two liganded molecules is a function only of the on- and off-rates, k_{on} and k_{off}, for their reaction. The on-rate is determined by long-range forces (electrostatic charge and dipole fields) and by hydrogen bonds, but the off-rate will be more a function of short-range forces, hydrophobic and van der Waals' forces. The physical dimensions of the buried surface in the complex thus play a role in determining K_d by influencing k_{off}.

C. Complementarity

Aside from the general roles of complementarity and hydrophobicity in determining the stability of protein–protein interactions, the stereochemical details of such interactions can also be analyzed for insights into fundamental properties of proteins that might influence

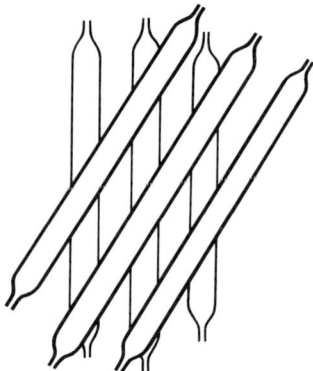

Fig. 1. En face schematic of sheet–sheet packing. In this figure, the aligned class is demonstrated by the angle of approximately $-30°$ between the strand direction in the two sheets.

their capacity to bind other proteins or ligands, in general. In particular, it is observed that there are a limited number of ways in which secondary structure elements of proteins can interact with each other. For example, several studies have addressed the question of how α-helices pack next to each other (Crick, 1953; Richmond and Richards, 1978; Efimov, 1979) or onto β sheets (Chothia et al., 1977).

For the purposes of reviewing antigen–antibody interactions, the most relevant class of secondary structure interaction between proteins is that of β sheet associations. The classic immunoglobulin fold is a pair of β sheets, comprising a so-called imunoglobulin domain, and the lateral noncovalent interactions between a pair of domains, which form the structural modules within an antibody molecule, are also examples of sheet–sheet packings.

Several studies have shown two common stereochemical themes for sheet–sheet interactions, one in which the strands of the two interacting sheets are nearly aligned (or more precisely inclined at about $-30°$) (Cohen et al., 1981; Chothia and Janin, 1981), and one in which they are approximately orthogonal (Chothia and Janin, 1982). The aligned packing mode is shown "en face" in Fig. 1, and the orthogonal packing is shown "edge on" in Fig. 2. The limited number of different sheet–sheet packings derives from restraints on the number of ways that side chains, associated with β sheets, can pack against side chains from other β sheets.

Examples of both the aligned and orthogonal sheet packing modes are found in immunoglobulin structures. The structures of both variable and constant domains (see Section III,B,1) conform to the

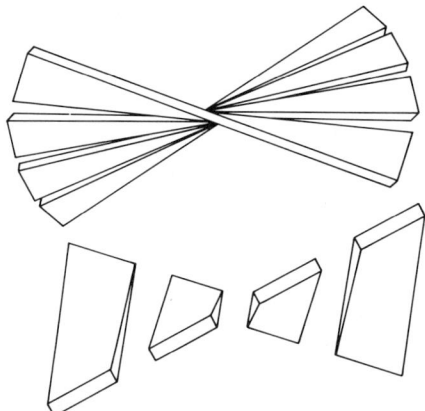

Fig. 2. Edge on schematic of sheet-sheet packing of the orthogonal class [adapted from Chothia and Janin (1982)]. This packing mode has been described for two sheets within the one polypeptide, but is also characteristic of the C_L-C_{H1} interface of the C module of Fab fragments.

aligned packing class, whereas the association of the constant light chain (C_L) and heavy chain (C_{H1}) domains and the dimerization of heavy chain C_{H3} domains demonstrate the orthogonal packing mode (see Section III,B,2).

In the protein structure data base (Bernstein et al., 1977), only one exception to this classification of β sheet associations is observed to date and that is in the association of V_L and V_H domains to form the V module in antibody molecules (Chothia et al., 1985). The importance of the V_L-V_H interface in determining, and possibly, modulating antibody specificity is discussed below (Sections III,B,2 and VI).

D. Conformational Changes

Many of the examples of protein–protein interaction that have been studied do not permit a comparison of the structures of the free and complexed components, which would, in turn, allow directly a determination of the extent to which the interacting structures are modified by the formation of a complex. Multimeric proteins often do not dissociate readily, and the structures of the free subunits are, therefore, not available.

Another approach is to look for proteins which can assemble in different multimeric forms, and to compare the structure of the monomer in each. Insulin, for example, can form two different hexameric assemblies, one with two Zn ions per hexamer (Blundell et al.,

1971) and the other with four (Bentley et al., 1976). The hexamers are, in each case, constructed from three identical dimers in which the structures of the two chemically identical monomers in each dimer are similar, but not identical. An analysis of the different structural forms of the monomer, found within and between the two hexamers (Chothia et al., 1983), reveals that helices packed in subunit interfaces can move as rigid bodies with respect to each other by about 1.5 Å. These movements are accommodated by torsion angle adjustments of up to 15° in the amino acid side chains at the interface.

Similar movements of helices (Chothia and Lesk, 1985) occur during the allosteric transition in hemoglobin (Fermi and Perutz, 1981) and following the binding of ligands by citrate synthase (Remington et al., 1982; Lesk and Chothia, 1984). In each of these examples, the movement of helices as semirigid bodies results in the propagation and, in the case of citrate synthase, amplification of conformational changes over large distances (20 Å or more).

Comparable studies of β sheet interfaces have not been reported, but it has been suggested that sheet structures, being less compact than helices, may be better able to absorb conformational changes locally (Chothia et al., 1983). Analysis of the structure of different immunoglobulin domains (Lesk and Chothia, 1982) shows that the two β sheets, which form the core of the domain structure, can adopt different positions with respect to each other, depending on the size and shape of the amino acids that are packed between them. In addition, recent data, from crystal structure studies of the Bence–Jones protein Loc in two different crystal forms, demonstrate that sheet structures can also move as semirigid bodies in accommodating different interface packing arrangements (Chang et al., 1987; see Section III,B,2).

The importance of these studies has been to reveal that two protein surfaces can adapt to each other, at least to a limited extent, and that it is not necessary for structures destined for the interface region to be rigidly and precisely complementary to each other. This is not to say that structural changes always occur in protein–protein complexes. The structure of the pancreatic trypsin inhibitor–trypsin complex (Huber et al., 1974) shows no significant changes in the structure of either protein, compared to their uncomplexed structures (Bode and Schwager, 1975). Similarly, the structure of one of the Fab–lysozyme complexes indicates no significant structural change in either protein from the unliganded state (Amit et al., 1986).

Structural complementarity is also observed in protein–crystal lattice interactions. These intermolecular contacts are characteristically tenuous, as evidenced by the mechanical fragility of many protein crystals, and frequently involve regions of the molecule that are not elements of regular secondary structure. Differences between the structure of a protein in two different crystal forms are small, but are usually largest in those regions engaged in lattice contacts. Above-average flexibility in chain segments outside of secondary structure regions is also commonly observed.

In coming to terms with structural complementarity, it is useful to distinguish between interactions engaging secondary structures and those involving unstructured surface loops. The types of structural alteration which can occur at an interface are clearly dependent on the underlying structures of the interacting surfaces.

Debate over the role of flexibility as an important aspect of, in particular, protein antigenicity (reviewed in Tainer et al., 1985) can be reduced to a consideration of the following fundamental question about immune recognition. To what extent do the complementary structures on antibody and antigen preexist before they encounter each other, or conversely, to what extent are they inducible as a result of encounter?

The functional importance of large-scale flexibility in protein molecules has been reviewed recently (Bennett and Huber, 1984) and encompasses ligand binding, zymogen activation, allostery, and viral capsid formation. One of the purposes of this review is to consider its role in immune recognition.

III. Antigen Receptor Structure

The humoral and cellular arms of the immune response apparently recognize antigen rather differently. Antibody, first as a receptor molecule on B cells and later as a soluble secreted molecule, binds antigen in an unprocessed or native state. T cell receptors (TCR), on the other hand, characteristically bind antigen only in association with histocompatibility proteins (Zinkernagel and Doherty, 1974), and furthermore, the protein antigen is typically processed to a small peptide prior to its presentation to TCR. It is emerging that this view of TCRs is now equally valid for both helper (Shimonkevitz et al., 1983) and cytotoxic (Townsend et al., 1986) T cells.

While there are now many data addressing the structure and chemistry of antibody molecules and their behavior toward antigen, TCRs are comparatively recently characterized (Meuer et al., 1983; Haskins et al., 1983). It is, therefore, not certain whether there are fundamental structural differences between antibodies and TCRs, or for that matter, between unprocessed antigen and peptides associated with histocompatibility molecules.

The balance of evidence seems to favor the reductionist view, that there are no fundamental distinctions between the two systems, at least, neither in terms of the genetic and structural basis of the diversity of the two types of antigen receptor nor indeed of the size of the epitopes to which these receptors bind.

A. DIVERSITY IN IMMUNE RESPONSES

There is thought to be sufficient coding capacity and genetic machinery in man to generate about 10 million different immunoglobin specificities (paratopes) (Tonegawa, 1983). This does not include somatic mutations, which accumulate as the immune response to a particular antigen matures (Berek and Milstein, 1987). A similar number of different human TCRs has recently been estimated (Klein *et al.*, 1987).

The question of preexisting versus inducible structural complementarity between antigen and antigen–receptor is somehow linked to this number. The immune system is required to economically generate enough receptors to counter an antigenic armada, the total size and individual shapes of which are unknown. One economy would be to relax the requirement for precise preexisting complementarity and to permit a degree of induced fit in the interaction. This would allow that one antibody might service two or more different antigens, and the balance that is required becomes one between economy and specificity. To retain an acceptable level of specificity, there should, therefore, be limits to the extent to which paratope structures are inducible by antigen.

Examples of antibodies cross-reacting with antigens apparently unrelated to the immunizing antigen are not uncommon (Srinivisappa *et al.*, 1986; Oldstone, 1987). Details of the antigenic epitopes involved in these studies are not available, but the prevailing view is that the immune response is not only heterogeneous (many antibodies responding to a particular antigen), but also degenerate (the same antibody responding to different antigens) (Richards and Konigsberg, 1973).

Exactly what mechanisms are operating to permit one antibody to bind to different antigens is not known. It is, of course, not necessary to demand that different antigens induce different paratope structures on a given antibody in order to explain degeneracy. These observations can be explained satisfactorily, by arguing that different regions of the paratope are involved in engaging different antigens. A partial contribution to this debate can now be put, based on the extent to which the antigens in the three known structures fill the antibody paratope (see below, Section V,A).

B. ANTIBODY STRUCTURE

It is not appropriate in this article to review antibody structure in detail. Several comprehensive analyses have been written recently, describing the domain structure and the nature of the combining site for antigen [for examples, see Amzel and Poljak (1979) and Davies and Metzger (1983)]. Fab structure studies, postdating these reviews, include Satow et al. (1986), Suh et al. (1986), Colman and Webster (1987), and Cygler et al. (1987). In this section, we shall concentrate on particular aspects of the structure that bear on the primary function of antibody, namely, the formation of complex with antigen.

1. Domain Structure

Antibody structures are modular. They appear as "**Y**"-shaped objects in the electron microscope, and the arms of the **Y** (Fab fragments) are removable by proteolysis and retain the capacity to interact with antigen. The Fab arm itself is composed of two polypeptide chains, heavy (H) and light (L), each of which is folded into two independent structural domains of approximately 100 amino acids, one variable (V) and one constant (C). The four domains are associated in such a way that V_L and V_H form a globular module, the V module, as do C_L and C_{H1} likewise form the C module. The C_{H2} and C_{H3} domains of the H chain comprise the Fc fragment of the antibody (the tail of the **Y**), which mediates various effector functions, recently reviewed by Morgan and Weigle (1987).

At the distal end of the V module, three hypervariable polypeptide loops from each of V_L and V_H are clustered in space and form the complementarity-determining regions (CDRs), or paratope, of the antibody. The CDRs occur approximately at residues 24–34, 50–56, and 89–97 in V_L domains and 31–35, 50–65, and 95–102 in V_H domains (Kabat et al., 1983). Between the CDRs are the more conserved regions of structure known as framework (FR), and the structure of a V domain may be read from N- to C-terminal as FR1, CDR1, FR2, CDR2, FR3, CDR3, and FR4. Because the CDRs are loop structures, they are more likely to be regions of above-average mobility, and in the McPC603 Fab, one of them remains "invisible," even in the refined structure (Satow et al., 1986). CDR loop flexibility may be an important feature of antigen binding, just as epitope flexibility is thought to facilitate antibody binding in some cases (see Tainer et al., 1985). Studies of CDR structures are revealing patterns which give some hope for future attempts to predict these structures from amino acid sequence data (Kabat, 1978; Padlan, 1979; Fine *et al.*, 1986; Chothia and Lesk, 1987).

Similarities in the structures of the CDRs from one antibody to another suggest that they might be held in fixed relationships in space to each other, at least within a domain. Using as reference points (see Chothia and Lesk, 1987) residues 29, 52, and 93 for the light chain CDRs (L1, L2, and L3) and 29, 51, and 94 for the heavy chain CDRs (H1, H2, and H3) (94 is really at the junction of FR3 and CDR3, but is used here, because H3 is the most structurally variable of the six CDRs), one can calculate the lengths of the sides of the two triangles formed by these markers within L1, L2, and L3 and H1, H2, and H3. For the three best-determined structures in the data base, McPC603 (Satow *et al.*, 1986), KO1 (Marquart *et al.*, 1980), and J539 (Suh *et al.*, 1986), the standard deviations of these lengths is, in all cases, less than 0.7 Å. This number is not increased by including the anti-neuraminidase Fab fragment NC41 (Colman *et al.*, 1987a) in the calculations. That structure is currently refined to a residual of 0.25 at 2.9 Å resolution. To a first approximation then, the CDR loops of these four structures are rigidly attached to the supporting framework of the domain. The spatial relationship between

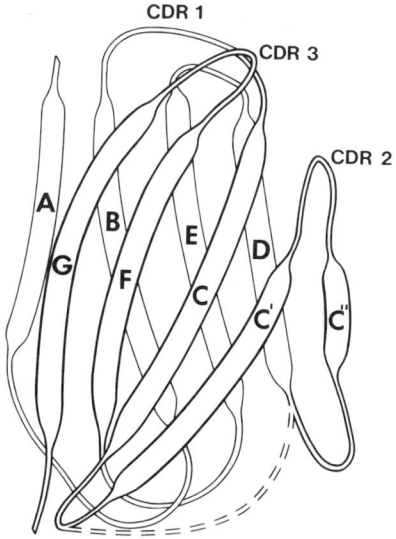

FIG. 3. Schematic of the immunoglobulin fold, showing the seven β strands characteristic of C domains and the additional two strands C' and C" found in V domains. [adapted from Richardson (1981)].

the H and L chain CDR triangles is discussed below (Section III,B,2).

The structures of V and C domains have long been recognized to be homologous (Edelman and Gall, 1969). The basic structure is one of seven β strands (Fig. 3), labeled A, B, C, D, E, F, and G from the N-terminus. V domains are characterized by two additional strands, C' and C", inserted after the third strand. The additional two strands in the V domain are joined by one connecting loop, which is the second of the three CDRs. Strands A and B constitute FR1, strands C and C' constitute FR2, strands C", D, E, and F constitute FR3, and strand G is FR4 (see Fig. 3). Each type of domain is a β sheet sandwich of the aligned packing class (Section II,C), with three and four strands per sheet in the basic C domain structure. C domains associate into a C module, through interactions between their A–B–E–D sheets, whereas the V module is formed by interactions between the C'–C–F–G sheets. Thus, not only do the additional strands of the V domain support CDR2, but one of them (C') is also a key element in the V_L–V_H interface.

One further significant difference between V and C domain structures is found in strand G, where V domains contain an additional residue in a β bulge that does not occur in C domains (Lesk and Chothia, 1982). This bulge is associated with Gly-X-Gly sequences in strand G and appears to play a role in the construction of the peculiar pairing of V domains (see Section III,B,2).

The variable domains are encoded by multiple germ-line gene segments, which rearrange during B cell ontogeny. For V_L, the V gene embraces amino acid residues 1 through 95 (i.e., FR1 through CDR3), and the J gene embraces the remaining 13 or so residues of the V domain (i.e., FR4 only). In V_H, three gene segments, V, D, and J, correspond approximately to residues in FR1 through FR3 (i.e., 1–94), CDR3, and FR4, respectively. The single strand of β structure that is FR4 in both V_L and V_H forms one of the edge strands in the interface between V_L and V_H (see Section III,B,2).

Although there are extensive amino acid sequence differences across different V_L and V_H domains, and particularly so within the three CDRs on each domain, the spatial arrangement of the six CDRs on different antibodies is remarkably similar. The surface, generated by the spatial clustering of the six CDRs, encodes the antigen specificity of the antibody. To a first approximation, the size of this surface is invariant among antibodies whose three-dimensional structures are known, being roughly circular with a diameter of about 30 Å. Therefore, the hypervariable surface of the V module has an area of about 700 Å^2 (Colman and Ward, 1985).

2. Module Structure

The noncovalent interactions between domains to form the V and C modules of the Fab fragment are further examples of protein–protein interactions or, more specifically, of sheet–sheet interactions. Consider first the C_L–C_{H1} domain association, because it is an example of a well-characterized packing mode for two β sheets, namely, the orthogonal packing class. Although this class of packing has only been described as a feature of sheet interactions within a polypeptide chain (Chothia and Janin, 1982), we note here that it describes quite precisely the interface between the C_L and C_{H1} domains of the C module, as in Fig. 2. A recent study of the packing of C_L and C_{H1} domains in the C module has revealed the existence of a cavity, whose function may be to permit some amino acid substitutions in the central interface residues, without disturbing the domain–domain packing arrangement (Padlan et al., 1986).

However, of particular interest to the question of antigen binding is the nature of the domain–domain interactions which generate the V module from the V_L and V_H domains. These interactions play a crucial role in determining the specificity of the antibody, because they control the relative spatial arrangements of the three CDRs of V_H with respect to the three CDRs of V_L.

The β strands from the interacting surfaces of V_L and V_H describe a nearly prefect, cylindrical surface (Novotny et al., 1983) shown schematically in Fig. 4. The radius of a least-squares-fitted cylinder to the interface of the myeloma proteins KO1 (Marquart et al., 1980),

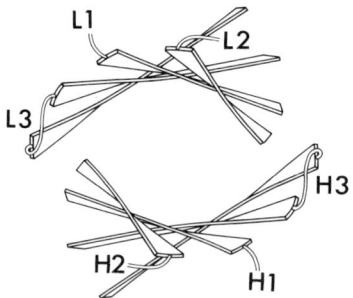

FIG. 4. Schematic of V_L–V_H interface, showing the view into the antigen-binding site. The interacting sheets are psuedosymmetrical. Unlike other sheet–sheet packings in which amino acids from the central layers of the sheets dominate the interactions, here amino acids from the edge strands fold into the center of the sandwich, forming a third layer separating residues from the central strands. The six CDR loops, L1 through H3, are labeled.

McPC603 (Satow *et al.*, 1986), and New (Saul *et al.*, 1978) are 8.2, 8.6, and 8.4 Å, respectively (Novotny *et al.*, 1983). More complex analytical surfaces have also been fitted to the V_L–V_H interface (Novotny *et al.*, 1984; Novotny and Haber, 1985).

It was noted above (Section II,C) that interacting β sheets in proteins conform to one of two packing classes in which the relative orientations of the two sheets are either $-30°$ (aligned packing) or $-90°$ (orthogonal packing). The values, in the case of the sheets in the V_L–V_H interface, of the three Fab structures referred to above are -52, -54, and $-52°$, respectively, and hence may be regarded as atypical. Slight differences in the packing of the V_L and V_H domains of different immunoglobulins have been attributed to the fact that amino acids in contact across the interface come from the variable CDRs, as well as from the conserved framework regions (Davies *et al.*, 1975). About one-quarter of the buried surface in the interface derives from the CDRs (Chothia *et al.*, 1985), and since CDR sequences vary considerably from one antibody to another, their capacity to form favorable contacts in the V_L–V_H interface might not only modulate the precise geometry of the V_L–V_H association, but also determine whether a particular V_L–V_H pairing is even possible.

Further study of the V module interface (Chothia *et al.*, 1985) confirms its unique place among protein structures analyzed to date. As a result of distortions in the two interacting sheets in this case, the most extensive interactions across the interface involve residues on the edge strands of the sheets, which fold into the center of the interface, forming a separating layer between the amino acid side chains of the central strands. This is to be compared with the more common packing modes in which amino acids from the central strands of the sheets dominate the packing interactions. In an attempt to rationalize the observations for V_L–V_H pairing, it has been suggested that conserved β bulges, in the edge strands of the V_L and V_H sheets in the interface, play a critical role in distorting the sheets to facilitate the observed packing (Chothia *et al.*, 1985).

Amino acid sequences (Kabat *et al.*, 1983) on the edge strands of the V_H interface sheet include the highly conserved G–L–E–W–L/I (44–48, strand C′) and W–G–X–G (103–106, strand G) segments. Similarly, in the V_L domain, P–X1–X2–L–X2 (44–48, where X1 is hydrophilic and X2 hydrophobic) and F–G–X–G (98–101) are strongly conserved or conservatively exchanged in immunoglobulin sequences. These amino acids participate in β bulges (Richardson *et al.*, 1978), which are conserved structural elements of strands C′ and G of variable domains (Chothia *et al.*, 1985). Strand G (FR4) is encoded by the J gene segment, which thus plays a crucial role in determining the peculiar structure of the V_L–V_H interface, and thereby the geometry of the CDR surface (Fig. 4).

One measure of the differences in V_L–V_H pairings among antibody structures is to align the (say) V_L domains from two different V modules and then observe the consequent misalignment of the two V_H domains. The additional rotation and translation required to align the V_H domains is a measure of the different interface packing within the V modules (Colman et al., 1987a). For the three myeloma protein structures KOl, McPC603, and New, such an analysis shows pair-wise differences of 4–6° (Colman et al., 1987a).

Although these angular differences are small, they result in significantly different arrangements of the CDRs in the three structures. This is demonstrated by extending the analysis of distances within the L1–L2–L3 and H1–H2–H3 triangles, referred to above (Section III,B,1), to include interdomain distances, L1–H1, L1–H2, etc. Using the same three Fab structures (McPC603, KOl, and J539), one finds that the standard deviation of the distribution of values for these distances is, in some cases, greater than 2 Å. In particular, the L3–H3 distance in KOl is 4 Å longer than in J539 or McPC603. This observation cannot be attributed to alterations in the structures of the CDRs within the V_L or V_H domains, as differences in these three Fabs at that level are minimal (Section III,B,1).

It would appear then that the part of the V_L–V_H interface deriving from invariant framework residues and comprising some 75% of the buried surface in the interface can accomodate slightly different packing modes, depending on the CDR sequences that contribute the remaining 25% of the interface. Whether this property can be directly linked to the particular and peculiar design of this interface has not yet been demonstrated.

Structure analyses of Bence–Jones proteins (Schiffer et al., 1973; Chang et al., 1985) and of isolated V_L–V_L dimers (Epp et al., 1974; Fehlhammer et al., 1975; Colman et al., 1977; Furey et al., 1983) have shown that V_L domains can associate in the same way as V_L–V_H domains. These structures also provide evidence that the geometry of V_L–V_L pairing is not affected by the covalent attachment of the C_L domains. For these structures, however, the picture is somewhat more confused, because two of them, Rhe (Furey et al., 1983), a V_L dimer, and Loc (Chang et al., 1985), a light chain dimer, have V module structures distinctly different from the other Bence–Jones proteins or Fab fragments. The more bizarre of these structures is Rhe. When it is measured against the canonical V_L pairing by aligning one domain and measuring the additional rotation required to align the second domain, a value of 61° is returned (Furey et al., 1983), compared with values of around 5° for the differences between the V modules of Fab fragments described above. Rhe differs least from the other V_L dimers on its back surface, i.e., where the switch peptides would connect to the C_L domains were they present. The relative positions of the switch peptides are altered only by 6–9 Å (Furey et al., 1983), implying that the C domains of Rhe could still associate in the normal way, despite the unusual V_L–V_L association. The amino acid sequence of Rhe remains uncertain on the question of acid–amide assignments, but if the usual immunoglobulin sequences are assigned in those places, the electrostatic interaction between the domains, as observed in the crystal structure, is one of repulsion (Novotny and Haber, 1985). This is not to say that the free energy of the interaction is unfavorable, but in other antibody V modules studied, the electrostatic potential is attractive in all cases (Novotny and Haber, 1985). No satisfactory explanation has yet been offered for this structural result.

If the V module interface can be modulated by the CDRs, then it is not unreasonable to expect that it might also be affected by crystal packing forces, at least where these involve other parts of the V domains. Whether lattice contacts or CDR sequences or both are responsible for the Rhe structure is not known, but a high-resolution NMR study of the Rhe V_L dimer in solution could shed some light on the issue. A recent result on the structure of a second crystal form of the Bence–Jones protein Loc establishes unequivocally that crystal field forces can influence the V module interface, and more importantly, that one Bence–Jones protein can have more than one V module structure (Chang et al., 1987).

The only Fab structure that has been studied in different crystal forms in the human myeloma protein KOl. The intact IgG (Colman et al., 1976; Marquart et al., 1980) and the free Fab structures (Matsushima et al., 1978) are both known. The differences between the V module structures in the two cases are "very small and probably insignificant" (Marquart et al., 1980), but the crystal packing contacts in the two structures are also rather similar. Thus, there are no direct structural data supporting the view that the V module of a Fab fragment can adopt more than one quaternary structure, but for V_L–V_L associations, the point is established.

3. Conformational Changes

During the 1970s, there was discussion over the question of conformational changes in antibody that were triggered by an encounter with antigen. The main issue was the mechanism for initiating antibody effector functions in the Fc portion of the antibody following antigen binding, and the problem was that the Fc region is distant from the antigen combining site. Much of the work contributing to this debate has been reviewed by Metzger (1978), who concludes that there are no unequivocal data in support of antigen-triggered meaningful conformational changes in antibodies. In contrast, some data indicate that aggregation of the Fc regions is a necessary and sufficient condition for switching on effector function (Metzger, 1978).

While the evidence for functionally relevant conformational change is marginal, a number of experiments do point toward some conformational change occurring. Altered circular dichroism (Holowka et al., 1972) and circular polarization of fluorescence (Schlessinger et al., 1975) on antigen binding indicate changes in the environment of aromatic residues, and kinetic data support a bimolecular reaction with distinct conformational states for free and bound Fab (Levison et al., 1975; Lancet and Pecht, 1976; Zidovetzki et al., 1981). The last of these studies has also shown that a recombinant antibody has two different conformations in solution and that the equilibrium between them is shifted on binding hapten, dinitrophenyllysine. Although the focus of all these experiments has been on their implication for effector functions, it might be better to view the conformational changes observed as structural adaptations of the antibody, which are important for securing a productive complex between antibody and antigen (Colman et al., 1987a). For antibodies whose paratopes match well the target epitope, little or no conformational change would be required [as, for example, in Amit et al. (1986)]. This point is further discussed in Sections V,D and VI.

There is some evidence for hapten-induced conformational change in light chain dimers (Ely et al., 1978; Edmundson et al., 1984). In that case, hydrophobic ligands appear to penetrate the V_L-V_L interface and signal their presence to the C module. Strain so-introduced into the C module can then be relaxed by reducing and reoxidizing the interchain disulfide bond at the C terminus of the two chains. There are no comparable results for hapten–Fab structures, where on the contrary, vitamin K_1 (Amzel et al., 1974) and phosphocholine (Segal et al., 1974) bind myeloma proteins without causing changes to their structure.

C. T Cell Receptors

The T cell receptor (TCR) for antigen is a heterodimer of disulfide-linked chains (Allison et al., 1982; Haskins et al., 1983; Meuer et al., 1983). Both the α and β chains contain a variable (V) domain, which encodes epitope specificity, and a constant (C) domain, the C-terminal end of which anchors the receptor to the T cell surface. The V and C domains show sequence homology with their counterparts in immunoglobulins (Yanagi et al., 1984; Saito et al., 1984), and there appears, therefore, to be a superficial similarity between TCR and the Fab portion of an antibody.

The V domains of both α and β chains display segments of hypervariability similar to that seen in immunoglobulins, and this variability, together with the sequence homology to immunoglobulins, suggests that the V domains of TCR can be partitioned into framework (FR) and complementarity-determining regions (CDR) in the same way as the immunoglobulin V_L and V_H domains.

The germ-line gene organization of the various gene segments, encoding the V regions of both α and β chains, has been determined (Hedrick et al., 1984; Arden et al., 1985). The α chain is the analog of the immunoglobulin light chain in this regard, since its gene is rearranged from two separated DNA segments, V_α and J_α, while the β chain gene is generated from three distinct genetic elements, V, D, and J, in the same way as the immunoglobulin V_H domain.

On the basis of many observations of three-dimensional structural homology between proteins of limited amino acid sequence homology, there can be little doubt that the V_α and V_β domains are very similar in structure to their antibody counterparts. Beyond this similarity in the gene and protein structure of the T and B cell antigen receptors and the underlying somatic mechanisms for generating receptor diversity in the two systems, there is one important structural difference. Whereas the J gene segment of the V_L domain encodes FR4 of that domain, in the V_α domain, it corresponds to CDR3 and FR4 (Fig. 3). FR4 of V_L plays little or no role in contacting antigen in the three-dimensional structures of the antibody–antigen complexes studied so far (see Section V,A), but in each case, CDR3 participates in contacting the epitope. Thus, the role of the J_α segment in generating functional (i.e., antigen specific) diversity is likely to be greater than that of the J gene in the immunoglobulin light chain. The V_β and V_H domains, on the other hand, are quite similar to each other in the respect of which structural elements of the domains are encoded by the V, D, and J gene segments.

In addition to the conservation of amino acids in these V domains that determine the domain structure itself, it is also striking that residues found in the V_L–V_H interface of the V module are also conserved in TCR sequences (Novotny et al., 1986; Colman et al., 1987a). In particular, the edge strands of the putative V–V interface in TCR show interesting parallels with antibody sequences. The F–G–X–G sequence of FR4, encoded by the J_α and J_β genes, is identical to a V_L

sequence in FR4 (strand G). The other edge strand of the interface, C′, also has interesting homologies between immunoglobulin and TCR. The sequence P–X1–X2–L–X2 (X1, hydrophilic; X2, hydrophobic), beginning nine residues after the conserved domain tryptophan in V_α (Arden *et al.*, 1985), is homologous to V_L sequences in strand C′ as discussed above. Some of the V_β sequences show good homology with the C′ strand V_H sequence of G–L–E/R–X2 [see table in Kronenberg *et al.* (1986)], but others are more similar to the analogous V_L sequence. The role of these sequences in twisting the sheets of the V–V interface in antibodies has been discussed above. It seems most likely, therefore, that not only are the domain structures of immunoglobulins and TCRs identical, but that the V–V pairings of the two molecules are also very similar. Indeed, the similarity extends to those features of the V–V interface that are believed responsible for its unique place among sheet–sheet interactions. This observation is important because, as discussed above, the V–V association determines the relative positions of the six CDRs in the paratope and, thereby, determines the antigen specificity of the receptor.

On the surface of T cells, the α–β antigen receptor dimer is normally associated with three invariable T3 membrane proteins, T3-γ, T3-δ, and T3-ε (Brenner *et al.*, 1986; Bank *et al.*, 1986; Van den Elsen *et al.*, 1984; Gold *et al.*, 1986). These proteins are thought not to function directly in antigen binding, but possibly to act as signal transducers, following encounter of antigen with the α–β dimer [reviewed in Weiss *et al.* (1986)]. The oligomeric state of these polypeptides on the cell surface is mostly unknown, although γ chains are reported to associate as homodimers (Alarcon *et al.*, 1987).

The T3 membrane proteins also appear to belong to the immunoglobulin supergene family (Novotny *et al.*, 1986; Gold *et al.*, 1987), but amino acid substitutions at certain critical residues suggest that they will not form dimers in the classic antibody V module mode. For example, the conserved F–G–X–G sequence in FR4 is absent in γ (Saito *et al.*, 1984) and ε (Gold *et al.*, 1987) chains, and the conserved glutamine, three residues beyond the conserved tryptophan in FR2, is absent in ε sequences, as is the entire β sheet strand C′, which forms the unusual edge strand of the V–V interface in antibody structures. Such sequence anomalies might be read as diagnostic for a function of these proteins other than antigen binding, as suggested above.

In summary, the TCR α–β dimer displays all of the expected characteristics of an antigen receptor molecule, namely, clonal variability, rearrangement of separate genetic elements during T cell ontogeny, and detailed amino acid sequence similarities with V region genes of immunoglobulins, including features which suggest that the pairing of V_α and V_β is similar to that of V_L and V_H. The consequence of this last observation is that the size of the TCR paratope will be similar to that of an antibody molecule and that epitope size (in this case, a complex of processed antigen and histocompatibility antigen) is also likely to be similar to that observed for antigen associated with antibody (Novotny et al., 1986).

Confirmatory evidence for this latter conclusion has now come from the structure analysis of a human class I histocompatibility molecule (Bjorkman et al., 1987a,b). Processed antigen is believed to bind between two α-helices of the histocompatibility antigen in a groove measuring 25 × 10 Å, forming an epitope for binding by TCR of overall dimensions of approximately 25 × 30 Å, i.e., around 750 Å^2.

IV. Antigens

Volumes of material have been written about antigens in general and, in particular, the antigenic structure of proteins (Sela, 1978). It is apparent enough, from a consideration of the structure of an antibody molecule, which part of it is involved in binding antigen, but the argument has no simple symmetry to indicate which parts of a protein antigen are involved in combining with antibody. Indeed, there are likely to be many sites on a given macromolecular antigen against which antibodies may be raised and to which antibodies may bind.

While many of these studies have been done on so-called model antigens, such as myoglobin, lysozyme, cytochrome, etc., the three-dimensional structures of a number of natural viral antigens have been determined recently. These include the influenza virus antigens, hemagglutinin (Wilson et al., 1981; Wiley et al., 1981) and neuraminidase (Varghese et al., 1983; Colman et al., 1983), and rhino (Rossman et al., 1985) and polio (Hogle et al., 1985) viruses. A distinguishing feature of these molecules is that the radius of curvature of their surfaces is large compared to that of an antibody, whereas the reverse is true for the low-molecular-weight protein antigens referred to above.

Among the advantages of studying antigens from natural pathogens is the capacity to select laboratory mutants under monoclonal antibody pressure (Gerhard and Webster, 1978). These mutants are typically distinguishable from the wild-type antigen by single amino acid substitutions, which are sufficient to, effectively, abolish the binding of the monoclonal antibody used for their selection. The site of the mutation is frequently read as a marker for the selecting eptiope, on the assumption that only local structural changes are associated with the mutation. In two cases, this has now been shown to be the case. A Gly-to-Asp mutant of the influenza hemagglutinin and a Lys-to-Glu mutant of the influenza neuraminidase both show strictly local structure perturbations with respect to wild type (Knossow et al., 1984; Varghese et al., 1988).

One issue, which has been widely debated, concerns the search for identifiable structural characteristics of antibody-binding sites (epitopes) on protein antigens. Whether an antibody can be raised against some particular antigenic shape is a function of many variables. Some of these, such as immune response genes, tolerance, and T cell help, are beyond the scope of this review, but the question of intrinsic structural characteristics of epitopes demands some attention.

Early studies on the antigenic structure of myoglobin (Atassi, 1975) and lysozyme (Atassi, 1978), using fractionated antisera, led to the view that there were on each of these proteins a small number of well-defined epitopes, with intrinsic "immunogenic" properties and which were independent of the host in which the antibody response was measured. These observations are not consistent with a body of more recent data, which has been reviewed by Benjamin et al. (1984). They have concluded that essentially the entire surface structure of a protein molecule can be antigenic, i.e., epitopes form a continuous and overlapping pattern on the surface of a protein. While the capacity of a given individual to respond to any particular one of these epitopes is determined by regulatory processes of the immune system, this conclusion implies that a principal structural property of an epitope is its surface accessibility.

Another property of protein molecules, which has been widely discussed in relation to antigenicity, is structural flexibility [reviewed in Tainer et al. (1985)]. X-Ray diffraction studies of protein molecules yield information about the positions of the atomic centers and also about the mean amplitude of oscillations about those centers. This latter parameter is easily confused for genuine structural disorder in which, say, two or three slightly different conformations of a structural segment are present. Although it was originally proposed that flexible epitopes provided a rationale for the binding of antipeptide antibodies to native protein molecules (Westhof et al., 1984; Tainer et al., 1984), the argument has now been extended to embrace immunizing epitopes on intact proteins (Geysen et al., 1987; Getzoff et al., 1987).

The approach used in these latter studies has been to probe antisera (raised by immunizing with myohemerythrin in complete Freund's adjuvant) with all of the five and six residue overlapping, homologous peptides from the protein sequence. Peptides which reacted with antisera in enzyme-linked immunosorbent assays (ELISA), were further analyzed by the systematic replacement of each residue in the peptide by every other of the 20 amino acids in an attempt to identify residues of the peptide that contact the antibody. Only two or three amino acids are judged by this criterion to be contact residues for each of the three peptides so analyzed. However, on the basis of the three-dimensional structures of complexes (see below), it is believed that epitopes on intact protein molecules extend over some 15 or so amino acids, and it could, therefore, be argued that the peptide data are fortuitously monitoring antibody binding at a site (or sites) on the protein unrelated to the site containing the homologous peptide. A careful study of the surface of myohemerythrin would no doubt clarify this point. Its importance is central to the identification of epitopes on proteins by this method and thus to conclusions that site mobility, convex shape, and high negative electrostatic potential are common themes among different epitopes (Geysen et al., 1987).

The separation of the two structural properties of surface accessibility and flexibility is not trivial, because the more-flexible regions of a protein are found on its surface. Some data purport to selectively imply the role of flexibility in antigenicity, but in many cases, the evidence is less than convincing [reviewed in Tainer et al. (1985)]. The best ground on this issue is probably somewhere in the middle. There may be cases in which either rigid or flexible epitopes engage antibodies. Indeed, if immune responses were limited to either type of structure, examples should exist of microorganisms exploiting this weakness in host defenses.

The functional relevance of epitope flexibility would be to facilitate better fit of complementary surfaces on antibody and antigen. The cost in binding energy enters through the loss of conformational entropy, resulting from the flexible epitope segments becoming rigid on the antibody. The gain comes from more favorable enthalpic terms, reflecting a higher degree of surface structure complementarity.

The argument reduces to a question of the degree of "preexisting verus inducible" complementarity of paratope and epitope. The demands on the immune system to detect an unknown number of foreign shapes suggest that elements of induced fit, so common in other protein–protein and protein–ligand interactions, are likely to be utilized.

V. Three-Dimensional Structures of Complexes

Early studies on antibody–hapten complexes have shown that small molecules, phosphocholine and vitamin K_1 hydroxide, bind to Fab fragments without altering the structure of the antibody (Padlan et al., 1973; Amzel et al., 1974). The main difference between that work and the contemporary studies of protein antigen–antibody complexes is the substantially larger surface area of interaction observed between macromolecular antigens and antibody.

Three three-dimensional structures of protein antigen–antibody complexes have now been reported, and more are anticipated in the near future. One of these complexes is with the influenza virus antigen neuraminidase of subtype N9 (Colman et al., 1987a, 1987b) and the other two are with hen egg-white lysozyme as antigen (Amit et al., 1986; Sheriff et al., 1987).

The published structures are at different levels of precision. The D1.3 Fab–lysozyme structure is determined at 2.8-Å resolution and refined to a crystallographic residual of 0.28 (Amit et al., 1986). The comparative numbers for the NC41 Fab–neuraminidase complex and the HyHEL5 Fab–lysozyme are 3.0 Å/0.35 and 2.5 Å/0.25, respectively. The structure of free lysozyme is particularly well characterized from several different crystal forms, whereas the structure of free N9 neuraminidase is still not completely refined (Baker et al., 1987). Details of the interaction between antigen and antibody in this latter case are still being analyzed. Statements about conformational changes in the neuraminidase upon antibody binding must be treated with more caution than those relating to lysozyme.

A. Epitope and Paratope Size

In all three cases, the sizes of the epitope (the contact surface on the antigen) and the paratope (the contact surface on the antibody) are very similar. Sixteen and thirteen amino acids, respectively, form the two lysozyme epitopes (Amit et al., 1986; Sheriff et al., 1987), and a similar number forms the epitope in the case of neuraminidase (Colman et al., 1987a). On the antibody, 17 amino acids contact the lysozyme in each case, and again a similar, but possibly smaller, number touches the neuraminidase.

In no case is the epitope formed from a continuous sequence of amino acids. The neuraminidase epitope involves at least four sequentially discontinuous surface loops in contact with the antibody, and the lysozyme structures use two and three discontinuous loops, respectively. This observation is consistent with the conclusions of Benjamin et al. (1984), who found that most epitopes "depend on the conformational integrity of the native molecule."

All six of the antilysozyme CDRs participate in binding the epitope in both of those complexes. In the neuraminidase complex, there is some doubt about the direct involvement of light chain CDR1 in forming bonds to the antigen. Whatever the final analysis there, it is clear that not all CDRs participate equally in forming contacts with the epitope. In the D1.3 antibody, heavy chain CDR3 interactions with lysozyme dominate, whereas in the HyHEL5 antibody, there are more contacting residues in light chain CDR3. These results suggest that one antibody might be able to service two somewhat different antigens by utilizing different portions of its CDRs as different, but overlapping, paratopes.

In none of the three structures do chain segments encoded by J genes or even the V–J or D–J junction points contact the antigen. V–J or D–J junctional variability might, nevertheless, influence antigen specificity indirectly by affecting the structures of neighboring elements of the CDR surface (Colman et al., 1977). One or two amino acids from the framework regions immediately adjacent to the CDRs of both D1.3 and HyHEL5 touch the lysozyme. This observation may be especially interesting in the context of somatic variation, since some of these mutants map to framework regions (Berek and Milstein, 1987).

The two Fab–lysozyme complexes both remove 750 $Å^2$ of lysozyme surface from contact with solvent to form the interface with the antibody. The exposed surface area of the two antilysozyme antibodies is reduced by 700 and 750 $Å^2$, respectively, when the lysozyme binds. All of these figures are consistent with data referred to earlier (Section II,A), which suggest that at least 600 $Å^2$ of surface area from each protein is buried in a protein–protein interaction, and in this respect, antibody–antigen interactions are not atypical.

B. Epitope Character

In the case of all three structures reported, numerous immunochemical studies have addressed the location of the epitope to which the antibody in question binds. For the lysozyme complexes, these studies have mostly taken the form of analyzing the binding to the antibody of lysozymes from different animal species (Harper et al., 1987; Smith-Gill et al., 1982). In the case of HyHEL5, a detailed model building of the complex was also undertaken (Smith-Gill et al., 1982), yielding results in striking agreement with X-ray analysis. For the anti-neuraminidase NC41 antibody, cross-reactions with laboratory-selected antigenic variants of neuraminidase are consistent with the observed epitope (Webster et al., 1987). In all cases, therefore, the X-ray-determined epitopes are consistent with serological studies.

The question of epitope mobility is not settled by these three structures. D1.3 binds to a region of low-thermal mobility on lysozyme, whereas HyHEL5 binds high-thermal-mobility regions. NC41 is also binding a less well-ordered region of neuraminidase, including part of one loop, whose structure has not yet been fully characterized in the free antigen. While epitope flexibility may be important in facilitating antibody binding in some cases, it would be surprising were it exclusively so.

C. Complementarity

The number and type of specific interactions between lysozyme and the two Fabs are similar. For D1.3, 12 hydrogen bonds, but no ion pairs, are found in the interface, and for HyHEL5, 10 hydrogen bonds and 3 ion pairs, contrary to some expectations (Roberts et al., (1987), provide the specificity. In addition, hydrophobic interactions contribute to the stabilization of the complexes.

Part of the specific hydrogen-bonded contact in the D1.3–lysozyme complex involves polypeptide main-chain atoms engaged in β sheet interactions. In addition, in other places, hydrogen bonds between the main chain of the antigen and the side chains of the antibody, as well as vice versa, are observed. It is important to recognize the potential for interactions of this type when assessing data on the peptide–antibody binding and on the consequences of amino acid substitution in the peptide. Loss of binding of antibody to a peptide, as a result of an amino acid substitution, might result from altered conformational properties of the peptide affecting peptide backbone–antibody interaction.

Antibody HyHEL5 binds lysozyme some 50 times more tightly than does antibody D1.3. It has been suggested (Sheriff et al., 1987) that, since the areas of interaction are similar in both cases, these data are evidence for the role of specific interactions in determining the binding energy and that this is not a simple function of buried surface area. Indeed, in the case of the HyHEL5 complex, the ion pairs in the interface might contribute significantly to the K_d, particularly if they are at all shielded from the solvent. Nevertheless, from the published data, HyHEL5 and lysozyme bury about 60 $Å^2$ more of their free surface area on complex formation than do D1.3 and lysozyme. In the simple model in which binding energy is only a function of buried surface area, this translates into 1.5 kcal/mol. It is not possible to convert this figure into a value for the difference in K_d of the two antibodies, because neither the relative free energies of the two antibodies in solution nor the relative differences in free energy of the free and complexed lysozymes is known. However, were those values similar, then 1.5 kcal/mol would equate to a difference in K_d of just one order of magnitude, close to the observed value of 50. However, the thermodynamic approximations used here might well be sufficiently inaccurate to reverse the sign of this prediction.

D. CONFORMATIONAL CHANGES

Of particular interest in studying these results is the question of conformational perturbation that might have occurred in either the antibody or the antigen. Such changes can occur on different levels; for example, they may involve only side-chain rotations, or main-chain conformational changes may be required. On the antibody, an additional level of structural rearrangement might relate to alterations in the pairing of the V_L and V_H domains of the V module.

1. Antigen

The first of these structures to be reported (Amit *et al.*, 1986) revealed a complex in which no conformational changes in antibody or antigen occurred on union. The data for the lysozyme part of that complex showed that there are conformational differences between different crystal forms of lysozyme that are as large as those occurring upon interaction with the D1.3 anti-lysozyme antibody.

The NC41 Fab–neuraminidase complex structure draws a rather different picture (Colman *et al.*, 1987a). When the structure of free (Baker *et al.*, 1987) and complexed N9 neuraminidase are compared, one of the surface loops (residues 367–370), engaged in contacts with the antibody, is seen to have moved the position of its backbone atoms by up to 1 Å. Other conformational changes in the antigen are still being analyzed, particularly in the region of residue 330 in which the model for the free antigen structure remains somewhat ambiguous. The difficulties in modeling this loop probably reflect elevated temperature or mobility factors for this part of the structure.

Small conformational changes in the antigen are also observed in the third structure (Sheriff *et al.*, 1987). The main chain at Pro-70 of the lysozyme has moved by 1.7 Å, and this may be correlated to the formation of a hydrogen bond of the peptide carbonyl to the hydroxyl of Tyr-97 in CDR3 of the heavy chain of the antibody. In addition, Trp-63, near but not in the epitope, appears to have flipped by 180° around its $C_\beta : C_\gamma$ bond.

The two structures demonstrating changes in antigen conformation are showing only small changes. Neither gross refolding of the polypeptide nor removal of side chain from the interior to the exterior (or vice versa) is observed. Although the movements detected are small, in energetic terms they are likely to be significant. These details can be evaluated only when the structure refinements in all cases are complete, and it should then be possible to calculate the additional binding energy of the complex that obtains as a result of the small conformational changes in the antigen. Whether we are sampling, in these three structures, conformational changes at the very low end of the antigen–antibody spectrum is a matter for speculation. Some data have been interpreted to imply that much larger structural changes than those reported in this section can occur (Djavadi-Ohaniance *et al.*, 1984; Geysen *et al.*, 1987; Getzoff *et al.*, 1987).

2. Antibody

Although the structure of the free Fab fragment is not known in any of the three structures reported to date, sufficient data on other Fab structures are available to allow a comparison of the complexed Fabs with the average structure of uncomplexed Fabs.

On this basis, it has been concluded that no conformational change has occurred in the D1.3 antibody on formation of complex with lysozyme (Amit *et al.*, 1986). The V_L-V_H association appears unremarkable, and the structures of four of the hypervariable loops (CDRs) were predicted correctly, based on principles derived from uncomplexed antibody structures (Chothia *et al.*, 1986). However, it is not possible to rule out conformational changes in CDRs H1 and L3, although the disagreement between structure and prediction, in those cases, has been ascribed to problems with the prediction, rather than to structural movement on complex formation (Chothia *et al.*, 1986). Sheriff *et al.* (1987) report a similar finding for the HyHEL5 Fab fragment complexed to lysozyme. The V module pairing is reported to be canonical, and statements about the CDR conformations are reserved, pending information about the structure of the free Fab fragment.

On the basis of the comparison of the Fabs McPC603, KO1, and J539 in Section III,B,2 and the differences observed there between the relative positions of the H and L chain CDRs, it seems that there is a family of so-called canonical V_L-V_H interfaces. It has been suggested (Colman *et al.*, 1987a) that the NC41 Fab complexed to neuraminidase is not a member of that family. The structure of the NC41–neuraminidase complex is now partially refined. The residual is 0.25 at 2.9-Å resolution, and shifts of tenths of angstroms only are anticipated in the final structure. When this structure is compared with the other Fabs by the angle analysis, the V module interface pairing of NC41 is found to differ from McPC603, KO1, and J539 by 4.0, 7.8, and 10.6°, respectively. The first two of these numbers were previously reported to be 8.8 and 12.1° on the basis of the unrefined NC41–neuraminidase complex structure (residual 0.35). In the present analysis, the difference between McPC603 and NC41 is not statistically significant. If, on the other hand, selected distances between CDRs are measured (as described in Sections III,B,1 and III,B,2), it is found, first, that NC41 in the complex has a constellation of L and H chain CDRs similar to other Fabs and, second, that between V_H and V_L there are large differences in the inter-CDR distances. For example, the distances between the chosen markers on L3 and H2 of McPC603 and J539 are 13.9 and 12.5 Å, respectively, but for NC41, this distance is 17.2 Å. While a difference of this magnitude is small in structural terms, it is very large in specificity terms. The distance between neighboring amino acids on a polypeptide chain is 4 Å, and any relocation of the CDRs by this amount would enable specific interactions with the antigen to be through amino acids adjacent to those that would otherwise be so engaged.

The amino acid sequences of the V_L and V_H domains (G. M. Air, unpublished) show that those residues which are normally conserved in the V module interface are indeed conserved in the NC41 sequence. Of course, the sequences of the CDRs of NC41 are different to sequences seen in other Fabs whose three-dimensional structures are known. The role of the CDRs in modulating the V_L–V_H pairing has been described above. It has now been suggested (Colman et al., 1987a) that antigen might also perturb the V_L–V_H interface by way of inducing a better fit between epitope and the six CDRs, as discussed below in Section VI.

The particular V module interface structure of the NC41 Fab–neuraminidase complex seems not to have affected the C module structure (Colman et al., 1987a). The pairing of C_L and C_{H1} domains is indistinguishable from that seen in the McPC603-free Fab structure (Satow et al., 1986). Thus, there is no suggestion from this work that perturbation of the V_L–V_H interface by antigen might trigger conformational changes through the C module to the hinge and Fc regions of the antibody.

VI. The Interface Adaptor Hypothesis

The hypothesis states that the V module interface is specifically designed to function as a flexible adaptor, allowing antigen to mold slightly different CDR surfaces by repositioning the V_L CDRs with respect to the V_H CDRs.

The basis for this hypothesis is illustrated in Fig. 5, which shows the interaction between antibody and antigen as a ternary complex of V_L, V_H, and antigen. Of the three interfaces that occur in this complex, one is largely conserved (V_L–V_H), although the contribution of the CDRs to this interaction is not insignificant. However, the constant part of the V_L–V_H interface, together with the V_L–antigen and the V_H–antigen interfaces, comprise the bulk of the buried surface area in the ternary

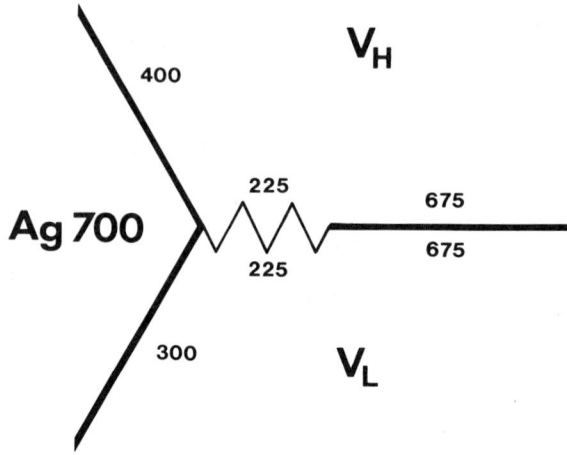

FIG. 5. Schematic of ternary complex of V_L, V_H, and antigen showing approximate sizes in squared angstroms of the various buried surfaces at each interface. The V_L–V_H interface has been partitioned to show contributions from the framework (invariant) and CDR (variable) amino acids, depicted here as straight and zigzag lines, respectively. Figures for the size of buried surfaces are derived from Marquart and Deisenhofer (1982), Chothia *et al.* (1985), and Amit *et al.* (1986).

complex. The evidence that the constant part of the V module interface can be modified by CDR sequences has been discussed above and by Colman *et al.* (1987a), suggesting that there is a family of different interface structures accessible to each other via low-energy transitions. If this surface is perturbable by interactions involving some 200 Å^2 of surface area (i.e., the CDRs), there is no conceptual problem with antigen similarly modifying the V module interface. Indeed, the simplified model of protein–protein interaction, which associates the binding energy with the buried surface area, would predict that antigen would have greater influence over the V module pairing than would the CDRs.

The dynamic aspect of the hypothesis requires the V_L–V_H interface to respond to an epitope structure by rearranging the domain pairing to optimize interactions with the epitope. There are kinetic data in support of such an isomerization and spectral data in support of structural changes affecting the environment of aromatic residues (discussed above in Section III,B,3). These latter data may well be measuring altered signals from the conserved tryptophan and tyrosine residues in the V module interface.

Affinity maturation of the immune response through somatic mutation has now been shown to involve, at least in some cases, the mutation of framework amino acids, and in one case, the substitution of a V_H–V_L interface residue, L-Tyr-36-to-Phe (Berek and Milstein, 1987). No change in the affinity of the antibody was observed for this substitution, but altered kinetics cannot be ruled out. It remains to be analyzed whether such a substitution might affect the capacity of the V_L–V_H interface to accommodate slightly different packing arrangements.

An antigen–receptor molecule, endowed with the properties suggested by this hypothesis, might seem contrary to the requirements of specificity of the immune system, but an important aspect of the hypothesis is that reorganization of the V module interface is limited. The very concept of induced fit implies a partial fit as the initiating step. Movement of the CDRs from their canonical positions by more than about 4 Å would require that the initial fit of antigen to antibody was sensing the capacity for improved contacts, not with neighboring amino acids on the antigen, but with next nearest neighbors. Small shifts (1–2 Å) in the positions of the CDRs may indeed have occurred in the two lysozyme–Fab complexes through interface adaptation, but these discrepancies would be interpreted in the first instance as deriving from particular CDR sequences. Shifts of this magnitude could, nevertheless, make a substantial contribution to the binding energy for antigen.

The most direct test of this hypothesis would come from studies of the three-dimensional structures of uncomplexed Fab fragments for comparison with the complexed structures. It is equally important that these studies be done on the Fabs whose structures have been interpreted as showing no interface rearrangement, so that the separate influences of antigen and the CDR sequences can be evaluated.

The detailed similarity between antibody and TCR sequences suggests that principles of immune recognition by both humoral and cellular arms of the immune system are likely to be similar.

The extent to which this hypothesis is consistent with established principles of protein–protein interactions, as discussed in Section II, can be summarized as follows.

1. Different oligomeric states of protein molecules exhibit different interface packing geometries. V_L–V_H–antigen can be considered as just another oligomeric form of V_L–V_H.

2. Structural adaptation by semirigid body movements of interface structural elements is well established.

3. Flexibility is important in other protein–ligand interactions and is likely to play a role in the most challenging of biological recognition phenomena, the immune response.

4. In the current protein structure data base, the V module interface stands alone as a unique structural feature. No other β sheet associations are organized in the same way. The structure of this interface directly controls antigen specificity suggesting, but not proving, an important dynamic role for the interface in antigen binding.

5. For V_L–V_L dimers, the principle of one sequence–two structures has been established.

Amit *et al.* (1986) conclude from their Fab–lysozyme structure that a "lock-and-key" description of that complex is adequate and that "somatic recombination of the germ-line gene repertoire provides all the complementary antibody templates necessary to bind all possible antigens." The interface adaptor hypothesis allows a structurally designed amplification of this repertoire, which helps account for degeneracy of immune responses and which endows the immune system with an even wider range of binding specificities. Conformational changes in antigen, as a result of antibody binding, have now been demonstrated in two of the three structures studied and, taken together with the possibility of changes in antibody structure during antigen binding, suggest a "handshake" view of immune recognition.

Acknowledgments

I thank Steven Sheriff and David Davies for communicating results in advance of this publication. Refinement of the NC41–neuraminidase complex is being done by William Tulip. Some results of that work have been included in this article ahead of its publication.

References

Alarcon, B., DeVries, J., Petty, C., Boylston, A., Yssel, H., Terhorst, C., and Spits, H. (1987). *Proc. Natl. Acad. Sci. U.S.A.* **84,** 3861–3865.
Allison, J. P., McIntyre, B. W., and Bloch, D. (1982). *J. Immunol.* **129,** 2293–2300.
Amit, A. G., Mariuzza, R. A., Phillips, S. E. V., and Poljak, R. J. (1986). *Science* **233,** 747–753.
Amzel, L. M., and Poljak, R. J. (1979). *Annu. Rev. Biochem.* **48,** 961–997.
Amzel, L. M., Poljak, R. J., Saul, F, Varga, J. M., and Richards, F. F. (1974). *Proc. Natl. Acad. Sci. U.S.A.* **71,** 1427–1430.
Arden, B., Klotz, J. L., Siu, G., and Hood, L. E. (1985). *Nature (London)* **316,** 783–787.
Atassi, M. Z. (1975). Immunochemistry **12,** 423–438.
Atassi, M. Z. (1978). Immunochemistry **15,** 909–936.
Baker, A. T., Varghese, J. N., Laver, W. G., Air, G. M., and Colman, P. M. (1987). *Proteins: Struct. Funct. Genet.* **2,** 111–117.
Bank, I., Dephino, R. A., Brenner, M. B., Cassimereis, J., Alt, F. W., and Chess, L. (1986). *Nature (London)* **322,** 179–181.
Bartlett, P. A., and Marlowe, C. K. (1987). *Science* **235,** 569–571.
Benjamin, D. C., Berzofsky, J. A., East, I. A., Gurd, F. R. N., Hannum, C., Leach, S. J.,

Margoliash, E., Michael, J. G., Miller, A., Prager, E. M., Reichlin, M., Sercarz, E. E., Smith-Gill, S. J., Todd, P. E., and Wilson, A. C. (1984). *Annu. Rev. Immunol.* **2,** 67–101.
Bennett, W., and Huber, R. (1984). *CRC Crit. Rev. Biochem.* **15,** 291–384.
Bentley, G. A., Dodson, E. J., Dodson, G. G., Hodgkin, D. C., and Mercola, D. A. (1976). *Nature (London)* **261,** 166–168.
Berek, C., and Milstein, C. (1987). *Immunol. Rev.* **96,** 23–41.
Bernstein, F. C., Koetzle, T. F., Williams, G. J. B., Meyer, E. F., Brice, M. D., Rodgers, J. R., Kennard, O., Shimanouchi, T., and Tasumi, M. (1977). *J. Mol. Biol.* **112,** 535–542.
Bjorkman, P. J., Saper, M. A., Samraoui, B., Bennett, W. S., Strominger, J. L., and Wiley, D. C. (1987a). *Nature (London)* **329,** 506–512.
Bjorkman, P. J., Saper, M. A., Samraoui, B., Bennett, W. S., Strominger, J. L., and Wiley, D. C. (1987b). *Nature (London)* **329,** 512–518.
Blundell, T. L., Cutfield, J. F., Cutfield, S. M., Dodson, E. J., Dodson, G. G., Hodgkin, D. C., Mercola, D. A., and Vijayan, M. (1971). *Nature (London)* **231,** 506–511.
Bode, W., and Schwager, P. (1975). *J. Mol. Biol.* **98,** 693–717.
Brenner, M. B., McLean, J., Dialynas, D. P., Strominger, J. L., Smith, J. A., Owen, F. L., Seidman, J. P., Ip, S., Rosen, F., and Krangel, M. S. (1986). *Nature (London)* **322,** 145–149.
Chang, C. H., Short, M. T., Westholm, F. A., Stevens, F. J., Wang, B. C., Furey, W., Solomon, A., and Schiffer, M. (1985). *Biochemistry* **24,** 4890–4897.
Chang, C. H., Carperos, W. E., Ainsworth, C. F., Olsen, K. W., and Schiffer, M. (1987). Fourteenth International Congress of Crystallography, Collected Abstracts, 02.1–5.
Chothia, C. (1974). *Nature (London)* **248,** 338–339.
Chothia, C. (1975). *Nature (London)* **254,** 304–308.
Chothia, C., and Janin, J. (1975). *Nature (London)* **256,** 705–708.
Chothia, C., and Janin, J. (1981). *Proc. Natl. Acad. Sci. U.S.A.* **78,** 4146–4150.
Chothia, C., and Janin, J. (1982). *Biochemistry* **21,** 3955–3965.
Chothia, C., and Lesk, A. M. (1985). *Trends Biochem. Sci.* **10,** 116–118.
Chothia, C., and Lesk, A. M. (1987). *J. Mol. Biol.* **196,** 901–917.
Chothia, C., Levitt, M., and Richardson, D. C. (1977). *Proc. Natl. Acad. Sci. U.S.A.* **74,** 4130–4134.
Chothia, C., Lesk, A. M., Dodson, G. G., and Hodgkin, D. C. (1983). *Nature (London)* **302,** 500–505.
Chothia, C., Novotny, J., Bruccoleri, R., and Karplus, M. (1985). *J. Mol. Biol.* **186,** 651–663.
Chothia, C., Lesk, A. M., Levitt, M., Amit, A. G., Mariuzza, R. A., Phillips, S. E. V., and Poljak, R. J. (1986). *Science* **233,** 755–758.
Cohen, F. E., Sternberg, M. J. E., and Taylor, W. R. (1981). *J. Mol. Biol.* **148,** 253–272.
Colman, P. M., and Ward, C. W. (1985). *Curr. Top. Microbiol. Immunol.* **114,** 117–255.
Colman, P. M., and Webster, R. G. (1987). In "Biological Organisation: Macromolecular Interactions at High Resolution" (R. Burnett, ed.) pp. 125–113. Academic Press, New York.
Colman, P. M., Deisenhofer, J., Huber, R., and Palm, W. (1976). *J. Mol. Biol.* **100,** 257–282.
Colman, P. M., Schramm, H. J., and Guss, J. M. (1977). *J. Mol. Biol.* **116,** 73–79.
Colman, P. M., Varghese, J. N., and Laver, W. G. (1983). *Nature (London)* **303,** 41–44.
Colman, P. M., Laver, W. G., Varghese, J. N., Baker, A. T., Tulloch, P. A., Air, G. M., and Webster, R. G. (1987a). *Nature (London)* **326,** 358–363.
Colman, P. M., Air, G. M., Webster, R. G., Varghese, J. N., Baker, A. T., Lentz, M. R., Tulloch, P. A., and Laver, W. G. (1987b). *Immunol. Today* **8,** 323–326.
Crick, F. H. C. (1953). *Acta Crystallogr.* **6,** 689–691.
Cygler, M., Boodhoo A., Lee, J. S., and Anderson, W. F. (1987). *J. Biol. Chem.* **262,** 643–648.

Davies, D. R., and Metzger, H. (1983). *Annu. Rev. Immunol.* **1,** 87–117.
Davies, D. R., Padlan, E. A., and Segal, D. M. (1975). *Annu. Rev. Biochem.* **44,** 639–667.
Djavadi-Ohaniance, L., Friguet, B., and Goldberg, M. (1984). *Biochemistry* **23,** 97–104.
Edelman, G. M., and Gall, W. E. (1969). *Annu. Rev. Biochem.* **38,** 415–466.
Edmundson, A. B., Ely, K. R., and Herron, J. N. (1984). *Mol. Immunol.* **21,** 561–576.
Efimov, A. V. (1979). *J. Mol. Biol.* **134,** 23–40.
Ehrlich, P. (1900). *Proc. R. Soc. London B*: **66,** 424–448.
Eisenberg, D., and McLachlan, A. D., (1986). *Nature (London)* **319,** 199–203.
Ely, K. R., Firca, J. R., Williams, K. J., Abola, E. E., Fenton, J. M., Schiffer, M., Panagiotopoulos, N. C., and Edmundson, A. B. (1978). *Biochemistry* **17,** 158–167.
Epp, O., Colman, P. M., Fehlhammer, H., Bode, W., Schiffer, M., Huber, R., and Palm, W. (1974). *Eur. J. Biochem.* **45,** 513–524.
Fehlhammer, H., Schiffer, M., Epp, O., Colman, P. M., Lattman, E. E., Schwager, P., Steigemnann, W., and Schramm, H. J. (1975). *Biophys. Struct. Mech.* **1,** 139–146.
Fermi, G., and Perutz, M. F. (1981). "Haemoglobin and Myoglobin: Atlas of Molecular Structures in Biology." Clarendon Press, Oxford.
Fine, R. M., Wang, H., Shenkin, P. S., Yarmush, D. L., and Levinthal, C. (1986). *Proteins: Struct. Funct. Genet.* **1,** 342–362.
Fischer, E. (1894). *Chem. Ber.* **27,** 2985–2993.
Furey, W., Wang, B. C., Yoo, C. S., and Sax, M. (1983). *J. Mol. Biol.* **167,** 661–692.
Gerhard, W., and Webster, R. G. (1978). *J. Exp. Med.* **148,** 383–392.
Getzoff, E. D., Geysen, H. M., Rodda, S. J., Alexander, H., Tainer, J. A., and Lerner, R. A. (1987). *Science* **235,** 1191–1196.
Geysen, H. M., Tainer, J. A., Rodda, S. J., Mason, T. J., Alexander, H., Getzoff, E. D., and Lerner, R. A. (1987). *Science* **235,** 1184–1190.
Gold, D. P., Puck, J. M., Pettey, C. L., Cho, M., Coligan, J., Woody, J. N., and Terhorst, C. (1986). *Nature (London)* **321,** 431–434.
Gold, D. P., Clevers, H., Alarcon, B., Dunlap, S., Novotny, J., Williams, A. F., and Terhorst, C. (1987). *Proc. Natl. Acad. Sci. U.S.A.* **84,** 7649–7653.
Harper, M., Lema, F., Boulot, G., and Poljak, R. J. (1987). *Mol. Immunol.* (in press).
Haskins, K., Dubo, R., White, J., Pigeon, M., Kappler, J., and Marrack, P. (1983). *J. Exp. Med.* **157,** 1149–1169.
Hedrick, S. M., Nielsen, E. A., Kavaler, J., Cohen, D., and Davis, M. M. (1984). *Nature (London)* **308,** 153–158.
Hogle, J. M., Chow, M., and Filman D. J. (1985). *Science* **229,** 1358–1365.
Holowka, A. D., Strosberg, A. D., Kimball, J. W., Haber, E., and Cathou, R. E. (1972). *Proc. Natl. Acad. Sci. U.S.A.* **69,** 3399–3403.
Huber, R., Kukla, D., Bode, W., Schwager, P., Bartels, K., Deisenhofer, J., and Steigemann, W. (1974). *J. Mol. Biol.* **89,** 73–101.
Kabat, E. A. (1978). *Adv. Protein Chem.* **32,** 1–75.
Kabat, E. A., Wu, T. T., Bilofsky, H., Reid-Miller, M., and Perry, H. (1983). "Sequences of proteins of immunological interest." (U.S. Dept. of Health and Public Services, Washington, D.C.).
Kauzmann, W. (1959). *Adv. Protein Chem.* **14,** 1–63.
Klein, M. H., Concannon, P., Everett, M., Kim, L. D. H., Hunkapillar T., and Hood, L. (1987). *Proc. Natl. Acad. Sci. U.S.A.* **84,** 6884–6888.
Knossow, M., Daniels, R. S., Douglas, A. R., Skehel, J. J., and Wiley, D. C. (1984). *Nature (London)* **311,** 678–680.
Kronenberg, M., Siu, G., Hood, L. E., and Shastri, N. (1986). *Annu. Rev. Immunol.* **4,** 529–591.

Lancet, D., and Pecht, I. (1976). *Proc. Natl. Acad. Sci. U.S.A.* **73,** 3549–3553.
Laver, W. G., Webster, R. G., and Colman, P. M. (1987). *Virology* **156,** 181–184.
Lesk, A. M., and Chothia, C. (1982). *J. Mol. Biol.* **160,** 325–342.
Lesk, A. M., and Chothia, C. (1984). *J. Mol. Biol.* **174,** 175–191.
Levison, S. A., Hicks, A. N., Portman, A. J., and Dandliker, W. B. (1975). *Biochemistry* **14,** 3778–3786.
Marquart, M., and Deisenhofer, J. (1982). *Immunol. Today* **3,** 160–166.
Marquart, M., Deisenhofer, J., Huber, R., and Palm, W. (1980). *J. Mol. Biol.* **141,** 369–391.
Matsushima, M., Marquart, M., Jones, T. A., Colman, P. M., Bartels, K., Huber, R., and Palm, W. (1978). *J. Mol. Biol.* **121,** 441–459.
Metzger, H. (1978). *Contemp. Top. Mol. Immunol.* **7,** 119–152.
Meuer, S. C., Fitzgerald, K. A., Hussey, R. E., Hodgdon, J. C., Schlossman, S. F., and Reinherz, E. L. (1983). *J. Exp. Med.* **157,** 705–719.
Miller, S, Lesk, A. M., Janin, J., and Chothia, C. (1987). *Nature (London)* **328,** 834–836.
Morgan, E. L., and Weigle, W. O. (1987). *Adv. Immunol.* **40,** 61–134.
Novotny, J., and Haber, E. (1985). *Proc. Natl. Acad. Sci. U.S.A.* **82,** 4592–4596.
Novotny, J., Bruccoleri, R., Newell, K., Murphy, D., Haber, E., and Karplus, M. (1983). *J. Biol. Chem.* **258,** 14433–14437.
Novotny, J., Bruccoleri, R. E., and Newell, J. (1984). *J. Mol. Biol.* **177,** 567–573.
Novotny, J., Tonegawa, S., Saito, H., Kranz, D. M., and Eisen, H. N. (1986). *Proc. Natl. Acad. Sci. U.S.A.* **83,** 742–746.
Oldstone, M. B. A. (1987). *Cell* **50,** 819–820.
Padlan, E. A. (1979). *Mol. Immunol.* **16,** 287–296.
Padlan, E. A., Segal, D. M., Spande, T. F., Davies, D. R., Rudikoff, S., and Potter, M. (1973). *Nature (London) New Biol.* **245,** 165–167.
Padlan, E. A., Cohen, G. H., and Davies, D. R. (1986). *Mol. Immunol.* **23,** 951–960.
Pauling, L. (1948). *Am. Sci.* **36,** 51–58.
Remington, S. J., Wiegand, G., and Huber, R. (1982). *J. Mol. Biol.* **158,** 111–152.
Richards, F. K., and Konigsberg, W. H. (1973). *Immunochemistry* **10,** 545–553.
Richardson, J. S. (1981). *Adv. Protein Chem.* **34,** 167–339.
Richardson, J. S., Getzoff, E. D., and Richardson, D. C. (1978). *Proc. Natl. Acad. Sci. U.S.A.* **75,** 2574–2578.
Richmond, T. J., and Richards, F. M. (1978). *J. Mol. Biol.* **119,** 537–555.
Roberts, S., Cheetham, J. C., and Rees, A. R. (1987). *Nature (London)* **328,** 731–734.
Rossman, M. G., Arnold, E., Erickson, J. W., Frankenberger, E. A., Griffith, J. P., Hecht, H.-J., Johnson, J. E., Kamer, G., Luo, M., Mosser, A. G., Rueckert, R. R., Sherry, B., and Vriend, G. (1985). *Nature (London)* **317,** 145–153.
Saito, H., Kranz, D. M., Takagaki, Y., Hayday, A. C., Eisen, H. N., and Tonegawa, S. (1984). *Nature (London)* **312,** 36–40.
Satow, Y., Cohen, G. H., Padlan, E. A., and Davies, D. R. (1986). *J. Mol. Biol.* **190,** 593–604.
Saul, F. A., Amzel, L. M., and Poljak, R. J. (1978). *J. Biol. Chem.* **253,** 585–597.
Schiffer, M., Girling, R. L., Ely, K. R., and Edmundson, A. B. (1973). *Biochemistry* **12,** 4620–4631.
Schlessinger, J, Steinberg, I. Z., Givol, I. D., Hochman, J., and Pecht, I. (1975). *Proc. Natl. Acad. Sci. U.S.A.* **72,** 2775–2779.
Segal, D. M., Padlan, E. A., Cohen G. H., Rudikoff, S., Potter, M., and Davies, D. R. (1974). *Proc. Natl. Acad. Sci. U.S.A.* **71,** 4298–4302.
Sela, M., ed. (1978). "The Antigens (Vol. 4)." Academic Press, New York.
Sheriff, S., Silverton, E. W., Padlan, E. A., Cohen, G. H., Smith-Gill, S. J., Finzell, B. C., and Davies, D. R. (1987). *Proc. Natl. Acad. Sci. U.S.A.* **84,** 8075–8079.

Shimonkevitz, R., Kappler, J., Marrack, P., and Grey, H. (1983). *J. Exp. Med.* **158,** 303–316.
Smith-Gill, S. J., Wilson, A. C., Potter, M., Prager, E. M., Feldman, R. J., and Mainhart, C. R. (1982). *J. Immunol.* **128,** 314–322.
Srinivisappa, J., Saegusa, J., Prabhakur, J. S., Gentry, M. K., Buchmeier, M. J., Wiktor, T. J., Koprowski, H., Oldstone, M. B. A., and Notkins, A. L. (1986). *J. Virol.* **57,** 397–401.
Suh, S. W., Bhat, T. N., Navia, M. A., Cohen, G. H., Rao, D. N., Rudikoff, S., and Davies, D. R. (1986). *Proteins: Struct. Funct. Genet.* **1,** 74–80.
Tainer, J. A., Getzoff, E. D., Alexander, H., Houghton, R. A., Olsen, A. J., Lerner, R. A., and Hendrickson, W. A. (1984). *Nature (London)* **312,** 127–134.
Tainer, J. A., Getzoff, E. D., Paterson, Y., Olson, A. J., and Lerner, R. A. (1985). *Annu. Rev. Immunol.* **3,** 501–535.
Tonegawa, S. (1983). *Nature (London)* **302,** 575–581.
Townsend, A. R. M., Rothbard, J., Gotch, F. M., Bahadur, G., Wraith, D., and McMichael, A. J. (1986). *Cell* **44,** 959–968.
Tronrud, D. E., Holden, H. M., and Matthews, B. W. (1987). *Science* **235,** 571–574.
Van den Elsen, P., Shepley, B.-A., Borst, J., Coligan, J. E., Markham, A. F., Orkin, S., and Terhorst, C. (1984). *Nature (London)* **312,** 413–418.
Varghese, J. N., Laver, W. G., and Colman, P. M. (1983). *Nature (London)* **303,** 35–40.
Varghese, J. N., Laver, W. G., Webster, R. G., and Colman, P. M. (1988). *J. Mol. Biol.* **200,** 201–203.
Webster, R. G., Air, G. M., Metzger, D. W., Colman, P. M., Varghese, J. N., Baker, A. T., and Laver, W. G. (1987). *J. Virol.* **61,** 2910–2916.
Weiss, A., Imboden, J., Hardy, K., Manger, B., Terhorst, C., and Stobo, J. (1986). *Annu. Rev. Immunol.* **4,** 593–619.
Westhof, E., Altschuh, D., Moras, D., Bloomer, A. C., Mondragon, A., Klug, A., and Van Regenmortel, M. H. V. (1984). *Nature (London)* **311,** 123–126.
Wiley, D. C., Wilson, I. A., and Skehel, J. J. (1981). *Nature (London)* **298,** 373–378.
Wilson, I. A., Skehel, J. J., and Wiley, D. C. (1981). *Nature (London)* **298,** 366–373.
Yanagi, Y., Yoshikai, Y., Leggett, K., Clark, S. P., Aleksander, I., and Mak, T. W. (1984). *Nature (London)* **308,** 145–149.
Zidovetzki, R., Blatt, Y., and Pecht, I. (1981). *Biochemistry* **20,** 5011–5018.
Zinkernagel, R. M., and Doherty, P. C. (1974). *Nature (London)* **248,** 701–702.

The $\gamma\delta$ T Cell Receptor

MICHAEL B. BRENNER,*,† JACK L. STROMINGER,* AND MICHAEL S. KRANGEL*

*Laboratory of Immunochemistry
Dana-Farber Cancer Institute,
Boston, Massachusetts 02115 and
†Department of Rheumatology and Immunology,
Brigham and Women's Hospital,
Harvard Medical School,
Boston, Massachusetts 02115

I. Introduction: Immunological Antigen Receptors

A focal point in immunity is the recognition of foreign antigens and the ability of the organism to distinguish such antigens from self. The molecules that mediate this distinction are the immunological antigen receptors on B lymphocytes and T lymphocytes. These receptors are unusual proteins, since they display remarkably extensive structural diversity generated by similar genetic mechanisms in both cell types. Along with important conceptual similarities, there are also fundamental differences in structure and in the mode of operation of the antigen receptors on B and T cells.

The B lymphocyte receptor is the immunoglobulin molecule. The large number of antibody molecules that can be produced by the organism result in part from somatic recombination of germ-line variable (V), diversity (D), joining (J), and constant (C) gene segments to form a contiguous (rearranged) gene encoding an immunoglobulin polypeptide chain (Dreyer and Bennett, 1965; Tonegawa, 1983). Antibody diversity is extensive and is contributed to by hundreds of V gene segments, dozens of D gene segments, and several J gene segments. Diversity is further increased by imprecise joining, by the addition of template-independent nucleotides at the joining ends of the rearranged segments, by somatic mutation during B cell replication, and by the combinatorial association of nonidentical light and heavy chain subunits. The paired light and heavy chains of each antibody molecule form a unique site that is involved in recognition of antigen; in turn, this unique site may be recognized by other antibodies as an idiotypic determinant.

When an antigen binds to an immunoglobulin molecule expressed on the surface of a B lymphocyte, that cell may be triggered to undergo cell division, resulting in clonal expansion (antigen-mediated clonal selection) (Burnet, 1957). Eventually secreted antibody molecules may directly bind to the cognate antigen, marking it for removal or destruction.

T cells differ from B cells in antigen recognition, since they often recognize different determinants, generally do not react with soluble or free antigen, and instead recognize antigen on the cell surface and only in conjunction with products encoded by self-major histocompatibility complex (MHC) genes (Shearer and Schmitt-Verhulst, 1977; Zinkernagel and Doherty, 1979). This requirement, referred to MHC-restricted recognition of antigen by T cells, is a hallmark of the T cell scheme [a few examples of direct antigen binding in the absence of MHC may represent exceptions (Rao et al., 1984; Siliciano et al., 1986)]. Recent studies have provided direct molecular evidence for the binding of antigenic peptides to MHC class II (Babbitt et al., 1985, 1986; Buus et al., 1986) and MHC class I molecules (Bjorkman et al., 1987a,b).

The development of monoclonal antibodies (mAb) (Kohler and Milstein, 1975) and the establishment of T cell clones and hybrids (Moller, 1981, 1983) set the stage for identifying the T cell antigen–receptor complex. Based on the concept that a T cell receptor, like surface immunoglobulin on B cells, might express an antigenic determinant (idiotype) unique to one clone of T cells, mAb were generated that recognized clone-specific cell surface molecules on T cell tumors, T cell clones expanded in vitro in the presence of interleukin 2 (IL-2), and cloned T cell hybrids (Allison et al., 1982; Meuer et al., 1983; Haskins et al., 1983; Samelson et al., 1983; Kaye et al., 1983). The molecules recognized by clone-specific mAb were heterodimers composed of a 40- to 50-kDa acidic TCR α-glycoprotein that was covalently associated by disulfide linkage to a 40- to 45-kDa basic or neutral TCR β-glycoprotein. Biochemical analysis revealed that the TCR α and β molecules varied among T cell clones having different antigenic specificities (Acuto et al., 1983; McIntyre and Allison, 1983; Kappler et al., 1983). Immobilized anti-TCR mAb were capable of activating the T cell clones to which they bound either to proliferate or to secrete lymphokines such as IL-2, suggesting that T cell receptors may mediate clonal expansion of T cells in response to antigen (Meuer et al., 1983; Kaye et al., 1983). Paradoxically, soluble mAb often blocked effector function such as the ability of a cytotoxic T lymphocyte clone to lyse a specific target cell (Meuer et al., 1983; Haskins et al., 1983; Samelson et al., 1983).

Even before identification of the clonotypic TCR $\alpha\beta$ protein complex, another T cell-specific surface structure was known to be functionally active. The CD3 complex (initially known as T3 or Leu-4) (Kung et al., 1979; Ledbetter et al., 1981; Borst et al., 1982) is expressed on all mature T cells. mAb directed against this structure were either potently mitogenic, or alternatively, were capable of blocking the effector capabilities of cytotoxic T cell clones (Van Wauwe et al., 1980; Chang et al., 1981; Platsoucas and Good, 1981; Biddison et al., 1981). Based on these activities, the CD3 molecule was referred to as a "T cell stimulation receptor" (Van Wuawe et al., 1980) or as a candidate for an "antigen–recognition structure on human T cells" (Chang et al., 1981). However, biochemical analyses showed that the CD3 molecules were not clonally variable (Borst et al., 1982; Spits et al., 1982). Based upon cocapping of CD3 and the TCR by mAb against either, Meuer et al. (1983) first claimed an association between the clonotypic, variable TCR $\alpha\beta$ molecule and the CD3 complex on human T cells. These results were supported by the observations that CD3 and TCR $\alpha\beta$ molecules can be coimmunoprecipitated from detergent lysates (Borst et al., 1982; Reinherz et al., 1983), are coordinately expressed on the cell surface (Weiss and Stobo, 1984), and can be chemically cross-linked on the cell surface, indicating that they are physically associated (Brenner et al., 1985; Allison and Lanier, 1985). The receptor is thus correctly known as the TCR $\alpha\beta$–CD3 complex (Fig. 1).

In parallel with the immunological–biochemical approach, molecular biologists assumed that like immunoglobulin (Ig) of B cells, the T cell receptor might be expressed specifically in T lymphocytes and might be encoded by gene segments that rearrange to form a functional gene. Subtractive hybridization (T minus B) was used to generate T cell-specific cDNA probes and libraries, and clones were isolated that corresponded to genes that underwent rearrangement specifically in T cells. Using this approach, two laboratories identified T cell-specific cDNA clones composed of V, D, J, and C regions, analogous to those of immunoglobulin genes (Yanagi et al., 1984; Hedrick et al., 1984a,b).

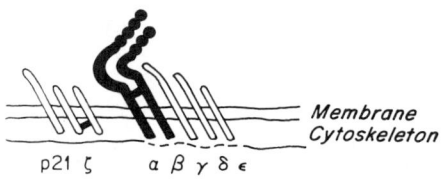

FIG. 1. TCR $\alpha\beta$–CD3 complex. The variable TCR α and β chains are disulfide linked to one another and associated with the nonpolymorphic CD3 subunits. The CD3 complex includes γ, δ, and ε subunits, as well as the disulfide-linked ζ homodimer. In addition, there exists a 21-KDa chain which may also be disulfide-linked ζ.

Based upon comparison with the partial amino acid sequence of purified TCR β chain protein, the first such genes characterized were found to encode the murine and the human TCR β chains (Acuto et al., 1984; Hannum et al., 1984). The TCR α gene was isolated soon thereafter, using both the subtractive library approach and protein sequence analysis in conjunction with synthetic oligonucleotide probes (Chien et al., 1984; Saito et al., 1984a; Sim et al., 1984; Hannum et al., 1984; Fabbi et al., 1984; Jones et al., 1985).

The TCR α and β genes encode transmembrane glycoproteins containing two extracellular domains (variable and constant) whose sequences are homologous to those of immunoglobulins, as well as a connecting segment, a transmembrane hydrophobic segment, and a short intracytoplasmic tail. In addition to intrachain loops in the V and C domains, each chain carries a cysteine residue in the membrane proximal connector that is thought to participate in interchain disulfide linkage.

Characterization of the murine and human TCR β loci has revealed 50–100 V_β gene segments. In addition, there are one D_β element and six to seven functional J_β segments situated upstream of each of two highly homologous C_β segments. The human and murine TCR α loci are composed of 50–100 V_α segments, as well as 30–100 J_α segments spread over more than 50 kb, upstream of a single C_α segment. To date, no evidence for D_α elements has been found. Rearrangement of V, D, and J elements appears to be mediated by heptamer and nonamer site-specific recombination sequences, similar to the signals used for immunoglobulin gene rearrangement in B cells. Detailed accounts of the structure and organization of these genes have recently been reviewed (Kronenberg et al., 1986; Marrack and Kappler, 1986; Toyonaga and Mak, 1987). In this article, we will discuss in detail the recent studies on the biochemistry, molecular genetics, and function of a distinct T cell receptor, TCR $\gamma\delta$.

II. T Cell Receptor (TCR) γ Genes

A. Murine TCR γ Genes

Following the cloning of TCR β, Saito et al. (1984b), in analyzing T cell-specific cDNA clones obtained from the murine CTL clone 2C, identified clones (pHDS4 and pHDS203) encoding a distinct, but related protein. Like TCR β, these clones detected genes that were rearranged and expressed specifically in T cells and were composed of·elements displaying homology to immunoglobulin variable (V), joining (J), and constant (C) gene segments. Homologies to TCR β and IgH, Igκ, and Igλ sequences ranged from 18 to 23% in the V region, from 21 to 50% in the J region, and from 16 to 22% in the C region, levels similar to those relating TCR β to Ig sequences. The V region sequence was preceded by a hydrophobic leader (L) or signal sequence, and the immunoglobulin-like portion of the C segment was followed by a short connecting peptide, as well as hydrophobic and hydrophilic segments, presumed to be transmembrane and intracytoplasmic portions of the polypeptide, respectively. Pairs of cysteine residues in the V and C regions predicted intradomain disulfide loops of 72 and 55 amino acids, respectively. In addition, as for TCR β, an extra cysteine was present in the connecting peptide, possibly to mediate interchain disulfide linkage, and a single positive charge interrupted the putative transmembrane hydrophobic segment. Further, the clone maintained an open reading frame across the V–J junction, indicating that it could encode a functional polypeptide in 2C cells. Based on these characteristics, the clones were concluded, mistakenly, to encode the α subunit of the TCR.

That these clones did not encode TCR α was suggested by the fact that, although the TCR α polypeptide was known to carry asparagine-linked carbohydrate, the predicted polypeptide sequence did not display potential sites for oligosaccharide addition. Subsequent amino acid sequence information for human TCR α (Hannum et al., 1984; Fabbi et al., 1984; Jones et al., 1985) and the isolation of murine and human cDNA clones consistent with that sequence (Chien et al., 1984; Saito et al., 1984a; Sim et al., 1984) confirmed that pHDS4 and pHDS203 defined a distinct TCR-like molecule, which was renamed TCR γ.

The TCR γ locus is found on murine chromosome 13 (Kranz et al., 1985a), distinct from the TCR β locus (chromosome 6) and the TCR α locus (chromosome 14). The organization of the locus (Fig. 2) is remarkably different from either TCR β or α (reviewed in Kronenberg et al., 1986; Toyonaga and Mak, 1987). Four distinct constant gene segments have now been identified (Hayday et al., 1985; Iwamoto et al., 1986), along with associated V and J segments. There is no evidence for distinct diversity (D) elements, as is the case for TCR β. Three of the constant segments are each associated with single V and J segments (Hayday et al., 1985; Iwamoto et al., 1986), whereas one constant segment is associated with four V segments and one J segment (Garman et al., 1986; Heilig and Tonegawa, 1986; Traunecker et al., 1986; Pelkonen et al., 1987). Only two of the V–J–C complexes have been directly linked; $V_\gamma 1.1$ (which rearranges to $J_\gamma 4$–$C_\gamma 4$) and $V_\gamma 1.2$ (which rearranges to $J_\gamma 2$–$C_\gamma 2$) lie only 2.5 kb apart in head-to-head orientation (Hayday et al., 1985). An unusual rearrangement of $V_\gamma 5$ to $J_\gamma 4$, with

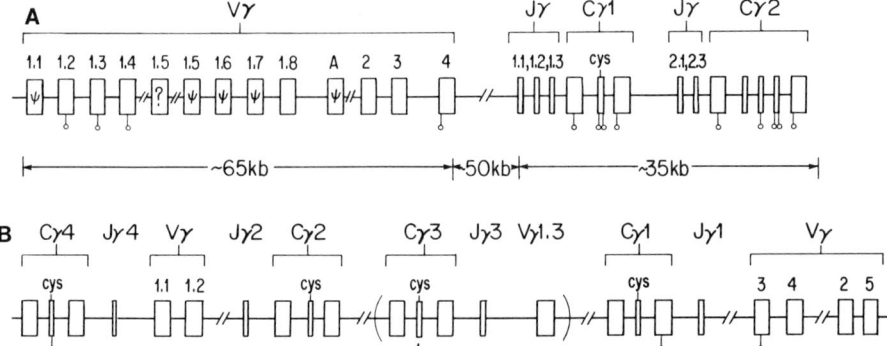

FIG. 2. Human (A) and murine (B) TCR γ gene organization. Human TCR γ gene organization is based on Fig. 7 of Forster *et al.* (1987) and Fig. 2B of Strauss *et al.* (1987) and on data from these references, as well as Lefranc *et al.* (1986a,b) and Quertermous *et al.* (1987). Nomenclature is according to Quertermous *et al.* (1987) and Strauss *et al.* (1987). An additional pseudogene ($V_\gamma B$) resides between $V_\gamma 3$ and $V_\gamma 4$ (Huck *et al.*, 1988). Murine TCR γ gene organization is based on Fig. 1 of Pelkonen *et al.* (1987) and on data from this reference, as well as Hayday *et al.* (1985), Garman *et al.* (1986), and Iwamoto *et al.* (1986). Nomenclature is according to Garman *et al.* (1986). The orientation of the $V_\gamma 1.3 - J_\gamma 3 - C_\gamma 3$ cluster relative to the other clusters is not known, and this is denoted by parentheses. Breaks in the solid lines connecting gene segments indicate that segments have not been linked on genomic clones. Pseudogenes are denoted by (ψ), and asparagine-linked glycosylation sites are denoted by (⌀). Cysteine residues encoded within CII exons are noted.

deletion of $C_\gamma 1$, $C_\gamma 2$, and $C_\gamma 3$, suggests the organization depicted in Fig. 2 (Pelkonen *et al.*, 1987). Heptamer and nonamer elements separated by 23 bp are found 3′ to the V segments, and similar elements separated by 12 bp are found 5′ to the J elements, suggesting V–J joining by a mechanism analogous to that of other TCR and immunoglobulin genes (Hayday *et al.*, 1985).

Each V gene is composed of two exons, one encoding most of a leader peptide and another encoding the main body of the V region (Hayday *et al.*, 1985). Using the nomenclature of Garman *et al.* (1986) (see Table I for other nomenclatures), three of the V segments, $V_\gamma 1.1$, $V_\gamma 1.2$, and $V_\gamma 1.3$, fall into a closely related family, displaying up to 88% homology at the amino acid level (Hayday *et al.*, 1985). By contrast, four other V segments ($V_\gamma 2$, $V_\gamma 3$, $V_\gamma 4$, and $V_\gamma 5$) are more distantly related both to $V_\gamma 1$ sequences and to each other, matching at 20–48% of their residues (Garman *et al.*, 1986; Heilig and Tonegawa, 1986; Traunecker *et al.*, 1986; Pelkonen *et al.*, 1987). Such a level of relatedness is similar to those among V_α subgroups and among V_β subgroups (reviewed in Kronenberg *et al.*, 1986; Toyonaga *et al.*, 1987). Strikingly,

TABLE I
HUMAN AND MURINE TCR γ NOMENCLATURE

Murine TCR γ nomenclature[a]					Human TCR γ nomenclature[b]	
V 1.1[c]	10.8B[d]	1[e]	1[f]	2[g]	V 1.1 pseudo[h]	1 pseudo[i]
V 1.2	10.8A	2	2	1	V 1.2	2
V 1.3	5.7	3	3	3	V 1.3	3
V 2			4	4.3	V 1.4	4
V 3			5	4.1	V 1.5	5
V 4				4.2	V 1.5 pseudo	5 pseudo
V 5				4.4	V 1.6 pseudo	6 pseudo
					V 1.7 pseudo	7 pseudo
C 1	13.4	1	2	4	V 1.8	8
C 2	10.5	2	1	1	V A pseudo	A pseudo
C 3	7.5	3	3	3	V 2	9
C 4			4	2	V 3	10
					V 4	11
					J 1.1	P1
					J 1.2	P
					J 1.3	1
					J 2.1	P2
					J 2.3	2

[a] Murine J segment nomenclature follows C segment nomenclature.
[b] Human C segments are universally denoted $C_\gamma 1$ and $C_\gamma 2$.
[c] Garman et al. (1986).
[d] Hayday et al. (1985).
[e] Reilly et al. (1986) and Heilig and Tonegawa (1986).
[f] Iwamoto et al. (1986).
[g] Pelkonen et al. (1987).
[h] Strauss et al. (1987) and Quertermous et al. (1987).
[i] Lefranc et al. (1986a), Forster et al. (1987), and Huck and Lefranc (1987).

$V_\gamma 5$ is much more closely related to human than to other murine V_γ sequences (Pelkonen et al., 1987). Whereas it displays only 32–38% nucleotide and 20–21% amino acid sequence identity with other murine V_γ segments, homology with human $V_\gamma 1$ subgroup segments ranges from 60 to 68% at the nucleotide level and from 42 to 48% at the amino acid level.

Each C_γ gene segment is composed of three exons, one encoding the immunoglobulin-like domain (CI exon), one encoding the connector peptide (CII exon), and one encoding the transmembrane and intracytoplasmic portions of the molecule (CIII exon) (Hayday et al., 1985; Garman et al., 1986). Of the four C_γ genes, $C_\gamma 1$, $C_\gamma 2$, and $C_\gamma 3$ are highly homologous. At the amino acid level, they differ in only 4–8% of their residues in the CI and CIII exons. The CII exons differ in size; those of $C_\gamma 1$ and $C_\gamma 3$ are 15 amino acids long, whereas that of $C_\gamma 2$ is only 10 amino acids long. By contrast, $C_\gamma 4$ differs at the amino acid level by 22% in the CI exon, 57% in the CIII exon, and displays a CII exon region which diverges in amino acid sequence almost completely from the CII exons of $C_\gamma 1$, $C_\gamma 2$, and $C_\gamma 3$, and encodes a peptide 33 amino acids long (Iwamoto et al., 1986). However, it is as yet unclear whether this results from a longer CII exon, or possibly the use of an additional exon. Sequence analysis indicates that $C_\gamma 3$ carries a defective splice signal at the end of the CII exon, suggesting that $C_\gamma 3$ is a nonfunctional pseudogene (Hayday et al., 1985). Furthermore, the $V_\gamma 1.3-J_\gamma 3-C_\gamma 3$ complex, although present in the BALB/c genome, is not found in the C57BL/10 genome (Iwamoto et al., 1986). The J regions display homology relationships analogous to their associated C regions; whereas $J_\gamma 1$ and $J_\gamma 2$ are identical at the amino acid level, $J_\gamma 4$ differs from them at 50% of the residues (Hayday et al., 1985; Iwamoto et al., 1986).

B. Human TCR γ Genes

Using probes derived from murine TCR γ sequences, human TCR γ genes were shown to constitute a T cell rearranging family of genes located on the short arm of chromosome 7 (Rabbitts et al., 1985; Lefranc and Rabbitts, 1985; Murre et al., 1985), distinct from the TCR β locus (long arm of chromosome 7) and the TCR α locus (chromosome 14). The locus is now well characterized, consisting of up to 14 V segments, 5 J segments, and 2 C segments (Lefranc and Rabbitts, 1985; Murre et al., 1985; Quertermous et al., 1986a,b 1987; Lefranc et al., 1986a,b,c; Dialynas et al., 1986; Forster et al., 1987; Huck et al., 1988) (Fig. 2). It spans roughly 160 kb, as judged by direct linkage of phage λ clones plus restriction mapping using both conventional and field-inversion gel electrophoretic techniques (Strauss et al., 1987). All gene segments lie in the same transcriptional orientation, the two C segments share the same pool of V segments, and rearrangement occurs by deletion of the intervening segment of DNA. This organization is, therefore, quite distinct from the organization of murine TCR γ genes, but is rather similar to that of human and murine TCR γ genes.

The characterized V segments fall into four subgroups, one of which consists of nine members, and three of which consist of one member each. The nomenclature of Strauss et al. (1987) ($V_\gamma 1.1$–1.8, $V_\gamma 2$, $V_\gamma 3$, and $V_\gamma 4$) will be used in this article, since it more closely approximates other TCR V gene nomenclature schemes in denoting different subgroups and members thereof. These V genes correspond to $V_\gamma 1-8$, $V_\gamma 9$, $V_\gamma 10$, and $V_\gamma 11$ in the nomenclature of Lefranc et al. (1986c) and Forster et al. (1987) (see Table I). $V_\gamma 1.1$ is considered a pseudogene due to a deletion between the heptamer and nonamer sequences at its 3' end, and $V_\gamma 1.5$, 1.6, and 1.7 are considered pseudogenes due to substitutions or frameshifts resulting in termination codons (Lefranc et al., 1986c). The status of $V_\gamma 4$ is uncertain, since it lacks one of the two cysteines involved in intradomain disulfide linkage, but is otherwise intact (Huck et al., 1988). Based upon cross-hybridization and deletion mapping, another $V_\gamma 1$ subgroup member has been suggested to lie between $V_\gamma 1.4$ and $V_\gamma 1.5$ (pseudo) (Forster et al., 1987). This gene has been denoted $V_\gamma 5$ or $V_\gamma 1.5$ (as distinguished from $V_\gamma 5$ or $V_\gamma 1.5$ pseudo), but since no other information is available, its status is unclear. In addition, a pseudogene ($V_\gamma A$) with a termination codon has been characterized (Forster et al., 1987; Huck et al., 1988). Another pseudogene ($V_\gamma B$) with numerous defects lies between $V_\gamma 3$ and $V_\gamma 4$ (Forster et al., 1987; Huck et al., 1988). Depending on the status of $V_\gamma 1.5$ and $V_\gamma 4$, there are likely to be six to eight functional V gene segments in the human TCR γ locus.

Comparisons among V sequences reveal 76–91% homology at the amino acid level among members of the $V_\gamma 1$ family (Lefranc et al., 1986c). Within this family, substitutions appear to cluster in three relatively discrete variable segments (Fig. 3). Homology between subgroups ranges from 20 to 44% (Lefranc et al., 1986a; Forster et al., 1987; Yoshikai et al., 1987; Huck et al., 1988). Polymorphism has been detected in the coding region of $V_\gamma 2$ (Forster et al., 1987).

```
V1.2  SQKSSNLEGRTKSVIRQTGSSAEITCDLAEGSNGY  IHWYLHQEGKAPQRLQYYDSYNSKVVLESGVSPGKYYTYASTRNNLRLILRNLIEN DSGVYYCATWDG
V1.3  -------------------------TVTNTF-     --------------L---VSTARD-----L-------HTPR-WSWI-R-Q----- ----------R
V1.4  ------------------------------T--    --------------L-----T-S------I-----D--G---K---M-------- ------------
V1.8  ----------------P-----V-----PVENAV-  T-------------L--------R------I-RE--H-----GKS-KF--E----R ----------R
V2    LCVYGAGH--QPQI-STKTLSKT-RLE-VVSGITISATSVY--RERPGEVIQF-VSI --DGT-RK---IPS--FEVDRIPETSTSTLTIHNV-KQ-IAT----L-EV
V3         QFQL-ISTEVKK-ID-P-KISSTRFETDV----RQKPNQ-LEH-I-IV-TK-AARRSM-KTSN-VEARKNSQTLTSILTIKSV-KE-MA-----A-WV
V4    ALGQ--QPEI-IS-PANK--H-SWKASIQGFSSKI----WQKPNTGLEY-LHVFLTI-AQDCSG-KT K-LEVSKNAHTSTSTLKIKFL-KE-EV--P--CQIL
```

FIG. 3. Human V_γ amino acid sequences. Deduced amino acid sequences encoded by the V exons of functional V_γ genes are aligned with each other. Nomenclature is according to Strauss et al. (1987). Identities are denoted by (-) and gaps by ( ). $V_\gamma 1$ subgroup V segments are known based on analysis of cDNA and unrearranged and rearranged genomic sequences (Lefranc et al., 1986c; Quertermous et al., 1986b; Dialynas et al., 1986; Littman et al., 1987). cDNA and unrearranged and rearranged genomic sequences are available for $V_\gamma 2$ (Lefranc et al., 1986a; Krangel et al., 1987a; Littman et al., 1987; Huck et al., 1988). Allelic variation in $V_\gamma 2$ has been reported (Forster et al., 1987). Unrearranged and rearranged genomic sequences are available for $V_\gamma 3$ (Forster et al., 1987; Huck et al., 1988). Thus, the 5' end is uncertain. The $V_\gamma 4$ sequence is from a rearranged genomic clone (Huck et al., 1988), and thus, both ends are uncertain.

```
Jγ1.3      NYYKKLFGSGTTLVVT
Jγ2.3      ----------------
Jγ1.1      TTGWF-I-AE--K-I--SP
Jγ2.1      SSDWI-T-AK--R-I--SP
Jγ1.2      GQELGKKI-V--P--K-II-
```

FIG. 4. Human J_γ amino acid sequences. Deduced amino acid sequences are from Lefranc et al. (1986a,c), Quertermous et al. (1986b, 1987), and Huck and Lefranc (1987). The 5' end of $J_\gamma 1.1$ is uncertain (Quertermous et al., 1987). Nomenclature is from Quertermous et al. (1987). Identities are denoted by (-).

Three J regions are situated 5' of $C_\gamma 1$, and two are situated 5' of $C_\gamma 2$ (Lefranc et al., 1986a,c; Quertermous et al., 1986b, 1987; Huck and Lefranc, 1987) (Fig. 2). Of these, $J_\gamma 1.1$ and 2.1 and $J_\gamma 1.3$ and 2.3 (nomenclature of Quertermous et al., 1987) form homologous pairs, whereas $J_\gamma 1.2$ has no counterpart associated with $C_\gamma 2$. $J_\gamma 1.3$ and 2.3 are identical to each other at the amino acid level, differing by only a single nucleotide, whereas $J_\gamma 1.1$ and 2.1 display roughly 70% nucleotide and amino acid sequence homology with each other. $J_\gamma 1.1$, 1.2, and 1..3 differ from each other in length, but share roughly 50% amino acid homology where they can be compared (Fig. 4).

The $C_\gamma 1$ gene, like its murine counterparts, is composed of three exons (Lefranc et al., 1986b). A CI exon encodes the immunoglobulin-like domain, a CII exon encodes a 16 amino acid connector peptide that includes a cysteine residue, which might be available for interchain disulfide bonding, and a CIII exon encodes the transmembrane and intracytoplasmic portions of the polypeptide (Fig. 5). Nucleotide homology with the murine exons is 74% in the CI exon, 64% in the CIII exon, but lower in the CII exon (Dialynas et al., 1986). Allelic variation in the CI exon has been described (Lefranc et al., 1986b). The $C_\gamma 2$ gene is composed of CI and CIII exons, which are virtually identical to their counterparts in $C_\gamma 1$. However, the $C_\gamma 2$ gene is considerably larger than $C_\gamma 1$ due to the duplication or triplication of a region encompassing the CII exon (Lefranc et al., 1986b; Littman et al., 1987; Krangel et al., 1987a; Pelicci et al., 1987; Hochstenbach et al., 1988). One polymorphic form of the $C_\gamma 2$ gene carries three copies of the 48-bp CII exon (copies a, b, and c), whereas another form carries two copies (copies b and c). The three $C_\gamma 2$ CII exons differ from each other by three to five nucleotides (two to four amino acids), and copy 3 of the $C_\gamma 2$ CII exon differs from the $C_\gamma 1$ form by only a single nucleotide. As a consequence of these substitutions, none of the CII exons in the $C_\gamma 2$ gene encodes the cysteine residue thought to be important for interchain disulfide linkage. As will be discussed later, differential use of $C_\gamma 1$ and allelic forms of $C_\gamma 2$, therefore, has the potential to encode polypeptides that vary in both size

FIG. 5. Human C_γ amino acid sequences. Deduced amino acid sequences for the $C_\gamma 1$ and $C_\gamma 2$ CI, CII, and CIII exons are shown (Dialynas et al., 1986; Lefranc et al., 1986b; Littman et al., 1987; Krangel et al., 1987b; Pelicci et al., 1987). Two polymorphic sequences of the $C_\gamma 1$ CI exon are shown, as well as three distinct sequences reported for the $C_\gamma 2$ CIII exon. As described in the text, the CII exon is present as a single copy in the $C_\gamma 1$ gene, but $C_\gamma 2$ genes carry $C_\gamma 2$ CII copies a, b, and c or copies b and c. Identities are denoted by (-).

and disulfide linkage (Krangel et al., 1987a; Hochstenbach et al., 1988). In addition, the CII exons carry multiple sites for asparagine-linked oligosaccharide addition, and hence, variable levels of glycosylation may occur as well.

Variation in C_γ gene structure in mouse and man presents a striking contrast to the situation for C_α and C_β genes. Only a single C_α gene has been identified in each species, and although two C_β genes exist, they are virtually identical (reviewed in Kronenberg et al., 1986; Toyonaga and Mak, 1987). However, in the mouse, of four C_γ genes, one is highly divergent, and in both mouse and man, variation in the size, sequence, and number of CII exons is extensive. Whether these differences will be reflected in the synthesis of functionally, as well as structurally, distinct proteins remains to be determined.

C. REARRANGEMENT AND DIVERSITY

The limited number of germ-line TCR γ V and J segments, as compared to TCR α and TCR β, indicates limited possibilities for TCR γ combinatorial diversity. In fact, initial studies in the murine system indicated that diversity was lower still, since a single predominant rearrangement ($V_\gamma 1.2$–$J_\gamma 2$–$C_\gamma 2$) was observed in a variety of CTL clones, as well as polyclonal adult spleen and thymus populations (Kranz et al., 1985b). However, rearrangement of $V_\gamma 2$ to $J_\gamma 1$–$C_\gamma 1$ was also demonstrated in adult thymocytes and $CD4^- 8^-$ (double negative) cells, and $V_\gamma 3$ and $V_\gamma 4$ rearrangements to $J_\gamma 1-C_\gamma 1$ were detected, but were more restricted (Garman et al., 1986; Heilig and Tonegawa, 1986). It should be noted that these early studies do not necessarily accurately reflect the rearrangement patterns observed in cells expressing cell surface TCR γ protein, which will be discussed below.

In man, a bias has been observed in J_γ segment usage in that rearrangement predominantly occurs to $J_\gamma 1.3$ and 2.3 (Quertermous et al., 1986a; Lefranc et al., 1986c). Since these J segments are of identical sequence, J usage contributes relatively little to human TCR γ diversity. Rearrangement is much less frequent to $J_\gamma 1.2$, possibly due to divergent heptamer and nonamer sequences (Lefranc et al., 1986c). Rearrangement to $J_\gamma 1.1$ and 2.1 are observed much more frequently in the thymus than in the periphery (Quertermous et al., 1987). However, these J_γ gene segment usage studies may or may not accurately reflect the usage of such gene segments from cells that express functional TCR $\gamma\delta$ receptors on their cell surfaces.

By far, the most significant contribution to TCR γ diversity occurs at the V–J junction (Kranz et al., 1985b; Quertermous et al., 1986b; Lefranc et al., 1986c). Comparsion of cDNA and genomic sequences reveals that, as for other TCR and immunoglobulin genes, this joining process may be imprecise, with the deletion of nucleotides at the 3' end of the V segment and/or the 5' end of the J segment, and new sequences may be inserted at the junction (Fig. 6). Since these sequences vary in random fashion from junction to junction and since there is no other evidence for distinct germ-line D elements or D–J rearrangements, they appear to represent N segments (Alt and Baltimore, 1982). Such N segments are thought to result from the activity of terminal transferase during the recombination process. However, an important by-product of this flexibility in joining at the V–J junction is that most joining events will result in a shift to an inappropriate translational reading frame. Indeed, until very recently, most attempts to identify T lymphocytes expressing TCR γ transcripts resulting from in-frame V–J joins had met with failure (Kranz et al., 1985b; Heilig et al., 1985; Reilly et al., 1986; Rupp et al., 1986; Heilig and Tonegawa, 1986; Dialynas et al., 1986; Quertermous et al., 1986b; Yoshikai et al., 1987).

D. TCR γ GENE PUZZLE

Although strides were made in delineating the structure and genomic organization of the TCR γ gene, it was unclear where to place this gene in a meaningful functional framework. Three facts dominated thinking about the TCR γ gene. First, its diversity was narrow, since there were only a limited number of V genes. Second, it was known to be expressed

```
A
GERMLINE   TGT GCC ACC TGG GAC AGG cacagtg.....      .....cactgtg AAT TAT TAT AAG AAA CTC

λF6        TGT GCC ACC TGG                                        AAT TAT TAT AAG AAA CTC
λS1        TGT GCC ACC TGG                  [CGGACG]              AAT TAT TAT AAG AAA CTC
pTγ1       TGT GCC ACC TGG GAC AGG          [CAAA]                AAT TAT TAT AAG AAA CTC




B
GERMLINE   TGT GCC TTG TGG GAG GTG cacagca.....      .....cactgtg AAT TAT TAT AAG AAA CTC

IDP2.11    TGT GCC TTG TGG GAG G            [G]                     T TAT TAT AAG AAA CTC
PBLC1.15   TGT GCC TTG TGG GAG G            [GAA]                    AT TAT AAG AAA CTC
Pγ12       TGT GCC TT                       [CCGGCCCG]                      AAG AAA CTC
pTγ10      TGT GCC TTG TGG GAG GTG                                   AT TAT TAT AAG AAA CTC
λK20       TGT GCC TTG                      [CGAGG]                   T TAT TAT AAG AAA CTC
λA6        TGT GCC TTG TGG GAG GTG          [C]                       T TAT TAT AAG AAA CTC
```

FIG. 6. Human TCR γ V–J junction nucleotide sequences. (A) Germ-line $V_\gamma 1.3$ (Lefranc et al., 1986c) and $J_\gamma 2.3$ (Quertermous et al., 1986b) sequences are shown, with coding nucleotides in capital letters and flanking heptamer sequences in small letters. Rearranged genomic sequences λF6 and λS1 (Lefranc et al., 1986c) and cDNA sequence pT$_\gamma$1 (Dialynas et al., 1986) utilize $V_\gamma 1.3$ and $J_\gamma 1.3$ or 2.3. N-region nucleotides are shown in brackets. (B) Germ-line $V_\gamma 2$ (Huck et al., 1988) and $J_\gamma 2.3$ (Quertermous et al., 1986a) sequences are shown. The structures of a variety of cDNA and rearranged genomic clones utilizing $V_\gamma 2$ and $J_\gamma 1.3$ or 2.3 (IDP2 and PBL C1, Krangel et al., 1987a: Pγ12, Littman et al., 1987; pTγ10, Quertermous et al., 1986b; λK20 and λA6, Lefranc et al., 1986c) are shown as well. N-region nucleotides are shown in brackets.

at high levels early in murine fetal thymic development (day 15–16) (Raulet et al., 1985; Snodgrass et al., 1985a) and before expression of TCR α on day 17 (Roehm et al., 1984). Following the early peak, TCR γ mRNA declined to very low levels by the time of birth and remained low in most mature cells, whereas TCR α and β levels remained high. Third, the vast majority of cloned TCR γ cDNAs corresponded to nonfunctional transcripts. These originated from diverse cell sources, including antigen-specific murine helper hybridomas (Helig et al., 1985), cloned murine cytotoxic T cell lines (Rupp et al., 1986; Reilly et al., 1986; Iwamoto et al., 1986), human thymus and peripheral blood (Yoshikai et al., 1987), and leukemia cell lines (Dialynas et al., 1986; Quertermous et al., 1986b).

These observations suggested a predominant role for TCR γ early in fetal life. Logical models incorporating this information were proposed, placing the γ chain of a hypothetical T cell receptor in fetal thymic development, possibly associated with the TCR β chain and participating only as an intermediate in the intrathymic process of selection of TCR αβ lymphocytes (Raulet et al., 1985; Snodgrass et al., 1985a; Pernis and Axel, 1985; Tonegawa, 1985; Yoshikai et al., 1986, 1987).

III. TCR $\gamma\delta$ Proteins

A. IDENTIFICATION OF TCR $\gamma\delta$

Recently, the TCR γ gene was found to encode part of a CD3-associated T cell receptor distinct from TCR $\alpha\beta$–CD3. Isolation of this receptor came from the observation that some CD3$^+$ human lymphocytes do not appear to express the TCR $\alpha\beta$ complex. These lymphocytes lack reactivity with mAbs βF1 (Brenner et al., 1986, 1987a) and WT31 (Tax et al., 1983; Oettgen et al., 1984; Spits et al., 1985a), which detect framework determinants present on all lymphocytes bearing $\alpha\beta$ T cell receptors. Based on the known association of CD3 with TCR $\alpha\beta$ (Borst et al., 1982; Meuer et al., 1983; Weiss and Stobo, 1984; Brenner, et al., 1985), it had been assumed that all T cells expressing CD3 would express TCR $\alpha\beta$. However, two-color cytofluorographic analyses of human peripheral blood revealed that, whereas 95–98% of CD3$^+$ cells were βF1$^+$ or WT31$^+$, 2–5% of CD3$^+$ cells failed to react with these mAb (Brenner et al., 1986; Weiss et al., 1986) (Fig. 7). It was, therefore, thought that these WT31$^-\beta$F1$^-$CD3$^+$ T cells might express a non-α, non-β CD3-associated T cell receptor. Consistent with this notion, the WT31$^-\beta$F1$^-$CD3$^+$ human peripheral blood cell line IDP2 lacked TCR α and TCR β mRNA (some lines expressed 1.0 kb of TCR β mRNA), yet expressed two cell surface polypeptides of 55 and 40 kDa, which could be chemically cross-linked to CD3. The 55-kDa species could be immunoprecipitated using antiserum made against synthetic peptides corresponding to the deduced amino acid sequence of a human TCR γ cDNA clone (Murre et al, 1985; Dialynas et al., 1986), suggesting that it was the elusive protein product of the TCR γ gene (Brenner et al., 1986). The 40-kDa species failed to react with anti-TCR γ sera and thus appeared to represent an additional component of the TCR γ–CD3 complex, which was termed TCR δ (Brenner et al., 1986). Thus, the TCR γ gene encoded part of a second T cell receptor present on T lymphocytes that expressed CD3, but lacked TCR $\alpha\beta$.

At the same time, three other laboratories found human CD3$^+$ T cell lines that expressed CD3-associated polypeptides other than TCR $\alpha\beta$. Bank et al. (1986) isolated a CD4$^-$8$^-$ thymus-derived T cell clone that lacked mature TCR α and β transcripts and lacked reactivity with βF1, but expressed abundant TCR γ transcripts and expressed a 42-kDa CD3-associated species. Weiss et al. (1986) observed that the T leukemic cell line PEER was WT31$^-$, lacked mature TCR α transcripts, and expressed a CD3-associated 55- to 60-kDa species that was considered a likely candidate for the TCR γ gene product. Similarly, Moingeon et al. (1986)

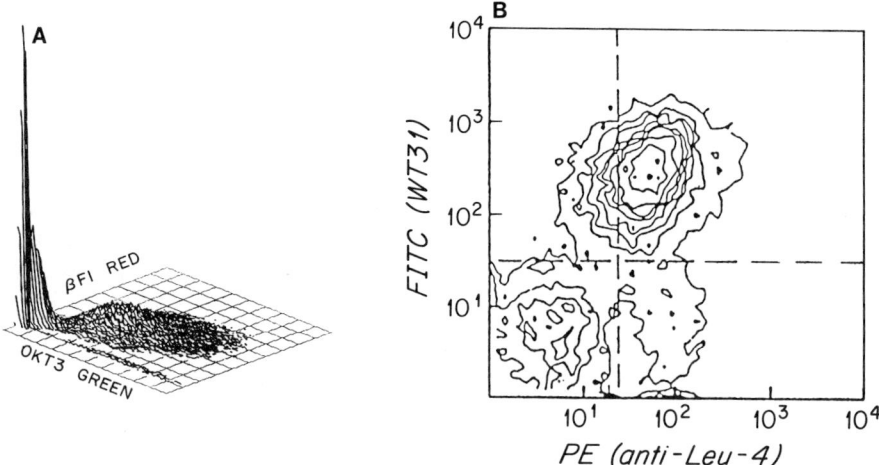

FIG. 7. Two-color cytofluorographic analysis of CD3 and TCR $\alpha\beta$ expression in human peripheral blood. (A) Two-color analysis staining of unseparated peripheral blood using mAb βF1 (red) compared to mAb OKT3 (green). The cells that do not react with either mAb are non-T cells (to the left). Of the T cells (defined as OKT3 or CD3$^+$), 95% also react with mAb βF1, which recognizes all TCR β chains. A small subpopulation corresponding to 5% of the cells is CD3$^+$, but unreactive with βF1 (cells along the x axis). These βF1$^-$CD3$^+$ peripheral blood lymphocytes correspond to TCR $\gamma\delta$ T lymphocytes. The lymphocytes were treated with 70% ethanol prior to staining with mAb βF1. [Modified from Brenner et al. (1986).] (B) Two-color analysis staining of peripheral blood using mAb WT31 (green) compared to mAb anti-Leu-4 (red). This experiment is similar to that in A. mAb WT31 brightly stains the surface of TCR $\alpha\beta$ lymphocytes, whereas anti-Leu-4 stains CD3$^+$ lymphocytes. The small subpopulation of WT31$^-$CD3$^+$ cells is seen along the x axis to the right. The reactivity of mAb WT31 is delineated further in the text. [Modified from Weiss et al. (1986).]

cloned 25-week human fetal T cells that were CD8$^+$ and expressed a CD3-associated disulfide-linked dimer, composed of 43-kDa subunits. These cells lacked TCR α mRNA, failed to react with mAbs WT31 and βF1, and thus expressed a CD3-associated complex, which could not represent the classical TCR $\alpha\beta$ complex. Ultimately, each of these cell lines was shown to express a TCR γ gene product after immunoprecipitation with anti-TCR γ peptide sera (Moingeon et al., 1987; Littman et al., 1987; Hochstenbach et al., 1988).

Reagents crucial to the identification of murine TCR $\gamma\delta$ were anti-TCR γ rabbit sera generated against synthetic peptides based on TCR γ-deduced amino acid sequences (Lew et al., 1986) and murine anti-CD3 mAb 145-2C11 (Leo et al., 1987). Based on the observations that TCR γ mRNA was highest in fetal thymus and double negative

(CD4⁻8⁻) adult thymocytes (Raulet et al., 1985; Snodgrass et al., 1985a) and that TCR γδ had been found on human CD4⁻8⁻ lymphocytes and thymocytes (above), murine adult double negative thymocytes were examined by immunoprecipitation using these reagents and were shown to express CD3-associated 35-kDa polypeptides specifically reactive with anti-TCR γ peptide sera (Lew et al., 1986). This TCR γ subunit was disulfide linked to a 45-kDa partner, TCR δ. Similarly, using a rabbit antiserum made against a TCR γ fusion protein (Nakanishi et al., 1987) or against synthetic peptides (Pardoll et al., 1987), the 35-kDa TCR γ subunit was found on 16-day fetal thymocytes, disulfide linked to a 45-kDa TCR δ partner chain.

B. Human TCR γδ Subunit Structure

Although TCR α and β chains are always disulfide linked and of similar size on virtually all T lymphocytes examined, the protein subunit structure of the TCR γδ has proved to be complex. In man, TCR γ subunits are found in disulfide-linked (Borst et al., 1987; Brenner et al., 1987b; Moingeon et al., 1987; Lanier et al., 1987; Alarcon et al., 1987; Van Dongen et al., 1988) and nondisulfide-linked forms (Brenner et al., 1986, 1987b; Weiss et al., 1986; Loh et al., 1987; Van Dongen et al., 1988; Hochstenbach et al., 1988). Further, the disulfide-linked form of TCR γ is 36–42 kDa, whereas the nondisulfide-linked forms are either 55 kDa or 40–44 kDa. Removal of asparagine-linked oligosaccharides reduces the disulfide-linked form to 32–34 kDa (Brenner et al., 1987b; Moingeon et al., 1987; Alarcon et al., 1987). By contrast, the large and small nondisulfide-linked forms are reduced to 40 and 35 kDa, respectively (Brenner et al., 1987b; Hochstenbach et al., 1988). Thus, human TCR γ peptides vary in disulfide linkage, polypeptide backbone size, and extent of glycosylation.

Characterization of cDNA clones reveals that the disulfide-linked TCR γ protein is encoded by the $C_\gamma 1$ gene, which contains a single CII exon that encodes a cysteine residue (Krangel et al., 1987a). The large nondisulfide-linked TCR γ protein is encoded by a $C_\gamma 2$ gene carrying three copies of the CII exon (copies a, b, and c), all lacking the cysteine residue (Littman et al., 1987; Krangel et al., 1987a). The small nondisulfide-linked TCR γ protein is encoded by a $C_\gamma 2$ gene carrying two copies of the CII exon (copies b and c) (Hochstenbach et al., 1988). These three forms have been denoted Form 1 (disulfide linked), Form 2abc (large, nondisulfide linked), and Form 2bc (small, nondisulfide linked) in accordance with $C_\gamma 1$ or $C_\gamma 2$ constant region and CII exon usage.

Due to variation in the number and kind of CII exons, these forms vary in the size and disulfide linkage of the membrane proximal

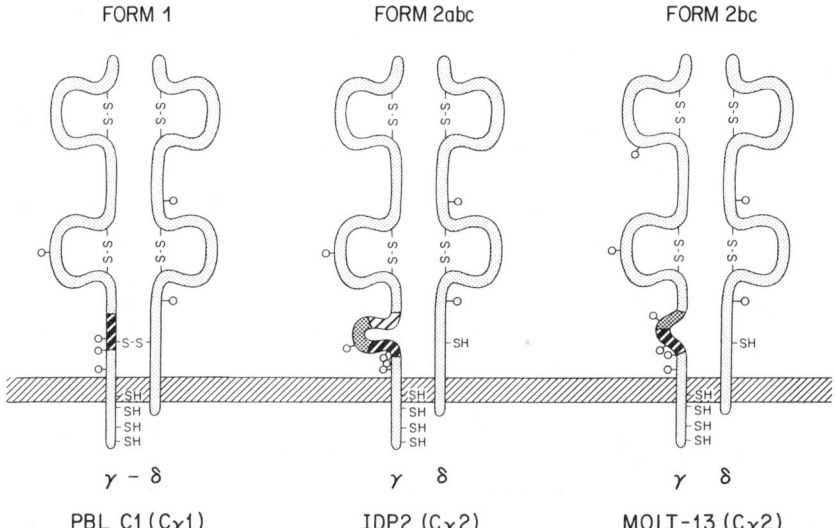

FIG. 8. Schematic representation of three forms of the human TCR $\gamma\delta$ complex. Form 1, represented by PBL C1 cells, displays a short (40 kDa) TCR γ polypeptide that is disulfide linked to TCR δ. The TCR γ chain is encoded by the $C_\gamma1$ gene that contains a single copy of the CII exon (hatched segment), bearing a cysteine in the connector region. Form 2abc, represented by IDP2 or PEER cell lines, displays a long (55 kDa) TCR γ polypeptide that is noncovalently associated with TCR δ. The TCR γ chain is encoded by the $C_\gamma2$ gene that contains three copies (copy a, copy b, copy c) of the CII exons (hatched segments), none of which bear a cysteine in the connector region. The TCR γ polypeptide encoded by this form is larger due to the CII exon repeats and the attachment of additional asparagine-linked carbohydrates. Form 2bc, represented by the MOLT-13 cell line, displays a short (40 kDa) TCR γ polypeptide noncovalently associated with TCR δ. The TCR γ chain is encoded by a version of the $C_\gamma2$ gene that contains two copies (copy b and copy c) of the CII exons, neither of which encodes a cysteine. In this form, less carbohydrate is attached to the potential asparagine-linked glycan acceptor sites that are located in the connector region, largely accounting for the smaller size of TCR γ. [Modified from Hochstenbach et al. (1988).]

connector region (Fig. 8). Further, this connector region carries most of the asparagine-linked oligosaccharide addition sites of the molecule. It is striking that Form 2bc carries significantly less carbohydrate than Form 2abc, even though they carry the identical number of acceptor sites in the connector region. This suggests that the conformation of the region dramatically affects the utilization of these sites (Hochstenbach et al., 1988). It is not known if use of these structurally distinct forms correlates with function; however, the preferential use of Form 1 in certain individuals has recently been observed (Hochstenbach et al., 1988).

The TCR δ subunit was initially distinguished as unreactive with serological reagents specific for the TCR γ chain (Brenner et al., 1986, 1987b; Borst et al., 1987). Peptide maps comparing TCR γ and TCR δ chains from the same cell line revealed these chains to be distinct (Krangel et al., 1987a). In man, the TCR δ subunit is 40 kDa and characteristically displays a marked shift in sodium dodecyl sulfate–polyacrylamide gel electrophoresis (SDS–PAGE) mobility when compared under reducing and nonreducing conditions (Brenner et al., 1987b; Band et al., 1987; Loh et al., 1987; Hochstenbach et al., 1988) (Fig. 9). When TCR δ is expressed with the 55-kDa TCR γ form, it is easily resolved by reducing SDS–PAGE, but when expressed with the shorter 40-kDa nondisulfide-linked TCR γ form, it is not resolved. However, due to the characteristic mobility shift, TCR γ and δ can be resolved in this circumstance under nonreducing conditions (Brenner et al., 1987b; Loh et al., 1987; Hochstenbach et al., 1988). The TCR δ polypeptide has a backbone size of 35 kDa and appears to carry two asparagine-linked glycans (Band et al., 1987; Hochstenbach et al., 1988). Consistent with sequence analysis (see below), biochemical studies of TCR δ chains from three independently derived TCR $\gamma\delta$ cell lines reveal highly related peptide map and glycosylation patterns (F. Hochstenbach and M. B. Brenner unpublished), indicating substantial relatedness among the TCR δ polypeptides. However, some differences in SDS-PAGE mobility have been reported. A TCR δ subunit of 43 kDa (Borst et al., 1987; Lanier et al., 1987) may represent a TCR δ subunit that carries more carbohydrate than the 40-kDa subunit and/or that utilizes a different V_δ gene segment.

In some instances, TCR γ subunits have been observed without detectable TCR δ species, and the existence of TCR γ–γ homodimers or γ–γ' heterodimers have been considered (Borst et al., 1987; Moingeon et al., 1987; Ang et al., 1987; Alarcon et al., 1987; Ioannides et al., 1987). In such cases, it is not clear whether TCR γ can be expressed without TCR δ, or whether TCR δ is not easily visualized due to poor radiolabeling or comigration with TCR γ. The recent development of mAb (Band et al., 1987) and antiserum (Loh et al., 1987) against TCR δ should allow a direct assessment of this issue.

In man, the CD3 polypeptides associated with TCR $\gamma\delta$ chains are subtly different from the CD3 polypeptides associated with TCR $\alpha\beta$ proteins (Brenner et al., 1986; Krangel et al., 1987b). This results from differential processing of the oligosaccharides of the CD3 δ chain. This chain normally carries one high-mannose (endoglycosidase H sensitive) and one complex asparagine-linked carbohydrate when expressed in association with human TCR $\alpha\beta$ (Borst et al., 1982, 1983; Kanellopoulos

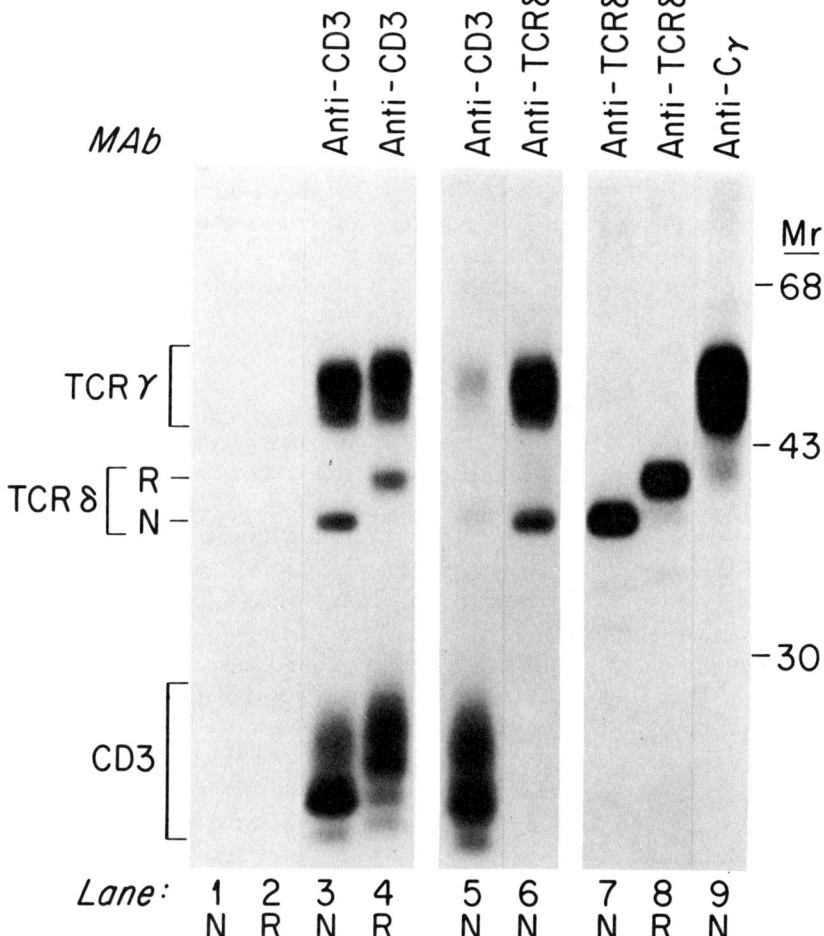

Fig. 9. Biochemical analysis of the TCR γδ–CD3 complex. TCR γδ cell line IDP2 was ^{125}I surface labeled, solubilized in the detergents noted, immunoprecipitated with specific mAb, and resolved by 10% SDS–PAGE under nonreducing (N) or reducing (R) conditions. The TCR γδ–CD3 complex remains associated when solubilized in 0.3% CHAPS (lanes 3 and 4, anti-CD3 mAb; lanes 1 and 2, control mAb). When 2% Triton X-100 detergent is added to the above, TCR γ and δ remain together, but they dissociate from the CD3 subunits (lane 5, anti-CD3 mAb; lane 6, anti-TCR δ1 mAb, which is specific for the TCR δ subunit). When anti-CD3 immunoprecipitates from CHAPS-solubilized cells (like those in lanes 3 and 4) are boiled in 1% SDS, all of the individual chains of the TCR γδ–CD3 complex dissociate. Immunoprecipitations with anti-TCR δ1 from such lysates results in isolation of the TCR δ chain alone (lanes 7 and 8), while immunoprecipitation with an antiserum specific for the TCR γ peptides results in isolation of the separated TCR γ chain (lane 9). Note that the TCR δ subunit displays a marked shift in SDS–PAGE mobility when resolved, comparing nonreducing and reducing conditions. This TCR γδ complex corresponds to the large nondisulfide-linked type, Form 2abc (see the text). [Modified from Band et al. (1987).]

et al., 1983; Van den Elsen *et al.*, 1984). However, on TCR $\gamma\delta$ T cells, the CD3 δ chain carries two complex asparagine-linked glycans, resulting in a slightly larger size when examined by SDS–PAGE and resulting in insensitivity to endoglycosidase H (Krangel *et al.*, 1987b; Alarcon *et al.*, 1987). No such differences have been reported in the mouse. The CD3 γ peptide from TCR $\gamma\delta$ lymphocytes may serve as a substrate for protein kinase C-mediated phosphorylation (Krangel *et al.*, 1987b), similar to the phosphorylation of this subunit on CD3-associated TCR $\alpha\beta$ lymphocytes (Cantrell *et al.*, 1985).

C. Murine TCR $\gamma\delta$ Subunit Structure

Like human TCR $\gamma\delta$ complexes, the murine receptor is CD3 associated. However, in contrast to the variation in covalent association of human TCR γ and δ chains, the murine TCR $\gamma\delta$ complex is always disulfide linked. The expressed TCR γ proteins have been characterized based on their reactivity with antisera that should recognize products of both $C_\gamma 1$ and $C_\gamma 2$, which are highly homologous, or antisera specific for $C_\gamma 4$, which is distinct. They have been further discriminated based on their molecular weight, as well as the presence or absence of asparagine-linked glycosylation. In both the BALB/c and C57BL/6 strains, $V_\gamma 1.2-J_\gamma 2-C_\gamma 2$ products are predicted to lack asparagine-linked carbohydrate, whereas $C_\gamma 1$ products should carry asparagine-linked carbohydrate. Fetal and adult CD4$^-$8$^-$ thymocytes and adult CD4$^-$8$^-$ splenocytes from these strains predominantly express TCR γ chains of 35 kDa, which reduce to 32 kDa on removal of asparagine-linked glycans, indicating that they are probably products of the $C_\gamma 1$ gene (Lew *et al.*, 1986; Pardoll *et al.*, 1987, 1988; Ioannides *et al.*, 1987). However, in other strains (for example C57BL/10), the predominant TCR γ peptides reacting with $C_\gamma 1$- and $C_\gamma 2$-specific antisera are 32 kDa and are not glycosylated (Maeda *et al.*, 1987; Pardoll *et al.*, 1988; Cron *et al.*, 1988). It has been argued that this results from a polymorphism in the $C_\gamma 1$ gene of this strain, resulting in the absence of an asparagine-linked oligosaccharide acceptor site (Maeda *et al.*, 1987; Pardoll *et al.*, 1988), although the $C_\gamma 1$ sequence from this strain is not known. This interpretation is supported by gene rearrangement data, which, coupled with protein data, suggest that the expressed TCR γ proteins are predominantly encoded by $V_\gamma 2-J_\gamma 1-C_\gamma 1$ (Pardoll *et al.*, 1988). Although TCR γ proteins encoded by the $C_\gamma 4$ gene were not detected previously in the BALB/c thymus (Lew *et al.*, 1986), $C_\gamma 4$-encoded TCR γ chains (41–45 kDa) have recently been found on the CD3$^+$4$^-$8$^-$ splenocytes of B10 mice (Cron *et al.*, 1988) and, to a lesser extent, on B10 adult and fetal thymocytes (J. Bluestone, unpublished).

Murine TCR γ is always disulfide linked to a TCR δ chain (45 kDa), which is reduced in size to 36–37 kDa by removal of asparagine-linked glycans (Lew et al., 1986; Pardoll et al., 1987). In contrast to the situation in man, the two chains can generally be resolved from one another (and to some extent, also from TCR α and β) on SDS–PAGE due to the larger size of the TCR δ subunit.

Recently, murine skin-derived Thy-1$^+$ dendritic epidermal cells (Thy-1$^+$ DEC) were shown to be of T cell lineage (see below). Several Thy-1$^+$ DEC cell lines were found to express 34-kDa TCR γ subunits (peptide core size 31 kDa) associated with 46-kDa putative TCR δ chains (peptide core size 34 kDa) (Koning et al., 1987a). Another Thy-1$^+$ DEC cell line was shown to carry a larger glycosylated TCR γ protein (41-kDa glycosylated, 34-kDa peptide core size) and a still larger TCR δ subunit (52-kDa glycosylated, 38-kDa core size), compared to the 35- and 45-kDa-glycosylated chains that are usually detected on murine thymocytes (Kuziel et al., 1987; Bonyhadi et al., 1987). Several DEC cell lines revealed unexplained biochemical features, namely, the possible expression of γ–γ homodimers and the apparent expression of two distinct heterodimers (thought to represent a TCR γ–γ homodimer and a TCR $\alpha\beta$-like heterodimer) on the same cell, although these analyses were preliminary (Stingl et al., 1987b). The possibility that Thy-1 DEC may use the distinct $C_\gamma 4$ gene was suggested by the presence of full-length $C_\gamma 4$ transcripts in one such cell line (Koning et al., 1987a).

IV. TCR δ Genes

A. Identification of TCR δ Genes

A better understanding of the structure of TCR $\gamma\delta$ has come from the recent identification and initial characterization of genomic clones and cDNA clones encoding the TCR δ polypeptide. In the mouse, Lindsten et al. (1987) initially identified a region 80 kb 5' to the C_α gene that was rearranged in adult dull Ly-1, CD4$^-$8$^-$ thymocytes and hybridomas. Based on the analysis of a panel of hybridomas, the rearrangements 5' to C_α appeared to just precede those at the TCR γ and TCR β loci. Similar results were obtained through the analysis of fetal thymocytes at different stages of development. Rearrangements were found to initiate as early as day 14, at or just prior to the time at which TCR β and γ rearrangements are first seen (Chien et al., 1987a). These rearrangements did not involve known J_α segments. Further, the rearranging region appeared to be deleted in most mature T cells, presumably as a result of V_α to J_α rearrangements. These observations suggested the possibility that an additional, previously undescribed rearranging gene was located in the region 5' to C_α, between V_α and C_α gene segments. Chien et al. (1987a) demonstrated that these rearrangements involved novel V, D, and J elements that were expressed in conjunction with a novel constant region (Cx). Transcripts of 2.0 and 1.55 kb (thought to originate from fully and partially rearranged genes, respectively) were detected at high levels in fetal thymocytes and in adult double negative thymocytes.

Human homologs of this rearranging gene were isolated soon thereafter. Hata et al. (1987) used a subtracted (T − B) cDNA probe generated from TCR $\gamma\delta$ cell line IDP2 RNA to obtain and characterize T cell-specific cDNA clones present in a human TCR $\gamma\delta$ cell line cDNA library. In this manner, they identified cDNA clones corresponding to transcripts specifically expressed in TCR $\gamma\delta$ T cells and to a gene that is rearranged in TCR $\gamma\delta$ T cells, but deleted in other T cell lines. Sequence analysis revealed elements homologous to TCR and immunoglobulin V, J, and C segments (Fig. 10) and revealed 79% nucleotide sequence identity between the constant region sequence and that of murine Cx. Transcript heterogeneity (2.2, 1.7, 1.3, and 0.8 kb) was shown to reflect both transcription from fully and partially or unrearranged genes, as well as differential polyadenylation. The high level of nucleotide sequence identity between the human and murine genes has also allowed the isolation of human cDNA clones from the PEER cell line by cross-hybridization with the murine cDNA clone (Loh et al., 1987).

The cDNA clones obtained by these approaches were presumed to encode TCR δ based on transcript expression and gene rearrangement data. Further, the human IDP2 TCR δ cDNA sequence predicted a

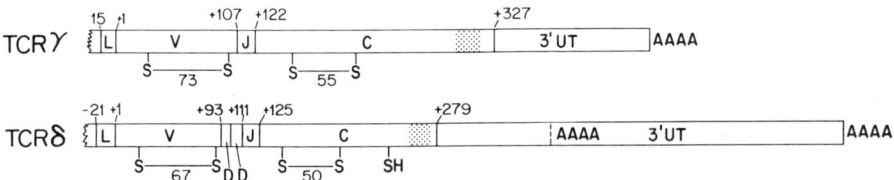

FIG. 10. Comparative organizations of the IDP2 TCR γ and TCR δ transcripts. Leader (L), variable (V), diversity (D), joining (J), constant (C), and 3'-untranslated (3'-UT) portions are noted (Krangel et al., 1987a; Hata et al., 1987). Amino acids are numbered from the presumed processing point between the leader and variable segments, and the locations and sizes of intrachain disulfide loops are noted. A free cysteine in TCR δ, thought to mediate interchain disulfide linkage with TCR γ peptides encoded by the $C_\gamma 1$ gene, is also shown. Presumed transmembrane regions are shaded. Polyadenylation is denoted by AAAA.

polypetide backbone size of 31.5 kDa, with two potential sites for N-linked glycosylation, in good agreement with the known properties of the TCR δ protein isolated from the same cell line (Band et al., 1987). Proof that the cDNA clones encoded TCR δ was obtained by Band et al. (1987), who showed by *in vitro* transcription and translation that the cDNA clones directed the synthesis of a polypeptide specifically recognized by a TCR δ-specific monoclonal antibody. Conversely, an antiserum raised against a synthetic peptide derived from a human cDNA clone-deduced amino acid sequence has been shown to specifically immunoprecipitate the human TCR δ peptide from the PEER and MOLT-13 cell lines (Loh et al., 1987), and an antiserum raised in a similar fashion against a murine sequence immunoprecipitated the murine TCR δ peptide from a dendritic epidermal cell line (Bonyhadi et al., 1987). These results are further confirmed by limited protein sequence information obtained for murine TCR δ purified from a fetal thymocyte × BW5147 hybridoma (Born et al., 1987).

B. Genomic Organization of TCR δ Gene Segments

There is evidence for only single cross-hybridizing TCR δ C segments in the murine and human genomes (Chien et al., 1987a; Hata et al., 1987; Loh et al., 1987). Initial characterization of the murine C_δ locus has been reported recently (Chien et al., 1987a,b; Korman et al., 1988) (Fig. 11). The C_δ gene segment lies roughly 70 kb 5′ to the C_α gene segment, but no more than 10–15 kb 5′ to the nearest J_α segments. The unusual location of C_δ, between V_α and J_α segments, results in the deletion of C_δ upon rearrangement of V_α to J_α. Two distinct J_δ segments ($J_\delta 1$ and $J_\delta 2$) lie 5′ of C_δ. In addition, two D_δ elements ($D_\delta 1$ and $D_\delta 2$) have been localized 5′ of the J segments. Heptamer and nonamer elements separated by 12–13 bp flank the 5′ ends of the D and J segments, whereas heptamer and nonamer elements separated by 23 bp flank the 3′ ends of the D elements.

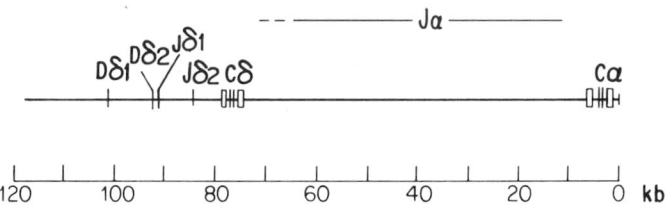

FIG. 11. Organization of the murine C_δ–C_α locus. Organization is based on Fig. 1a of Chien et al. (1987b) and on data from this reference, as well as Chien et al. (1987a) and Lindsten et al. (1987). Approximate distances in kilobases (kb) are noted.

This organization should permit the incorporation of zero, one, or two D elements into rearranged TCR δ genes, a situation without precedent in other TCR genes. Indeed, murine cDNA and rearranged genomic clones reflecting $V-D_\delta 2-J_\delta 1$ and $V-D_\delta 1-D_\delta 2-J_\delta 1$ joining events have been observed (Chien et al., 1987a,b).

Although analysis of the human C_δ locus is less complete, its organization appears to be quite similar. The C_δ gene segment lies roughly 85 kb 5' of C_α (Boehm et al., 1988). Two J_δ segments have been localized 5' to C_δ, and analysis of the junctional sequences of cDNA clones suggests the use of at least two human D_δ elements in tandem (see below; Hata et al., 1988; K. Satyanarayana, S. Hata, P. Devlin, J. L. Strominger, and M. S. Krangel, unpublished).

C. C_δ Gene Segment

The genomic organization of C_δ exons has not yet been reported. However, cDNA sequence analysis reveals that the TCR δ C region, like those of other TCR chains, is composed of an immunoglobulin-like domain, a connector peptide, and transmembrane and intracytoplasmic segments (Fig. 12). The immunoglobulin-like domain shares roughly 20% homology with the corresponding regions of other TCR and immunoglobulin chains. This domain carries two cysteine residues separated by 50 amino acids that are presumably involved in intradomain disulfide bonding. The distance between these cysteines is similar to that found in C_α (Chien et al., 1984; Saito et al., 1984a; Sim et al., 1984), but is shorter than that found in C_β (Yanagi et al., 1984; Hedrick et al., 1984b) or C_γ (Saito et al., 1984b). Alignments of the C region sequences support the notion that C_δ and C_α may encode immunoglobulin-like domains with structural features similar to each other, but somewhat distinct from those of C_β and C_γ. The connector region carries an additional cysteine residue, which would most likely mediate disulfide linkage with the TCR γ peptide. It is notable that, although this cysteine residue is available for disulfide linkage, it is apparently not utilized if the TCR δ peptide is expressed as part of a nondisulfide-linked complex along with the product of the human $C_\gamma 2$ gene, which lacks the corresponding cysteine residue (Fig. 8). Thus, TCR γ constant region usage appears to be the sole factor regulating interchain disulfide linkage of human TCR $\gamma\delta$, a notion that has been confirmed by transfection experiments (H. Band, M. S. Krangel, and M. B. Brenner, unpublished).

The putative TCR δ transmembrane and intracytoplasmic regions display remarkable sequence similarity with the analogous regions of TCR α (Fig. 12). Both sequences display a pair of basic residues spaced

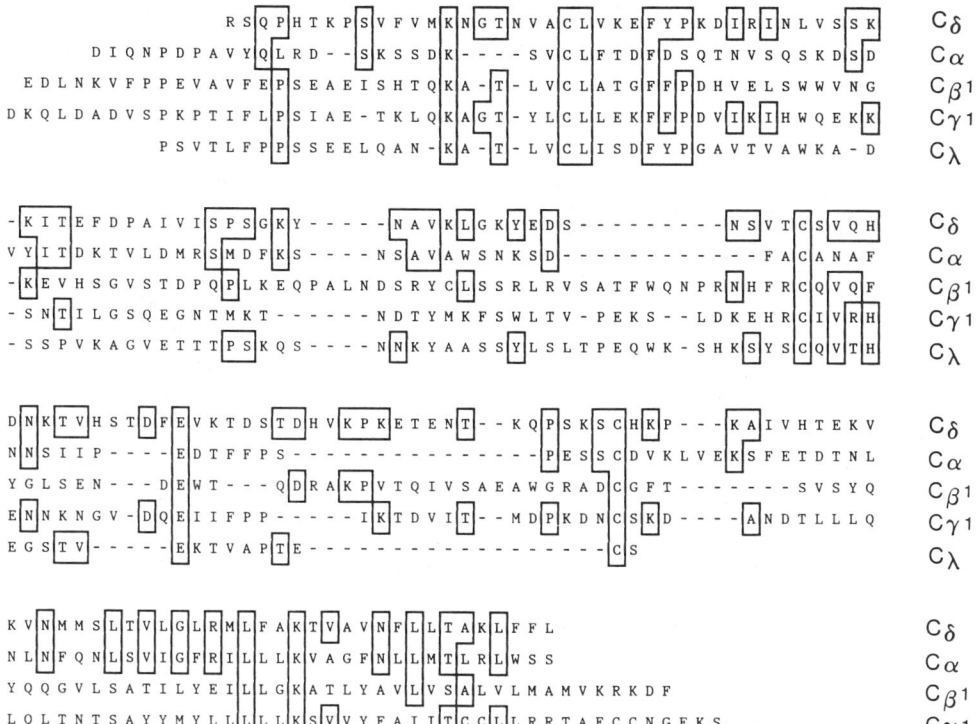

FIG. 12. Comparison of the human C_δ amino acid sequence with those of C_α, C_β, C_γ, and C_λ. The deduced amino acid sequence of the human TCR δ C segment (Hata et al., 1987; Loh et al., 1987) is aligned and compared with the sequence of C_α (Sim et al., 1984; Yoshikai et al., 1985), $C_\beta1$ (Yanagi et al., 1984; Toyonaga et al., 1985), $C_\gamma1$ (Lefranc et al., 1986b), and C_λ (Kabat et al., 1987). Identities are boxed. Gaps are denoted by (-).

by four amino acids that are probably buried within the membrane, followed by a ten amino acid spacer, an additional basic residue, and a short (three or four residues), uncharged carboxy-terminal tail. This contrasts with the analogous regions of TCR β and TCR γ in that these polypeptides display only a single basic residue within the transmembrane segment and display longer and more highly charged intracytoplasmic tails. Further, overall sequence identity within this region is 30–35% between C_δ and C_α, but much lower (10–15%) between C_δ and either C_β or C_γ. These relationships may indicate that this portion of the C_δ polypeptide is involved in interactions with other polypeptides (for example, CD3 components), which are analogous to those of C_α.

D. Rearrangement and Diversity of V_δ, D_δ, and J_δ Segments

Seven murine and one human TCR δ V sequences have been reported (Chien et al., 1987a,b; Hata et al., 1987; Loh et al., 1987; Korman et al., 1988). Most display considerably more relatedness to TCR α V sequences than to TCR β or TCR γ V sequences (Fig. 13). However, most TCR δ V sequences characterized to date have not been largely found to be utilized by TCR α, and thus, the two V repertoires may be distinct, despite the fact that they are all presumed to be located 5' of the C_δ gene. It is possible that V segment usage by C_δ and C_α might be controlled by chromosomal location; for example, all V_δ segments may lie between V_α and D_δ segments, but this remains to be seen.

Initial studies suggest that the TCR δ V gene pool may be limited. For example, five different human TCR γδ peripheral blood cell lines and leukemias utilize a single V_δ segment and a single J_δ segment (Hata et al., 1987, 1988; Loh et al., 1987;) Analysis of an expanded panel of cell lines has identified only two additional human V_δ segments (M. S. Krangel, S. Hata, and J. de Vries, unpublished). Furthermore, the murine V_δ segments already identified appear to account for a large proportion of the rearrangements in TCR γδ cells in fetal and adult thymus (Chien et al., 1987a,b; Korman et al., 1988). Although other human and murine V_δ segments may be identified in future studies, this initial work certainly suggests that the size of the TCR δ V repertoire will be much more like that of TCR γ than that of TCR α or TCR β.

The murine and human $J_\delta 1$ segments are highly homologus to each other, matching at 13/16 amino acids. Similarly, the murine and human $J_\delta 2$ segments form a closely related pair, matching at 18/19 amino acids (Chien et al., 1987a,b; Hata et al., 1987, 1988; Loh et al., 1987; Korman

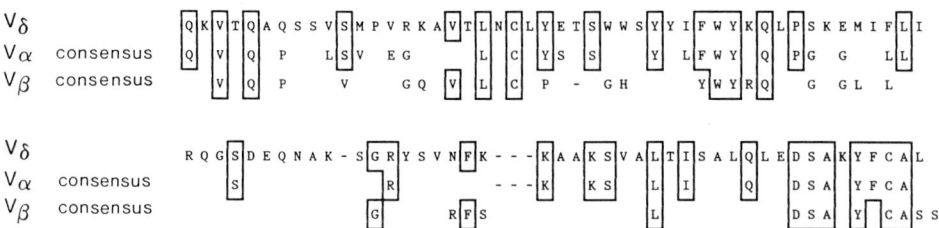

FIG. 13. Comparison of the human V_δ amino acid sequence with V_α and V_β consensus sequences. The deduced amino acid sequence of the human TCR δ V segment (starting at the presumed amino-terminus of the mature protein) is compared with V_α and V_β subgroup consensus sequences, as in Fig. 4A of Hata et al. (1987). V_α and V_β consensus residues were assigned on the basis of their appearance in 50% or more V_α or V_β subgroups, using data compiled in Toyonaga and Mak (1987). Blanks indicate no consensus assignment, gaps are denoted by (-), and identities are boxed.

	V		J	C
IDP2	YFCAL	AVRGKLLERNGGYAVFPS	DKLIFGKGTRVTVEP	RSQPQHT
PBL C1	-----GE	PGSLQWGWGRGIGG	--------------	-------
MOLT-13	-----GE	PGGY	T--------------	-------
MOLT-13t			SWDTRQMF--T-IKLF---	-------
PBL L1a	-----GE	SQPPYWGIRRILY	T--------------	-------
PBL L1b	-----GE	KWTDFPKTYWGIRN	T--------------	-------
PEER	-----G	TGVRGLQD	T--------------	-------

FIG. 14. Human TCR δ V–D–J–C junctional amino acid sequences. Deduced amino acid sequences of six functional transcripts (IDP2, PBL C1, MOLT-13, PBL L1a, PBL L1b, and PEER) and one truncated transcript (MOLT-13t) are presented (Hata et al., 1987, 1988; Loh et al., 1987). V, J, and C segment borders are based on V_δ and J_δ unrearranged genomic sequences (K. Satyanarayana, S. Hata, P. Devlin, J. Strominger, and M. S. Krangel, unpublished). A single J segment ($J_\delta 1$) is present in all cDNA clones derived from functionally rearranged genes. A second J segment ($J_\delta 2$) is found in a truncated cDNA from the MOLT-13 cell line, which appears to result from transcription initiating upstream of the J segment. Identities are denoted by (-).

et al., 1988). $J_\delta 1$ is highly homologous to consensus J sequences of other TCR chains and is particularly closely related to the J_α consensus. $J_\delta 2$ shares less than 50% nucleotide and amino acid sequence identity with $J_\delta 1$ and is only rarely utilized (Fig. 14).

By contrast to the low levels of diversity introduced by an apparently limited number of germ-line V_δ and J_δ segments, junctional diversity is apparently extremely high (Chien et al., 1987a,b; Hata et al., 1987, 1988; Loh et al., 1987; Korman et al., 1988; Fig. 14). Murine fetal thymocyte cDNA and rearranged genomic clones have been shown to incorporate $D_\delta 2$ alone or $D_\delta 1$ and $D_\delta 2$ in tandem (Chien et al., 1987a,b). However, N-region nucleotides are notably absent from the junctional regions of rearranged murine fetal thymocyte TCR δ genes (Chien et al., 1987b). Strikingly, V–D, D–D, D–J, and D–D–J rearrangements were all observed in fetal thymocyte genomic clones. This stands in sharp contrast to data for TCR β and IgH, in which D–J rearrangements invariably precede V–D rearrangements (Yaoita et al., 1983; Born et al., 1986).

The analysis of a panel of human TCR δ cDNA clones isolated from peripheral blood and leukemic cell lines has revealed dramatic variability in the junctional region (Fig. 14). As many as 51 and as few as 9 nucleotides separate the ends of the germ-line V and J segments. As in the murine system, there is evidence for the use of two D_δ elements that may be incorporated in tandem into the junctional region (Hata et al., 1988; K. Satyanarayana, S. Hata, P. Devlin, J. L. Strominger, and M. S. Krangel, unpublished). One of these D elements is clearly homologous to the murine $D_\delta 2$ sequence. In addition, there appears to be extensive incorporation of template-independent N-region nucleotides at each site of recombination within the junctional region, resulting in the structure V–N–D–N–D–N–J.

This preliminary work suggests that, as a result of the use of two (or perhaps more) D elements arranged in tandem, imprecise joining at the V, D, and J ends, and the polymerization of template-independent *N*-nucleotides, the capacity for diversity in the TCR δ V–J junctional region is tremendous. Thus, although germ-line TCR δ diversity is rather limited, a high degree of diversity is introduced into a small segment of the polypeptide in the junctional region. TCR γ diversity follows a similar pattern, although junctional diversity is less dramatic. The contrasts between these patterns of diversity and those of TCR α and β may serve as hints as to the nature of antigens recognized by TCR $\gamma\delta$.

E. Chromosomal Translocations

A limited number of recurrent translocations involving the TCR loci have been described in human T cell leukemias and lymphomas. Most commonly, these rearrangements involve the TCR α locus at 14q11 (Caccia *et al.*, 1985; Croce *et al.*, 1985; Collins *et al.*, 1985). A paracentric inversion of chromosome 14, inv(14)(q11;q32), involving the TCR α and IgH loci, and a related reciprocal translocation, t(14;14)(q11;q32) are typically observed in T-CLL (Zech *et al.*, 1984; Hecht *et al.*, 1984; Ueshima *et al.*, 1984; Baer *et al.*, 1985; Denny *et al.*, 1986; Mengle-Gaw *et al.*, 1987). In addition, reciprocal translocations t(11;14)(p13;q11) (Williams *et al.*, 1984; Lewis *et al.*, 1985) and t(8;14)(q24;q11) (Williams *et al.*, 1984; Shima *et al.*, 1986; Finger *et al.*, 1986; Mathieu-Mahul *et al.*, 1985, 1986; McKeithan *et al.*, 1986; Erikson *et al.*, 1986) are commonly found in T-ALL, and t(10;14)(q24;q11) is found in T cell lymphoma (Hecht *et al.*, 1984; Kagan *et al.*, 1987). Another T-ALL translocation involving chromosomes 14 and 11 t(11;14)(p15;q11) has been observed (LeBeau *et al.*, 1986).

Based on *in situ* hybridization, restriction mapping, and, in some cases, molecular cloning, many breakpoints in 14q11 occur between V_α and C_α (Lewis *et al.*, 1985; Erikson *et al.*, 1986; Kagan *et al.*, 1987). In some cases, the breakpoints have been demonstrated to occur within the J_α region (Shima *et al.*, 1986; McKeithan *et al.*, 1986), apparently mediated by J_α recombination signals (Finger *et al.*, 1986; Mengle-Gaw *et al.*, 1987). Further, at least one instance of inv(14)(q11;q32) has been mediated by site-specific recombination between V_H and J_α sequences, resulting in a productive rearrangement (Baer *et al.*, 1985; Denny *et al.*, 1986). In the case of t(8;14)(q24;q11) translocations, the reciprocal breakpoints occur 3' to c-*myc* (Mathieu-Mahul *et al.*, 1985, 1986; Erikson *et al.*, 1986; Shima *et al.*, 1986; McKeithan *et al.*, 1986; Finger *et al.*, 1986). Breakpoints within 14q32 may vary (Baer *et al.*, 1985; Denny *et al.*, 1986; Kennaugh *et al.*, 1986; Mengle-Gaw *et al.*, 1987)

Given the location of the TCR δ gene within the TCR α locus, between V_α and C_α, it seems likely that some of the translocations involving breakpoints that split the TCR α locus might directly involve TCR δ. Indeed, at least one example of t(11;14)(p15;q11) (LeBeau *et al.*, 1986) has recently been shown to result from site-specific recombination between chromosome 11 sequences and D_δ heptamer and nonamer sequences (Boehm *et al.*, 1988). The relative frequency of translocations into the TCR δ locus remains to be established.

V. Phenotype and Distribution of TCR $\gamma\delta$ Lymphocytes

A. Peripheral Blood and Thymic TCR $\gamma\delta$ Lymphocytes

One of the phenotypic hallmarks of TCR $\gamma\delta$ cells is the lack of reactivity with mAb directed against TCR $\alpha\beta$ (Brenner *et al.*, 1986; Weiss *et al.*, 1986; Moingeon *et al.*, 1986). These mAb include βFramework 1 (βF1), which detects a determinant on the TCR β chain and immunoprecipitates TCR $\alpha\beta$, but not TCR $\gamma\delta$ (Brenner *et al.*, 1987a). However, mAb βF1 does not bind to the surface of living T cells. mAb WT31 displays a less clearly defined reactivity, but is, nevertheless, extremely useful (Tax *et al.*, 1983). This antibody binds strongly to the surface of TCR $\alpha\beta$ cells and thus has been used to identify and to isolate brightly stained (TCR $\alpha\beta$) and unstained (TCR $\gamma\delta$) cells. However, under certain conditions (such as high antibody concentration or pretreatment of lymphocytes with neuraminidase), WT31 can bind at low levels to TCR $\gamma\delta$ cells (Van de Griend *et al.*, 1988). Very weak binding can also be demonstrated on fresh peripheral blood lymphocytes of some individuals (S. Porcelli and M. B. Brenner, unpublished). Further, although WT31 is a poor immunoprecipitator, appropriate solubilization conditions (using digitonin or low detergent concentrations) allow immunoprecipitation of either TCR $\alpha\beta$ (Oettgen *et al.*, 1984; Spits *et al.*, 1985a; Brenner *et al.*, 1986; Littman *et al.*, 1987) or TCR $\gamma\delta$–CD3 (Van de Griend *et al.*, 1988). Thus, the epitope recognized by WT31 is unclear, but may be TCR $\alpha\beta$ or a combinatorial determinant formed with CD3, with weak, but detectable, cross-reaction on TCR $\gamma\delta$–CD3. Although strong staining by WT31 suggests TCR $\alpha\beta$ expression and weak or no staining by WT31 on CD3$^+$ cells suggests TCR $\gamma\delta$ expression, corroboration as to which receptor is expressed should be obtained using either TCR $\gamma\delta$-specific mAb or biochemical analysis.

The WT31⁻CD3⁺ cells in human peripheral blood were found to constitute about 3% of peripheral blood lymphocyte (PBL) T cells (range 0.5–10%) (Brenner et al., 1986; Weiss et al., 1986; Lanier and Weiss, 1986; Lanier et al., 1987). Recently, mAb that react specifically with human TCR γδ cells were made. mAb anti-Ti-γA (Jitsukawa et al., 1987) reacts with about 3% (range 1–15%) of CD3⁺ lymphocytes in human peripheral blood. However, since it failed to react with two of seven TCR γδ clones propagated *in vitro*, it does have framework reactivity on all TCR γδ lymphocytes. This mAb may recognize a particular TCR γ V gene-encoded determinant (T. Hercend, personal communication). Band et al. (1987) characterized a mAb, anti-TCR δ1, that reacts with the TCR δ chain. It stains from <0.5 to 12% of peripheral blood T cells and all WT31⁻CD3⁺ T cells in the peripheral blood of several adults examined (L. Lanier, S. Porcelli, and M. B. Brenner, unpublished). It is a good candidate to be a framework anti-TCR δ mAb. In addition, a mAb, δTCS1, that is specific for the TCR δ chain reacts with a subset of TCR γδ lymphocytes and may detect a $V_δ$ uncoded determinant (Wu et al., 1988; W. T. Tian, S.-L. Ang, and S. H. Ip, unpublished).

The use of these mAbs has revealed no evidence for cell surface expression of TCR αβ and TCR γδ on the same cells. Similarly, there is no evidence for expression of mixed T cell receptors (i.e., βγ or αγ). Even in the PEER cell line in which TCR γδ is expressed on the cell surface and TCR β chains are synthesized and associate with CD3 intracellularly, no evidence for cell surface expression of TCR βγ or βδ could be detected (Koning et al., 1987b). The phenotypic expression of the two receptors thus appears to be mutually exclusive, in general, TCR α appears only to interact stably with TCR β and TCR γ appears only to interact stably with TCR δ.

A further striking feature to emerge from the identification of lymphocytes expressing TCR γδ was that many of the cells displayed the double negative (CD4⁻8⁻) phenotype (Brenner et al., 1986; Bank et al., 1986; Borst et al., 1987). WT31⁻CD3⁺ cells in human peripheral blood are CD4⁻8⁻ (Lanier and Weiss, 1986; Lanier et al., 1987). Two- and three-color cytofluorographic analyses using TCR γδ-specific mAbs, anti-Ti-γA and anti-TCR δ1, revealed that CD8 was found on greater than 20% of the TCR γδ in peripheral blood (but at levels approximately one-tenth that found on CD8⁺ TCR αβ cells), while the remainder of the TCR γδ cells were CD4⁻8⁻ (Jitsukawa et al., 1987; L. Lanier, S. Porcelli, and M. B. Brenner, unpublished). In addition, several cultured TCR γδ lines and clones have been found which are CD4⁻8⁺ (single positive) (Brenner et al., 1986; Moingeon et al., 1986, 1987). The low levels of CD8 expression on many resting TCR γδ lymphocytes makes quantitation of

CD8$^+$ and CD8$^-$ subsets difficult and may account for the failure to observe expression of this antigen in some studies. In contrast, CD4 was not detected on any TCR $\gamma\delta$ cells in peripheral blood.

The TCR $\gamma\delta$ population in human peripheral blood are either negative or express low levels of CD5 (T1) (approximately one-fifth the level found on TCR $\alpha\beta$ cells) (Lanier and Weiss, 1986; Jitsukawa et al., 1987; L. Lanier and M. B. Bernner, unpublished). CD2 (T11) was expressed at levels comparable to that on TCR $\alpha\beta$ cells (Lanier and Weiss, 1986; Jitsukawa et al., 1987; L. Lanier and M. B. Brenner, unpublished), but, in one individual, was absent from all the TCR $\gamma\delta$ cells in peripheral blood (Jitsukawa et al., 1987). Interestingly, Borst et al. (1987) found that many cultured TCR $\gamma\delta$ clones express the IgG Fc receptor (CD16), using the mAb VD2, although the majority of TCR $\gamma\delta$ cell lines and resting peripheral blood $\gamma\delta$ cells do not react with other anti-IgG Fc receptor mAb such as B73.1 (anti-Leu-11c) (Van de Griend et al., 1987). Resting TCR $\gamma\delta$ cells lack CD19, but express CD7 (Lanier and Weiss, 1986; L. Lanier and M. B. Brenner, unpublished). CD11 (OKM1) was found to be present on peripheral blood WT31$^-$CD3$^+$ lymphocytes (Moretta et al., 1988). Several TCR $\gamma\delta$ lymphocyte clones express NKH1 [present on many peripheral blood lymphocytes displaying natural killer (NK)-like function], but resting TCR $\gamma\delta$ cells in peripheral blood are NKH1$^-$ (Jitsukawa et al., 1987). WT31$^-$CD3$^+$ PBL were found to lack CD28 (T44), compared to the expression of CD28 on virtually all WT31$^+$ cells in peripheral blood (Poggi et al., 1987), although recently CD28$^+$ TCR $\gamma\delta$ cells have been found (L. Lanier, unpublished). In fresh thymus from 2- to 6-year old donors, 0.8% of total thymocytes showed the WT31$^-$CD3$^+$ phenotype and were otherwise similar in phenotype to the counterpart cells in PBL, except that about 20–40% of these cells were weakly CD1$^+$ (T6) (Lanier and Weiss, 1986). Direct staining with TCR $\gamma\delta$-specific mAb confirm the existence of this population (L. Lanier, V. Groh, and M. B. Brenner, unpublished). Recent studies using the anti-TCR δ1 mAb reveal that TCR $\gamma\delta$ cells are found in human lymph node, spleen, and tonsillar tissue at levels typical of those for peripheral blood ($<1–10\%$ of the lymphocytes) (V. Groh, L. Picker, R. Warnke, and M. B. Brenner, unpublished).

In the mouse, double negative (CD4$^-$8$^-$) adult thymocytes constitute 3–5% of the total and appear similar to day 15–16 fetal thymocytes, which contain precursors of the mature single positive peripheral T cells (Fowlkes et al., 1985). About 10% of double negative thymocytes express CD3, and such cells include TCR $\gamma\delta$ lymphocytes (Lew et al., 1986; Bluestone et al., 1987; Pardoll et al., 1987; Maeda et al., 1987). Recently, CD3$^+$4$^-$8$^-$ lymphocytes have also been found on murine peripheral blood lymphocytes in the spleen (Maeda et al., 1987;

Pardoll et al., 1988; Cron et al., 1988). Given the lack of framework mAb against either TCR $\alpha\beta$ or TCR $\gamma\delta$, direct phenotypic characterization of TCR $\gamma\delta$ cells in polyclonal populations has not been possible in the mouse. CD8 expression on murine TCR $\gamma\delta$ cells was not observed initially, but, in virtually all instances, murine TCR $\gamma\delta$ cells have been isolated using methods to specifically obtain double negative cells, prior to phenotypic analysis. However, in a recent study, low levels of CD8 expression on cultured TCR $\gamma\delta$ cell lines was observed, although it was thought to have been induced by *in vitro* culture (Cron et al., 1988).

B. Skin TCR $\gamma\delta$ Lymphocytes

The mammalian epidermis is composed of heterogeneous cell types including ectodermal keratinocytes, which comprise the vast majority (90%) of epidermal cells (EC), as well as less-abundant cell types with striking dendritic morphology, i.e., neuroectodermal melanocytes and Merkel cells and bone marrow-derived leukocytes. In the murine system, these epidermal leukocytes encompass two separate populations that differ in lineage, phenotype, and function. Antigen-presenting Langerhans cells are Ia^+, IgG Fc receptor$^-$, C3 receptor$^+$, and $F4/80^+$ (macrophage-specific antigen) and are members of the mononuclear–phagocyte or dendritic cell series (Schuler and Steinman, 1985). The second dendritic cell population residing within murine epidermis, termed Thy-1$^+$ dendritic epidermal cells (Thy-1$^+$ DEC), was originally identified by their abundant expression of Thy-1 antigens (also found on murine thymocytes, peripheral blood T cells, and neuronal cells), suggesting they were of T lymphocyte lineage (Tschachler et al., 1983; Bergstresser et al., 1983). Histochemical and phenotypic studies on Thy-1$^+$ DEC *in situ* and on freshly isolated cell suspensions have shown that, while expressing the Thy-1 antigen, these cells lacked other markers considered typical for T cells, namely, they were $CD5^-$ (Lyt 1$^-$), $CD4^-$ (L3T4$^-$), and $CD8^-$ (Lyt 2$^-$) (Romani et al., 1985). In contrast to Langerhans cells, Thy-1$^+$ DEC were Ia^- (Tschachler et al., 1983; Bergstresser et al., 1983; Romani et al., 1985). DEC express asialo-G_{M1}, have arylsulfatase (found in large granular lymphocytes), and contain membrane-bound granules, similar to cells having natural killer activity (Romani et al., 1985).

Recently, more compelling evidence that Thy-1$^+$ DEC are specialized T lymphocytes has emerged from analyses of their T cell receptors. The expression of a TCR complex on DEC was suggested by staining of all DEC with anti-CD3 mAb, 145-2C11 (Stingl et al., 1987b). Staining epidermal skin sheets (Stingl et al., 1987b) or staining and immunoprecipitation from interleukin 2-dependent *in vitro* cultured Thy-1$^+$ DEC cell lines (Stingl et al., 1987a; Koning et al., 1987a; Kuziel et al., 1987) revealed the cell surface expression of the TCR $\gamma\delta$–CD3 complex. The expression of mRNA transcripts for TCR α were typically absent from

TCR $\gamma\delta$ DEC cell lines, similar to the circumstances found in thymocytes and peripheral blood TCR $\gamma\delta$ lymphocytes. Consistent with their T lineage, DEC were able to be propagated *in vitro* in the presence of concanavalin A and conditioned media containing interleukin 2 activity. Under such conditions, they expressed interleukin 2 receptors and sometime lost expression of asialo-G_{M1} (Nixon-Fulton *et al.*, 1986; Stingl *et al.*, 1987a). Strikingly, one such *in vitro* cultured cell line reacted both by surface staining and in immunoprecipitation analyses with mAb F23.1, indicating that the DEC cell line expressed TCR $\alpha\beta$ ($V_\beta 8$) on the cell surface (Stingl *et al.*, 1987a). However, immunostaining of epidermal skin sheets with the F23.1 mAb failed to reveal positively stained DEC cells (Stingl *et al.*, 1987a), suggesting that, while the studied DEC cell line may express TCR $\alpha\beta$, TCR $\alpha\beta$ expression may be a rare occurrence *in situ* in skin. In fact, immunostaining of epidermal sheets *in situ* revealed that 40–60% of the DEC cells stain with an antiserum-reactive to $C_\gamma 1$-, 2-, and 3-encoded determinants (Stingl *et al.*, 1987b). It is possible that many of the DEC unreactive with this antiserum may utilize the $C_\gamma 4$ gene segment.

While the DEC lineage is still poorly understood, it appears that such cells are of bone marrow, rather than of thymic origin, based on the study of radiation chimeras reconstituted with allogeneic bone marrow cells from congenic mice differing only at the *Thy-1* locus (Breathnach and Katz, 1984; Bergstresser *et al.*, 1984). Thus, Thy-1$^+$ DEC may represent a prethymic or extrathymic pathway for development of cells expressing TCR $\gamma\delta$ (and possibly TCR $\alpha\beta$). Clearly Thy-1$^+$ DEC may represent specialized T cells in the skin that primarily utilize the TCR $\gamma\delta$ in immune functions.

In man, a cell population strictly analogous to the murine DEC has not been described, since Thy-1 antigens are not expressed on human epidermal cells or on most other human T cells (Cooper *et al.*, 1985; Cohen *et al.*, 1986). However, using TCR $\gamma\delta$ specific mAB, CD3$^+$ TCR $\gamma\delta^+$ lymphocytes are regularly found scattered within the most basal keratinocyte layer, as opposed to the suprabasal location of the murine highly dendritic Thy-1$^+$ CD3$^+$ TCR $\gamma\delta^+$ DEC. Human intraepidermal TCR $\gamma\delta^+$ lymphocytes comprise between 0.5 and 10% of the CD3$^+$ lymphocytes within the basal epidermis (V. Groh, unpublished). The relationship between human and murine TCR $\gamma\delta$ expressing cells in the skin is currently under investigation.

VI. Ontogeny of TCR $\gamma\delta$

At the time of thymic colonization, precursor thymocytes are not clonally precommitted. During the course of thymic T cell development, lymphocytes destined to express $\alpha\beta$ T cell receptors proliferate, differentiate, and are selected as appropriate clonally committed T-cells expressing unique TCRs (reviewed in Adkins et al., 1987; Stutman, 1985; Mathieson and Fowlkes, 1984; Fowlkes and Mathieson, 1985; Scollay et al., 1984; Rothenberg and Lugo, 1985; Acuto and Larocca, 1987). Evidence that TCR $\gamma\delta$-bearing lymphocytes also undergo an important stage of thymus-influenced development is emerging (Lew et al., 1986; Maeda et al., 1987; Pardoll et al., 1987, 1988). The process of T cell ontogeny is complex and is only partially understood. In this article, selected aspects are summarized only briefly, and the reader is referred to comprehensive reviews devoted to this topic (see above). We will focus on the timing and relationships among TCR α, β, γ, and δ gene rearrangements and expression in populations of developing T lymphocytes in the thymus.

Early in fetal development, the thymus is organized into cortical and medullary thymic microenviroments, including both epithelial and bone marrow-derived stromal cells. The subcapsular cortex, cortex, and medulla are histologically descrete thymic subregions having distinct stromal elements. T cells are hypothesized to pass through these regions in sequential order during various stages of development, although the pathway is uncertain.

The early precursor thymocytes (Pgp-1^+ in mouse or CD7^+ in man) colonize the thymus at d11 (fetal day 11) in the mouse (Moore and Owen, 1967; Fontaine-Persu et al., 1981; Trowbridge et al., 1985) or w8–9 (fetal week 8–9) in man (reviewed in Stutman, 1985; Lobach and Haynes, 1987). In the d14 murine thymus, the predominant thymocytes have a phenotype of Ly-1^+ (dull), Thy-1^+, Lyt 2^- (CD8^-), and L3T4^- (CD4^-) (Ceredig et al., 1983; Mathieson et al., 1981; Van Ewijk et al., 1982); while, in man, cells that may be their counterparts are CD$1^-3^-4^-8^-$ and CD2^+7^+ (Lobach and Haynes, 1987; Kamps and Cooper, 1984; Acuto and Larocca, 1987). These double negative (CD4^-8^-) thymocytes include the precursors of all mature T cell subsets. Two or three days later, "double positive" cells expressing Lyt 2 (CD8) and L3T4 (CD4), as well as CD1 (T6) and CD2 (T11), appear and then account for most of the cells that die within the thymus. This latter population may or may not include precursors of "single positive" CD4^+8^- or CD4^-8^+ mature thymocytes. CD3 expression, believed to correspond to associated cell surface TCR ($\alpha\beta$ or $\gamma\delta$) expression, is observed on a relatively large subpopulation of double positive (CD4^+8^+) thymocytes, although it is expressed at lower levels on mature (single positive) cells (Bluestone et al., 1987; Acuto and Larocca, 1987). It is now known that, although many double negative thymocytes represent immature precursors of more mature T cell subsets, a subpopulation of

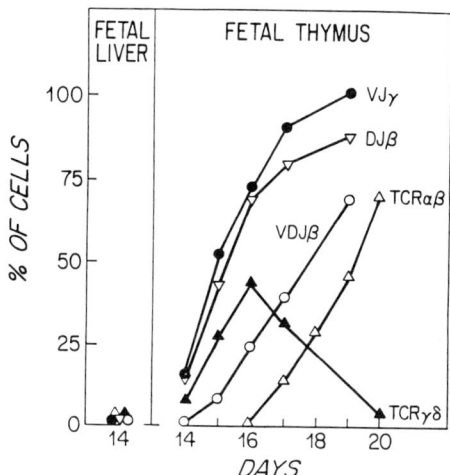

FIG. 15. Time course of rearrangement and expression of TCR subunits during fetal murine thymic development. An estimate of the percentage of cells in the fetal murine thymus, showing various TCR gene rearrangement, is shown from d14 of development to birth. TCR γ rearrangements attain substantial levels by d15–16. Nonproductive TCR β gene rearrangements follow a similar time course, but functional (V_β–D_β–J_β) rearrangements do not reach substantial levels until after d17. Cell surface expression of the TCR proteins ($\alpha\beta$ and $\gamma\delta$) are shown during the same time. TCR $\gamma\delta$ is expressed early and is the only TCR expressed at d15–16 of fetal thymic development. After this peak, the proportion of thymocytes expressing the TCR $\gamma\delta$ decreases. TCR $\alpha\beta$ is first expressed at low levels on d17 and, thereafter increases through the time of birth. Data are derived from Born *et al.* (1986) and Pardoll *et al.* (1987) and others (see the text).

double negative thymocytes appear not to be immature, but rather represent the mature phenotype of T cells that express TCR $\gamma\delta$.

It is assumed that TCR genes are unrearranged in the first lymphoid cells to reach the thymus, but by d14 of murine fetal development, TCR gene rearrangement and transcription is observed (Fig. 15). TCR γ mRNA transcripts can be detected as early as d14, increase rapidly to peak at d15, and subsequently decline until birth (Raulet *et al.*, 1985; Snodgrass *et al.*, 1985a; Haars *et al.*, 1986; Born *et al.*, 1986). Similarly, TCR γ gene rearrangements are first observed in d14 fetal thymocytes and peak on d15. TCR δ transcripts can also be detected by d14 (Chien *et al.*, 1987a). TCR δ gene rearrangements occur at least as early as d14 in fetal thymocytes, initiating at the same time or possibly just prior to the time TCR γ gene rearrangements are first detected (Chien *et al.*, 1987a,b). Thus, TCR γ and TCR δ genes are both rearranged and expressed in parallel fashion early in fetal thymic ontogeny.

By contrast, functional TCR β and TCR α gene rearrangements and transcription lag behind and then largely replace those of TCR γ and δ in the latter part of thymic development in the mouse. TCR β chain rearrangements are first detected at d14 (although most are still in germ-line form at that time) and increase markedly on d15 through d18 (Born et al., 1985; 1986; Snodgrass et al., 1985b; Haars et al., 1986). However, virtually no full-length TCR β transcripts (1.3 kb, corresponding to $V_\beta-D_\beta-J_\beta$) are present during the first few days in which rearrangements are detected (d14–15). By d16, full-length TCR β transcripts are present at high levels, and, after d17, TCR β transcription plateaus (Snodgrass et al., 1985a,b; Raulet et al., 1985; Haars et al., 1986). TCR α transcription lags until d16, at which time the transcripts are probably largely nonfunctional (1.4 kb). Then full-length (1.7 kb) transcripts increase from d16 onward to reach maximal levels in adult thymocytes (Raulet et al., 1985; Haars et al., 1986).

The recent availability of mAb and antiserum against the different TCRs made it possible to correlate the observations on gene rearrangement and transcription with the appearance of cell surface TCR proteins during development. Pardoll et al., (1987) showed that TCR $\gamma\delta$–CD3 protein complexes were expressed on the surface of ^{125}I-labeled d15 fetal thymocytes by using TCR γ-specific antiserum. No cell surface TCR $\alpha\beta$–CD3 complexes could be detected in these samples. By d17.5 of fetal thymic development, the radiolabeled TCR $\gamma\delta$ species were more abundant, and were shown to reside largely (if not totally) on $CD3^+4^-8^-$ thymocytes. TCR $\alpha\beta$ complexes were detectable by d17.5, and they became more abundant through d20. TCR $\gamma\delta$–CD3 complexes remained detectable through birth, but at later times (d17.5–20), they represented a minor species relative to the more abundant TCR $\alpha\beta$. These results are consistent with those of Roehm et al. (1984), who first detected TCR $\alpha\beta$ cells at d17 using an anti-TCR $V_\beta 8$-specific mAb. Thus, TCR $\gamma\delta$ rearrangement, transcription, and expression represents a first wave of T cell receptor development, which is followed by a second wave, TCR $\alpha\beta$. A possible third wave (birth–d5), consisting of $CD4^-8^-$ TCR $\alpha\beta$ lymphocytes expressing a single V_β segment ($V_\beta 8$), has recently been identified (Crispe et al., 1987; Fowlkes et al., 1987).

Consistent with the above data implying a thymic-influenced developmental pathway for TCR $\gamma\delta$ lymphocytes, recent studies in nude mice reveal that thymic engraftment results in a marked increase in the appearance of TCR $\gamma\delta$ lymphocytes in the periphery (Pardoll et al., 1988). However, TCR $\gamma\delta$-bearing lymphocytes are found in nude mice (even without thymus engraftment), and Thy-1$^+$ DEC in skin express TCR $\gamma\delta$ (Stingl et al., 1987b; Koning et al., 1987a; Kuziel et al., 1987;

Bonyhadi et al., 1987). Since Thy-1$^+$ DEC are thought to be of bone marrow origin, these results suggest that an extrathymic or thymic-independent pathway for production of TCR $\gamma\delta$-bearing lymphocytes may exist as well.

In man, analyses of T cell development are limited by the difficulty in obtaining tissue samples from fetal sources. Studies primarily employed staged human T lineage tumor cell lines, although aberrant and unrepresentative expression of differentiation antigens can limit the usefulness of such studies (Picker et al., 1987). Nevertheless, some generalizations have emerged from these studies, as well as limited analysis of human fetal thymocytes. In human cells that lack surface CD3–TCR, TCR β and TCR γ gene rearrangement and transcription are frequently observed. CD3 expression (CD3δ and CD3ε were tested) occurs even earlier in development, since these genes are transcribed and proteins are expressed intracellularly even in thymocytes displaying germ-line TCR β chain genes. In contrast, TCR α transcription appears to be a late event, since it characterizes the most mature cells (Royer et al., 1984, 1985; Collins et al., 1985; Furley et al., 1986; Sangster et al., 1986; van Dongen et al., 1987).

The ontogenetic history of TCR $\gamma\delta$ and TCR $\alpha\beta$ lymphocytes in mammals now appears to have received remarkable confirmation in birds. Molecular complexes that appear to be similar to human and murine CD3, CD4, and CD8 have been defined in the chicken and are referred to as CT3, CT4, and CT8, respectively (Chen et al., 1986; Chan et al., 1988). In addition, a mAb called TCR1 is reactive with a CT3-associated disulfide-linked heterodimer composed of 40- and 50-kDa subunits (Sowder et al., 1988). Among the 85% of blood lymphoid cells that are CT3$^+$, 16% are TCR1$^+$. Strikingly, the TCR1$^+$ cells are CT4$^-$8$^-$, whereas the remaining peripheral blood T cells are either CT4$^+$8$^-$ (49%) or CT4$^-$8$^-$ (19%). TCR1 almost certainly defines the chicken equivalent of human and murine TCR $\gamma\delta$. Ontogenetic analyses reveal that TCR1 is expressed on a small percentage of thymocytes by embryonic d11, increases to 30% of thymocytes by d15, and then declines to about 5% of the cells by hatching. TCR1$^-$CT3$^+$ thymocytes, which presumably express the chicken equivalent of TCR $\alpha\beta$, appear after d15 of fetal development and then quickly increase in number to exceed the level of TCR1$^+$ cells (Sowder et al., 1988). These studies suggest a remarkable evolutionary conservation of the developmental pattern of T cell subpopulations and their T cell receptors. The high levels of TCR1$^+$ lymphocytes (up to 30% of splenic T cells) in the chicken supports the view that these lymphocytes play an important functional role in the periphery.

As might be expected from the temporal relationships of TCR rearrangements outlined above, TCR $\gamma\delta$ lymphocytes typically display partial TCR β gene rearrangements and truncated TCR β transcripts (Brenner et al., 1986, 1987b; Bank et al., 1986; Borst et al., 1987; Lanier et al., 1987; Alarcon et al., 1987), and TCR $\alpha\beta$ lymphocytes typically display nonfunctional TCR γ gene rearrangements (Heilig et al., 1985; Rupp et al., 1986; Reilly et al., 1986; Iwamoto et al., 1986; Dialynas et al., 1986; Quertermous et al., 1986b; Sangster et al., 1986; Yoshikai et al., 1987). Due to selectivity of chain pairing, functional rearrangement of TCR α, β, and γ genes in the same cell results only in the assembly of a cell surface TCR $\alpha\beta$–CD3 complex (Nakanishi et al., 1987), whereas functional rearrangement of TCR γ, δ, and β genes results only in the assembly of a cell surface TCR $\gamma\delta$–CD3 complex (Koning et al., 1987b). The deletion of the C_δ locus in all TCR $\alpha\beta$ lymphocytes that have been examined (Chien et al., 1987a; Hata et al., 1987) makes it unlikely that lymphocytes carrying four functionally rearranged TCR genes will be found. It would appear unlikely (although it has certainly not been proved) that lymphocytes carrying functionally rearranged and expressed TCR γ, δ, and β genes can go on to rearrange TCR α at significant frequency (if at all), thereby moving from the expression of TCR $\gamma\delta$–CD3 to the expression of TCR $\alpha\beta$–CD3. If this does not occur, then TCR $\gamma\delta$ lymphocytes and TCR $\alpha\beta$ lymphocytes would represent distinct sublineages within the T cell lineage, related to each other by a common precursor.

The branchpoint of these presumed sublineages is uncertain. It is possible that thymocytes may attempt TCR γ and δ (and TCR β) rearrangement, and failing functional TCR γ and δ rearrangement, may then move on to attempt TCR α rearrangement, thereby deleting the C_δ locus. Alternatively, commitment to one or the other sublineage may occur at an earlier stage, such that cells may attempt TCR δ or TCR α rearrangement, but not both. In this regard, de Villartay et al. (1987) have described a probe (TEA, for "T early α") within the TCR α locus that is in the germ-line configuration in TCR $\gamma\delta$ lymphocytes, is deleted in TCR $\alpha\beta$ lymphocytes, but detects discrete rearrangement(s) within the thymus. Also, some human leukemic cell lines appear to have deleted the C_δ locus on both chromosomes, yet retain V_δ on at least one chromosome (M. S. Krangel, unpublished). Since V_δ is deleted in lymphocytes expressing TCR $\alpha\beta$ (and, therefore, presumably lies proximal to C_δ, between V_α and J_α segments), these observations suggest the possibility of a rearrangement deleting the C_δ locus that may be distinct from V_α to J_α rearrangement. Such a deletional event might initiate divergence of the sublineages, committing the cell to either attempt

TCR δ or TCR α rearrangement, but not both. The key to understanding these lineage relationships may be the reciprocal joint products of V_α to J_α rearrangement, which have recently been shown to occur as extrachromosomal circular DNAs (Fujimoto and Yamagishi, 1987). Structural analysis of these DNA fragments may reveal the history of rearrangement events on an individual chromosome.

VII. Functional Capabilities of TCR $\gamma\delta$ Lymphocytes

A. ACTIVATION

Activation of TCR $\alpha\beta^+$ lymphocytes results from perturbations of the TCR–CD3 complex on the cell surface (Meuer et al., 1983; Haskins et al., 1983). This is assumed to occur by interaction of the T cell receptor with a ligand composed of self-MHC class I or II molecules with bound peptide antigen (Buus et al., 1986; Babbitt et al., 1985, 1986). In many eukaryotic cell systems, transmembrane signaling results in the release of two second messengers, diacylglycerol, which activates protein kinase C, and inositol trisphosphate, which functions in mobilizing intracellular Ca^{2+} (Nishizuka, 1984; Berridge and Irvine, 1984; Bell, 1986). It appears likely that lymphocytes may be activated via these same second messengers (Tsien et al., 1982; Michell et al., 1982). Recently, perturbations of the TCR–CD3 complex have been shown to result in early changes in cytoplasmic free Ca^{2+} and inositol trisphosphate, suggesting that this pathway mediates the cascade of activation in T cells (Weiss and Stobo, 1984; Weiss et al., 1984; Imboden and Stobo, 1985; Oettgen et al., 1985; O'Flynn et al., 1985). Evidence is accumulating to suggest that many processes, similar to those observed for the activation and proliferation of TCR $\alpha\beta$ cells, may operate in TCR $\gamma\delta$ cells. In model systems, mAb against the CD3 complex have been used to mimic perturbation of the TCR–CD3 complex, in the absence of a cognate antigen. As in the case of TCR $\alpha\beta$ lymphocytes, it appears that signal transduction involves increases in cytoplasmic calcium and inositol phosphates, since anti-CD3 mAb (Krangel et al., 1987b; Pantaleo et al., 1988) and anti-CD2 mAb (Pantaleo et al., 1988) induce abrupt increases in the levels of intracellular free Ca^{2+} and inositol phosphate in TCR $\gamma\delta$ T cells. Once stimulated by mAb, TCR $\gamma\delta$ cells lines may secrete lymphokines. For example, Bank et al. (1986) found that human thymic-derived clones secreted IL-2 in response to anti-CD3 mAb in the presence of phorbol esters. Similarly, anti-CD3 mAb, phytohemagglutinin (PHA), or an appropriate combination of anti-CD2 mAb induces IL-2 production from human WT31$^-$CD3$^+$ peripheral blood-derived cell lines (Ferrini et al., 1987; Moretta et al., 1988). These studies also point out that it may be possible to trigger human TCR γδ cells through alternative pathways of activation such as via the CD2 molecule as reported earlier for TCR αβ cells (Meuer et al., 1984). Interestingly, in contrast to TCR αβ cells, TCR γδ lymphocytes appear to lack CD28 (T44), an additional molecule capable of mediating activation on TCR αβ cells (Poggi et al., 1987).

Anti-CD3 mAb also induce proliferation of human thymic-derived TCR γδ lymphocytes (Bank et al., 1986), of murine $CD3^+4^-8^-$ fetal thymocytes (Bluestone et al., 1987; Tentori et al., 1988), and of murine splenic $CD3^+4^-8^-$ T cells (Cron et al., 1988). In one case, a clonotypic anti-TCR γ mAb coupled to solid beads was shown to stimulate proliferation of a human TCR γδ fetal cord blood-derived T cell clone (Moingeon et al., 1987). In addition, allogeneic cell simulators were capable of inducing proliferation of specific TCR γδ nu/nu mouse cells (Matis et al., 1987). Recent studies show that anti-Thy-1 mAb stimulate murine fetal TCR γδ thymocytes to proliferate (Tentori et al., 1988; MacDonald et al., 1988), similar to the ability of these mAb to stimulate TCR αβ T cells (Gunter et al., 1984; MacDonald et al., 1985; Pont et al., 1985). The subsequent proliferation was IL-2 dependent and could be augmented by interleukin 1 (IL-1) (MacDonald et al., 1988). Moreover, stimulation of early fetal thymocytes (day 13–15) with anti-CD3 or anti-Thy-1 mAb resulted in secretion of interleukin 4 (IL-4) (BSF-1) in addition to IL-2 (Tentori et al., 1988). Similarly, anti-CD3 and anti-Thy-1 mAb induced lymphokine secretion, including IL-2 and IL-4 by murine $CD3^+4^-8^-$ splenocytes (Cron et al., 1988). These results raise the possibility that lymphokines secreted by fetal TCR γδ thymocytes may have an important role in development (Tentori et al., 1988).

B. CYTOLYSIS

One of the most dramatic functions observed for TCR γδ lymphocytes is their ability to carry out cell-mediated lysis of target cells. This cytolysis can be divided into three types, namely, redirected lysis, tumor lysis, and MHC-linked lysis. In redirected lysis, anti-CD3 mAb bind to and presumably signal effector cells, which then conjugate in part through the anti-CD3 mAb binding IgG Fc receptors on appropriate target cells (Spits et al., 1985b; Leeuwenberg et al., 1985; Mentzer et al., 1985). Through this mechanism, a variety of target cells expressing IgG Fc receptors were lysed by TCR γδ lymphocytes, indicating that the effector cells have cytolytic machinery (Bank et al., 1986; Borst et al., 1987; Brenner et al., 1987b). Such experiments did not yield information on the nature of the antigens that might be recognized on target cells, but they first showed the potent effector capability of these cells.

In addition to mAb-mediated redirected lysis, *in vitro* cultured TCR $\gamma\delta$ cells of various tissue origins, including PBL (Borst *et al.*, 1987; Brenner, 1987b; Ferrini *et al.*, 1987; Moretta *et al.*, 1988), pleural effusion (Borst *et al.*, 1987), fetal blood (Moingeon *et al.*, 1986, 1987; Alarcon *et al.*, 1987), cerebrospinal fluid (Ang *et al.*, 1987), and thymus (Lanier *et al.*, 1987), were found to exhibit spontaneous cytolytic activity against tumor targets. In some cases, lysis of tumor targets revealed broad reactivities, while in other cases, selective lysis of one or only a few tumor cell lines was observed. Lysis of tumor cells that lacked MHC class I and class II antigens such as K562 was observed, and tumor cell lysis was not inhibited by mAb to classical MHC class I and II molecules, when such molecules were expressed on target cells. These results, coupled with the lack of CD4 and CD8 expression on the effector cells, suggested that recognition of tumor targets by *in vitro* cultured TCR $\gamma\delta$ cells was not restricted by classical MHC class I and II molecules. In several examples, mAb to CD3 (Borst *et al.*, 1987; Brenner *et al.*, 1987b; Moingeon *et al.*, 1986; Alarcon *et al.*, 1987; Ang *et al.*, 1987; Van de Griend *et al.*, 1987), as well as to TCR $\gamma\delta$ (Moingeon *et al.*, 1986), blocked killing, while in other cases, anti-CD3 mAb did not block lysis of tumor targets (Lanier *et al.*, 1987). CD16$^+$ TCR $\gamma\delta$ cell lines were also capable of antibody-dependent cellular cytotoxicity (ADCC) *in vitro* against antibody-coated P815 cells (Van de Griend *et al.*, 1987).

The meaning of the cytotoxicity noted in the above studies is unclear at present. Although all examples of human *in vitro* cultured TCR $\gamma\delta$ cell lines display non-MHC-restricted tumor lysis, at least some *in vitro* cultured T cell lines, which presumably express TCR $\alpha\beta$ (Hercend *et al.*, 1983; Ritz *et al.*, 1985; Schmidt *et al.*, 1986; David *et al.*, 1987) and/or rearrange TCR $\alpha\beta$ genes (Yanagi *et al.*, 1985; Ikuta *et al.*, 1986), display similar activity. Furthermore, Van de Griend *et al.* (1987) have shown that molecules other than the TCR may mediate the triggering of TCR $\gamma\delta$ lymphocytes in the lysis of tumor targets. For example, mAb against the IgG Fc receptors on the effector cells appeared to trigger redirected lysis against target cells, similar to the manner in which anti-CD3 mAb trigger such lysis. Moreover, unlike traditional natural killer cells, it is likely that fresh TCR $\gamma\delta$ cells do not display tumor-specific lysis, since this is clearly the case for the CD3$^+$4$^-$8$^-$ PBL examined in one study, which should include TCR $\gamma\delta$ cells (Lanier *et al.*, 1986). Thus, whereas it is striking that virtually all *in vitro* cultured human TCR $\gamma\delta$ cell lines display spontaneous tumor lysis, the nature of the ligand(s) recognized and the role of the TCR $\gamma\delta$ in the observed tumor lysis remains to be elucidated.

Recently, Matis *et al.* (1987) reported evidence for TCR $\gamma\delta$ recognition of allogeneic target cells. BALB/c nu/nu mice (H-2^d) were immunized in vivo and restimulated in vitro with irradiated B10.BR spleen cells (H-2^k). A TCR $\gamma\delta^+$ cell line obtained in this way both proliferated to and lysed concanavalin A T cell blasts from several H-2^k strains of mice. Moreover, recognition of targets in cytolysis experiments using recombinant strains of mice correlated with the presence of the k haplotype at the H-2D end of the MHC. However, the pattern of reactivity was broad (including H-2^b and H-2^q targets), and the specificity appears to correlate with the recognition of nonclassical MHC-linked antigens, possibly Tla (J. Bluestone and L. Matis, personal communication). This report demonstrates allogeneic reactivity that appeared to be MHC-linked, suggesting that TCR $\gamma\delta$ may be capable of self–non-self discrimination. Maeda et al. (1987) also have found evidence that TCR $\gamma\delta$ cells may respond in allogeneic mixed lymphocyte cultures.

Although the physiological role played by TCR $\gamma\delta$ cells is not known, the functional studies suggest that these cells are activated by mechanisms and display effector capabilities that are similar to those of TCR $\alpha\beta$ cells. TCR $\gamma\delta$ cells can be activated using calcium and inositol phosphate-mediated pathways. Since this occurs through perturbation of their TCR–CD3 complexes, TCR $\gamma\delta$ cells may be capable of undergoing clonal proliferation in an antigen-specific manner. The studies on cell-mediated cytotoxicity suggest that TCR $\gamma\delta$ lymphocytes may function as effector cytotoxic lymphocytes that recognize antigens on other cell surfaces. In vitro cultured TCR $\gamma\delta$ cell lines lyse tumor targets, although it is not clear if tumor lysis is a T cell receptor-mediated, antigen-specific interaction. TCR $\gamma\delta$ cells may recognize (or be restricted by) MHC-encoded molecules, although data from several studies suggests that, in at least some cases, the molecules recognized may not be classical MHC class I and II proteins. Clearly, the molecular nature of the TCR $\gamma\delta$ ligand (such as physiological antigen and antigen-presenting structure, if analogous to those used by TCR $\alpha\beta$ cells) remains to be defined.

VIII. Summary and Hypothesis

One may begin to understand the TCR $\gamma\delta$ by comparing it to TCR $\alpha\beta$, but it may be understood completely only through an appreciation of its novelty (Table II).

A. Protein Structure

Like TCR $\alpha\beta$, TCR $\gamma\delta$ is coexpressed with CD3. TCR $\gamma\delta$ is not expressed on the cell surface as part of TCR $\alpha\beta$, or on cells that display TCR $\alpha\beta$. This may result from the inherent inability of TCR γ to pair with any

TABLE II
Comparison of TCR $\alpha\beta$ and TCR $\gamma\delta$ Lymphocytes and Receptors

	TCR $\alpha\beta$	TCR $\gamma\delta$
Structure	Disulfide-linked heterodimer	Disulfide-linked heterodimer (in mouse)
		Disulfide linked or unlinked (in man)
	CD3-associated	CD3-associated
Diversity	50–100 V_α and V_β and 2 D_β, used singly; 50–100 J_α and 12–13 J_β, throughout V domain	6–8 V_γ, 3–7 V_δ, and 2 D_δ, used in tandem; 3–5 J_γ and 2 J_δ, primarily junctional
Distribution	Peripheral blood (60–70% of PBL)	Peripheral blood (1–10% of PBL)
	Peripheral lymphoid organs ??	Peripheral lymphoid organs
		Epidermis (Thy-1$^+$ DEC) in mouse
Phenotype	CD4$^+$CD8$^-$ (60%)	CD4$^+$CD8$^-$ (0%)
	CD4$^-$CD8$^+$ (35%)	CD4$^-$CD8$^+$ dim (20–50%)[a]
	CD4$^-$CD8$^-$ (<1%)	CD4$^-$CD8$^-$ (50–80%)
	CD2$^+$ (100%)	CD2$^+$ (100%)
	CD5$^+$ (>95%)	CD5$^-$ or CD5$^+$ dim
Ontogeny	Thymic (second wave, later)	Thymic (first wave, earlier)
	Minimal extrathymic pathway	Probable extrathymic pathway for DEC cells

[a] CD8 staining of resting cells is very weak and may be classified as negative.

chain other than TCR δ and is further assured by the required deletion of TCR δ in the process of TCR α rearrangement.

TCR α and β polypeptides are always disulfide linked and of similar constant domain structure, since $C_\beta 1$ and $C_\beta 2$ are virtually identical. However, TCR $\gamma\delta$ exists in different forms in both man and mouse due to variation in TCR γ constant region structure. TCR $\gamma\delta$ may be either disulfide linked or nonlinked in man, and although always disulfide linked in mouse, divergent C_γ segments ($C_\gamma 1$ and 2 versus $C_\gamma 4$) may be used. These different forms may be selectively expressed in different circumstances or in particular locations, suggesting the possibility that they may represent functionally distinct isotypes, without counterpart in TCR $\alpha\beta$.

B. Diversity

At first glance, diversity of TCR $\gamma\delta$ is less than that of TCR $\alpha\beta$, but closer examination reveals that this may not be accurate. Clearly, there are only a limited number of germ-line elements, with only six to eight functional V_γ segments and fewer J_γ segments. Germ-line TCR δ diversity will be similar, if not more limited. TCR γ diversity is increased by N nucleotide incorporation at the V–J junction. However, TCR δ junctional diversity is unprecedented, with the use of at least two tandemly arranged D_δ elements and extensive N nucleotide incorporation. While it is not possible to quantitatively estimate the resulting junctional diversity of TCR $\gamma\delta$, it is clear that significant diversity exists, which might rival that of TCR $\alpha\beta$. The striking feature of TCR $\gamma\delta$ diversity is that it is focused at the V–J junction, in a single hypervariable region for each chain, rather than throughout the V domain.

It seems likely that TCR $\gamma\delta$ recognizes antigen on the surface of other cells, although this has so far been demonstrated only in one instance (Matis et al., 1987). By analogy with TCR $\alpha\beta$, we suppose that specific antigen is presented on the cell surface in association with a restricting element. The limited TCR $\gamma\delta$ V segment repertoire might suggest that such a restricting element, if it exists, would not be as polymorphic as MHC class I and class II antigens. Such a restricting element might be a relatively nonpolymorphic MHC class I-like or class II-like antigen or a determinant thereof. Alternatively, a set of distinct cell surface molecules unrelated to MHC class I and class II molecules may be recognized. The most attractive hypothesis now is that TCR $\gamma\delta$ does not see antigen on classical MHC class I and II molecules, but may instead recognize less polymorphic MHC-linked molecules of related structure. Presumably, TCR $\gamma\delta$ cells are $CD4^-8^-$ or weakly $CD8^+$, since these cell interaction molecules are not used in recognition of the restricting element. This may occur because the restricting element is not classical MHC class I or class II and/or because the presence of CD4 or CD8 would lead to excessive self-reactivity. The extensive TCR $\gamma\delta$ diversity in the junctional region might suggest that these portions of the V domains may form part of the combining site which is important in specific antigen recognition, and that a large antigenic universe may exist for TCR $\gamma\delta$, despite limited polymorphism of the restricting element.

C. Is It a Receptor?

Like TCR $\alpha\beta$, TCR $\gamma\delta$ forms part of a cell surface complex that can transduce an activation signal, resulting in cell proliferation, lymphokine secretion, and effector cytolysis. If TCR $\gamma\delta$ can recognize specific antigen, T cells bearing this receptor should be able to undergo clonal expansion in an antigen-specific manner. In this case, like B lymphocytes and TCR $\alpha\beta$ lymphocytes, TCR $\gamma\delta$ lymphocytes would qualify as an arm of the antigen-specific immune response.

D. Thymic Ontogeny

Recent studies suggest strongly that most TCR $\gamma\delta$ cells are thymus dependent. Since TCR $\alpha\beta$ lymphocytes probably acquire a preference for MHC reactivity during thymic development, TCR $\gamma\delta$ cells may similarly encounter and be selected based on appropriate reactivity with their putative restricting elements. However, the Thy-1$^+$ dendritic epidermal cells described in mice offer evidence for the existence of an additional extrathymic pathway. Such cells may or may not undergo similar selection processes.

TCR $\gamma\delta$ and TCR $\alpha\beta$ lymphocytes are thought to represent terminally differentiated phenotypes of separate branches or sublineages of T cell development, likely arising from common precursor thymocytes. TCR γ and δ genes undergo rearrangement and expression first, under developmental controls that are not known. Thus, TCR $\gamma\delta$ lymphocytes appear before TCR $\alpha\beta$ lymphocytes during thymic ontogeny. Some of the cells that successfully rearrange and express TCR γ and δ genes and proteins subsequently exit the thymus to the periphery. Despite often lacking CD4 and CD8, they represent mature, functional, and stably differentiated cells.

Rearrangement of TCR α would necessarily result in the deletion of TCR δ. Cells displaying successful TCR β and TCR α rearrangements may leave the thymus, following a still poorly understood selection process. It is not known whether cells which have unsuccessfully (or successfully) rearranged TCR δ can go on to attempt TCR α rearrangement, or whether the cells are committed to attempt either TCR δ or TCR α rearrangement at an earlier stage.

E. Is TCR $\gamma\delta$ Important?

It is possible that TCR $\gamma\delta$ lymphocytes are functionally similar to TCR $\alpha\beta$ lymphocytes, only present in much smaller numbers and displaying more limited diversity (i.e., "Who needs more?"). Even given this trivial explanation, TCR $\gamma\delta$ appears to represent the first CD3-associated T cell receptor in ontogeny and might represent the ancestral T cell receptor in evolution. It will be intriguing to assess such a relationship between TCR $\alpha\beta$ and TCR $\gamma\delta$ by analysis of the TCRs of more primitive species. However, it seems likely that this receptor is much more than an evolutionary relic or just a miniature TCR $\alpha\beta$. Rather, it is our view that TCR $\gamma\delta$ is a functionally distinct T cell receptor. Its displays a pattern of diversity distinct from TCR $\alpha\beta$. It appears early in development and may play a role in the thymus prior to the availability of TCR $\alpha\beta$. It may be the principal TCR in selected sites, such as the skin or other locations. TCR $\gamma\delta$ lymphocytes display cytotoxic activity and secrete lymphokines, suggesting that they may have most (or all) of the effector capabilities of TCR $\alpha\beta$ lymphocytes. Given all of these features, we suggest that TCR $\gamma\delta$ cells are likely to fill certain distinct and selective roles. The short time period since the discovery of this receptor has seen dramatic progress at the level of protein structure and molecular genetics. However, the key for the future will clearly be an understanding of how this receptor and the lymphocytes that bear it fit into the grand scheme of immunity.

Acknowledgments

We would like to thank Veronica Groh for helpful comments on the manuscript and Gail Dudley for secretarial assistance. In addition, we are deeply grateful to the many investigators who graciously provided us with unpublished data. This work was supported in part by grants ACS-NY-IM-425, NIH-1-KO8-AM01598, and CA47724 to M. B. B. and NSF-DCB-8617540 to M. S. K.

References

Acuto, O., Meuer, S. C., Hodgdon, J. C., Schlossman, S. F., and Reinherz, E. L. (1983). Peptide variability exists within α and β subunits of the T cell receptor for antigen. *J. Exp. Med.* **158**, 1368–1373.

Acuto, O., Fabbi, M., Smart, J., Poole, C. B., Protentis, J., Royer, H. D., Schlossman, S. F., and Reinherz, E. L. (1984). Purification and NH_2-terminal amino acid sequencing of the β subunit of a human T-cell antigen receptor. *Proc. Natl. Acad. Sci. U.S.A.* **81**, 3851–3855.

Acuto, O., and Larocca, L. M. (1987). T-cell antigen receptor expression in the thymus. *Hum. Immunol.* **18**, 93–109.

Adkins, B., Mueller, C., Okada, C. Y., Reichert, R. A., Weissman, I. L., and Spangrude, G. J. (1987). Early events in T-cell maturation. *Annu. Rev. Immunol.* **5**, 325–365.

Alarcon, B., DeVries, J., Pettey, C., Boylston, A., Yssel, H., Terhorst, C., and Spits, H. (1987). The T-cell receptor γ chain-CD3 complex: Implication in the cytotoxic activity of a $CD3^+CD4^-CD8^-$ human natural killer clone. *Proc. Natl. Acad. Sci. U.S.A.* **84**, 3861–3865.

Allison, J. P., McIntyre, B. W., and Bloch, D. (1982). Tumor-specific antigen of murine T-lymphoma defined with monoclonal antibody. *J. Immunol.* **129**, 2293–2300.

Allison, J. P., and Lanier, L. L. (1985). Identification of antigen receptor-associated structures on murine T cells. *Nature (London)* **314**, 107–109.

Alt, F. W., and Baltimore, D. (1982). Joining of immunoglobulin heavy chain gene segments: Implications from a chromosome with evidence of three $D-J_H$ fusions. *Proc. Natl. Acad. Sci. U.S.A.* **79**, 4118–4122.

Ang, S.-L., Seidman, J. G., Peterman, G. M., Duby, A. D., Benjamin, D., Lee, S. J., and Hafler, D. A. (1987). Functional γ chain-associated T cell receptors on cerebrospinal fluid-derived natural killer-like T cell clones. *J. Exp. Med.* **165**, 1453–1458.

Babbitt, B. P., Allen, P. M., Matsueda, G., Haber, E., and Unanue, E. R. (1985). Binding of immunogenic peptides to Ia histocompatibility molecules. *Nature (London)* **317**, 359–361.

Babbitt, B. P., Matsueda, G., Haber, E., Unanue, E. R., and Allen, P. M. (1986). Antigenic competition at the level of peptide-Ia binding. *Proc. Natl. Acad. Sci. U.S.A.* **83**, 4509–4513.

Baer, R., Chen, K.-C., Smith, S. D., and Rabbitts, T. H. (1985). Fusion of an immunoglobulin variable gene and a T cell receptor constant gene in the chromosome 14 inversion associated with T cell tumors. *Cell* **43,** 705–713.
Band, H., Hochstenbach, F., McLean, J., Hata, S., Krangel, M. S., and Brenner, M. B. (1987). Immunochemical proof that a novel rearranging gene encodes the T cell receptor δ subunit. *Science* **238,** 682–684.
Bank, I., DePinho, R. A., Brenner, M. B., Cassimeris, J., Alt, F. W., and Chess, L. (1986). A functional T3 molecule associated with a novel heterodimer on the surface of immature human thymocytes. *Nature (London)* **322,** 179–181.
Bell, R. M. (1986). Protein kinase C activation by diacylglycerol second messengers. *Cell* **45,** 631–632.
Bergstresser, P. R. Tigelaar, R. E., Dees, J. H., and Streilein, J. W. (1983). Thy-1 antigen-bearing dendritic cells populate murine epidermis. *J. Invest. Dermatol.* **81,** 286–288.
Bergstresser, P. R., Tigelaar, R. E., and Streilein, J. W. (1984). Thy-1 antigen-bearing dendritic cells in murine epidermis are derived from bone marrow precursors. *J. Invest. Dermatol.* **83,** 83–87.
Berridge, M. J., and Irvine, R. F. (1984). Inositol trisphosphate, a novel second messenger in cellular signal transduction. *Nature (London)* **312,** 315–321.
Biddison, W. E., Shearer, G. M., and Chang, T. W. (1981). Regulation of influenza virus-specific cytotoxic T cell responses by monoclonal antibody to human T cell differentiation antigen. *J. Immunol.* **127,** 2236–2240.
Bjorkman, P. J., Saper, M. A., Samraoui, B., Bennett, W. S., Strominger, J. L., and Wiley, D. C. (1987a). Structure of the human class I histocompatibility antigen, HLA-A2. *Nature (London)* **329,** 506–512.
Bjorkman, P. J., Saper, M. A., Samraoui, B., Bennett, W. S., Strominger, J. L., and Wiley, D. C. (1987b). The foreign antigen binding site and T cell recognition regions of class I histocompatibility antigens. *Nature (London)* **329,** 512–518.
Bluestone, J. A., Pardoll, D., Sharrow, S. O., and Fowlkes, B. J. (1987). Characterization of murine thymocytes with CD3-associated T-cell receptor structures. *Nature (London)* **326,** 82–84.
Boehm, T., Baer, R., Lavenir, I., Forster, A., Waters, J. J., Nacheva, E., and Rabbitts, T. H. (1988). The mechanism of chromosomal translocation t(11;14) involving the T cell receptor Cδ locus on human chromosome 14q11 and a transcribed region of chromosome 11p15. *EMBO J.* **7,** 385–394.
Bonyhadi, M., Weiss, A., Tucker, P. W., Tigelaar, R. E., and Allison, J. P. (1987). Delta is the C_X-gene product in the γ/δ antigen receptor of dendritic epidermal cells. *Nature (London)* **330,** 574–576.
Born, W., Yague, J., Palmer, E., Kappler, J., and Marrack, P. (1985). Rearrangement of T-cell receptor β-chain genes during T-cell development. *Proc. Natl. Acad. Sci. U.S.A.* **82,** 2925–2929.
Born, W., Rathbun, G., Tucker, P., Marrack, P., and Kappler, J. (1986). Synchronized rearrangement of T-cell γ and β chain genes in fetal thymocyte development. *Science* **214,** 479–482.
Born, W., Miles, C., White, J., O'Brien, R., Freed, J. H., Marrack, P., Kappler, J., and Kubo, R. T. (1987). Peptide sequences of T cell receptor δ and γ chains are identical to predicted X and γ proteins. *Nature (London)* **330,** 572–574.
Borst, J., Prendiville, M. A., and Terhorst, C. (1982). Compexity of the human T lymphocyte-specific cell surface antigen T3. *J. Immunol.* **128,** 1560–1565.
Borst, J., Alexander, S., Elder, J., and Terhorst, C. (1983). The T3 complex on human T lymphocytes involves four structurally distinct glycoproteins. *J. Biol. Chem.* **258,** 5135–5141.

Borst, J., van de Griend, R. J., van Oostveen, J. W., Ang, S.-L., Melief, C. J., Seidman, J. G., and Bolhuis, R. L. H. (1987). A T-cell receptor γ/CD3 complex found on cloned functional lymphocytes. *Nature (London)* **325**, 683–688.

Breathnach, S. M., and Katz, S. I. (1984). Thy-1$^+$ dendritic cells in murine epidermis are bone marrow-derived. *J. Invest. Dermatol.* **83**, 74–77.

Brenner, M. B., Trowbridge, I. S., and Strominger, J. L. (1985). Cross-linking of human T cell receptor proteins: Association between the T cell idiotype beta subunit and the T3 glycoprotein heavy subunit. *Cell* **40**, 183–190.

Brenner, M. B., McLean, J., Dialynas, D. P., Strominger, J. L., Smith, J. A., Owen, F. L., Seidman, J. G., Ip, S., Rosen, F., and Krangel, M. S. (1986). Identification of a putative second T-cell receptor. *Nature (London)* **322**, 145–149.

Brenner, M. B., McLean, J., Scheft, H., Warnke, R. A., Jones, N., and Strominger, J. L. (1987a). Characterization and expression of the human alpha beta T cell receptor by using a framework monoclonal antibody. *J. Immunol.* **138**, 1502–1509.

Brenner, M. B., McLean, J., Scheft, H., Riberdy, J., Ang, S.-L., Seidman, J. G., Devlin, P., and Krangel, M. S. (1987b). Two forms of the T-cell receptor gamma protein found on peripheral blood cytotoxic T lymphocytes. *Nature (London)* **325**, 689–694.

Burnet, F. M. (1957). A modification of Jerne's theory of antibody production using the concept of clonal selection. *Aust. J. Sci.* **20**, 67.

Buus, S., Colon, S., Smith, C., Freed, J. H., Miles, C., and Grey, H. M. (1986). Interaction between a "processed" ovalbumin peptide and Ia molecules. *Proc. Natl. Acad. Sci. U.S.A.* **83**, 3968–3971.

Caccia, N., Bruns, G. A. P., Kirsch, I. R., Hollis, G. F., Bertness, V., and Mak, T. W. (1985). T-cell receptor α chain genes are located on chromosome 14 at 14q11-14q12 in humans. *J. Exp. Med.* **161**, 1255–1260.

Cantrell, D. A., Davies, A. A., and Crumpton, M. J. (1985). Activators of protein kinase C down-regulate and phosphorylate the T3/T-cell antigen receptor complex of human T lymphocytes. *Proc. Natl. Acad. Sci. U.S.A.* **82**, 8158–8162.

Ceredig, R., Dialynas, D. P., Fitch, F. W., and MacDonald, H. R. (1983). Precursors of T cell growth factor producing cells in the thymus. *J. Exp. Med.* **158**, 1654–1671.

Chan, M. M., Chen, C-L. H., Ager, L. L., and Cooper, M. D. (1988). Identification of the avian homologues of mammalian CD4 and CD8 antigens. *J. Immunol.* **140**, 2133–2138.

Chang, T. W., Kung, P. C., Gingras, S. P., and Goldstein, G. (1981). Does OKT3 monoclonal antibody react with an antigen-recognition structure on human T cells? *Proc. Natl. Acad. Sci. U.S.A.* **78**, 1805–1808.

Chen, C-L. H., Ager, L. L., Gartland, G. L., and Cooper, M. D. (1986). Identification of a T3/T cell receptor complex in chickens. *J. Exp. Med.* **164**, 375–380.

Chien, Y., Becker, D. M., Lindsten, T., Okamura, M., Cohen, D. I., and Davis, M. M. (1984). A third type of murine T-cell receptor gene. *Nature (London)* **312**, 31–35.

Chien, Y., Iwashima, M., Kaplan, K. B., Elliott, J. F., and Davis, M. M. (1987a). A new T-cell receptor gene located within the alpha locus and expressed early in T-cell differentiation. *Nature (London)* **327**, 677–682.

Chien, Y., Iwashima, M., Wettstein, D. A., Kaplan, K. B., Elliott, J. F., Born, W., and Davis, M. M. (1987b). T cell receptor δ gene rearrangements in early thymocytes. *Nature (London)* **330**, 722–727.

Cohen, R. L., Crawford, J. M., and Chambers, D. A. (1986). Thy-1$^+$ epidermal cells are not demonstrable in rat and human skin. *J. Invest. Dermatol.* **87**, 30–32.

Collins, M. K. L., Tanigawa, G., Kissonerghis, A.-M., Ritter, M., Price, K. M., Tonegawa, S., and Owen, M. J. (1985). Regulation of T-cell receptor gene expression in human T-cell development. *Proc. Natl. Acad. Sci. U.S.A.* 82:4503–4507.

Cooper, K. D., Breathnach, S. M., Caughman, S. W., Palini, A. G., Waxdal, M. J., and Katz, S. I. (1985). Fluorescence microscopic and flow cytometric analysis of bone marrow-derived cells in human epidermis: A search for the human analogue of the murine dendritic Thy-1$^+$ epidermal cell. *J. Invest. Dermatol.* **85**, 546–552.

Crispe, I. N., Moore, M. W., Husmann, L. A., Smith, L., Bevan, M. J., and Shimonkevitz, R. P. (1987). Differentiation potential of subsets of CD4$^-$CD8$^-$ thymocytes. *Nature (London)* **329**, 336–339.

Croce, C. M., Isobe, M., Palumbo, A., Puck, J., Ming, J., Tweardy, D., Erikson, J., Davis, M., and Rovera, G. (1985). Gene for α-chain of human T-cell receptor: Location on chromosome 14 region involved in T-cell neoplasms. *Science* **227**, 1044–1047.

Cron, R., Koning, F., Maloy, W. L., Pardoll, D., Coligan, J. E., and Bluestone, J. A. (1988). A functional subpopulation of peripheral murine T lymphocytes which express a novel T cell receptor structure. *J. Immunol.* In press.

David, V., Bourge, J.-F., Guglielmi, P., Mathieu-Mahul, D., Degos, L., and Bensussan, A. (1987). Human T cell clones use a CD3-associated surface antigen recognition structure to exhibit both NK-like and allogeneic cytotoxic reactivity. *J. Immunol.* **138**, 2831–2836.

Denny, C. T., Yoshikai, Y., Mak, T. W., Smith, S. D., Hollis, G. F., and Kirsch, I. R. (1986). A chromosome 14 inversion in a T-cell lymphoma is caused by site-specific recombination between immunoglobulin and T-cell receptor loci. *Nature (London)* **320**, 549–551.

de Villartay, J-P., Lewis, D., Hockett, R., Waldmann, T. A., Korsmeyer, S. J., and Cohen, D. I. (1987). Deletional rearrangement in the human T-cell receptor α-chain locus. *Proc. Natl. Acad. Sci. U.S.A.* **84**, 8608–8612.

Dialynas, D. P., Murre, C., Quertermous, T., Boss, J. M., Leiden, J. M., Seidman, J. G., and Strominger, J. L. (1986). Cloning and sequence analysis of complementary DNA encoding an aberrantly rearranged human T-cell γ chain. *Proc. Natl. Acad. Sci. U.S.A.* **83**, 2619–2623.

Dreyer, W. J., and Bennett, J. C. (1965). The molecular basis of antibody formation: A paradox. *Proc. Natl. Acad. Sci. U.S.A.* **54**, 864–869.

Erikson, J., Finger, L., Sun, L., Ar-Rushdi, A., Nishikura, K., Minowada, J., Finan, J., Emanuel, B. S., Nowell, P. C., and Croce, C. M. (1986). Deregulation of c-myc by translocation of the α-locus of the T-cell receptor in T-cell leukemias. *Science* **232**, 884–886.

Fabbi, M., Acuto, O., Smart, J. E., and Reinherz, E. L. (1984). Homology of Ti α-subunit of a T-cell antigen/MHC receptor with immunoglobulin. *Nature (London)* **312**, 269–271.

Ferrini, S., Bottino, C., Biassoni, R., Poggi, A., Sekaly, R. P., Moretta, L., and Moretta, A. (1987). Characterization of CD3$^+$, CD4$^-$, CD8$^-$ clones expressing the putative T cell receptor γ gene product. *J. Exp. Med.* **166**, 277–282.

Finger, L. R., Harvey, R. C., Moore, R. C. A., Showe, L. C., and Croce, C. M. (1986). A common mechanism of chromosomal translocation in T- and B-cell neoplasia. *Science* **234**, 982–985.

Fontaine-Perus, J. C., Calman, F. M., Kaplan, C., and LeDouarin, N. M. (1981). Seeding of the 10-day mouse embryo thymic rudiment by lymphocyte precursors *in vitro*. *J. Immunol.* **126**, 2310–2316.

Forster, A., Huck, S., Ghanem, N., LeFranc, M.-P., and Rabbitts, T. H. (1987). New subgroups in the human T cell rearranging V_γ gene locus. *EMBO J.* **6**, 1945–1950.

Fowlkes, B. J., and Mathieson, B. J. (1985). Intrathymic differentiation: Thymocyte heterogeneity and the characterization of early T-cell precursors. *Surv. Immunol. Res.* **4**, 96–109.

Fowlkes, B. J., Edison, L., Mathieson, B. J., and Chused, T. M. (1985). Early T lymphocytes: Differentiation *in vivo* of adult intrathymic precursors. *J. Exp. Med.* **162**, 802–822.

Fowlkes, B. J., Kruisbeek, A. M., Ton-That, H., Weston, M. A., Coligan, J. E., Schwartz, R. H., and Pardoll, D. M. (1987). A novel population of T-cell receptor αβ-bearing thymocytes which predominantly expresses a single Vβ gene family. *Nature (London)* **329**, 251–254.

Fujimoto, S., and Yamagishi, H. (1987). Isolation of an excision product of T-cell receptor α-chain gene rearrangements. *Nature (London)* **327**, 242–243.

Furley, A. J., Mizutani, S., Weilbaecher, K., Dhaliwal, H. S., Ford, A. M., Chan, L. C., Molgaard, H. V., Toyonaga, B., Mak, T., can den Elsen, P., Gold, D., Terhorst, C., and Greaves, M. F. (1986). Developmentally regulated rearrangement and expression of genes encoding the T cell receptor-T3 complex. *Cell* **46**, 75–87.

Garman, R. D., Doherty, P. J., and Raulet, D. H. (1986). Diversity, rearrangement, and expression of murine T cell gamma genes. *Cell* **45**, 733–742.

Gunter, K. C., Malek, T. R., and Shevach, E. M. (1984). T cell-activating properties of an anti-Thy-1 monoclonal antibody. *J. Exp. Med.* **159**, 716–730.

Haars, R., Kronenberg, M., Gallatin, W. M., Weissman, I. L., Owen, F. L., and Hood, L. (1986). Rearrangement and expression of T cell antigen receptor and γ genes during thymic development. *J. Exp. Med.* **164**, 1–24.

Hannum, C. H., Kappler, J. W., Trowbridge, I. S., Marrack, P., and Freed, J. H. (1984). Immunoglobulin-like nature of the α-chain of a human T-cell antigen/MHC receptor. *Nature (London)* **312**, 65–67.

Haskins, K., Kubo, R., White, J., Pigeon, M., Kappler, J., and Marrack, P. (1983). The major histocompatibility complex-restricted antigen receptor on T cells. *J. Exp. Med.* **157**, 1149–1169.

Hata, S., Brenner, M. B., and Krangel, M. S. (1987). Identification of putative human T cell receptor δ complementary DNA clones. *Science* **238**, 678–682.

Hata, S., Satyanarayana, K., Devlin, P., Band, H., McLean, J., Strominger, J. L., Brenner, M. B., and Krangel, M. S. (1988). Extensive junctional diversity of rearranged human T cell receptor receptor δ genes. *Science.* In press.

Hayday, A. C., Saito, H., Gillies, S. D., Kranz, D. M., Tanigawa, G., Eisen, H. N., and Tonegawa, S. (1985). Structure, organization, and somatic rearrangement of T cell gamma genes. *Cell* **40**, 259–269.

Hecht, F., Morgan, R., Hecht, B. K. M., and Smith, S. D. (1984). Common region on chromosome 14 in T-cell leukemia and lymphoma. *Science* **226**, 1445–1447.

Hedrick, S. M., Cohen, D. I., Nielsen, E. A., and Davis, M. M. (1984a). Isolation of cDNA clones encoding T cell-specific membrane-associated proteins. *Nature (London)* **308**, 149–153.

Hedrick, S. M., Nielsen, E. A., Kavaler, J., Cohen, D. I., and Davis, M. M. (1984b). Sequence relationships between putative T-cell receptor polypeptides and immunoglobulins. *Nature (London)* **308**, 153–158.

Heilig, J. S., Glimcher, L. H., Kranz, D. M., Clayton, L. K., Greenstein, J. L., Saito, H., Maxam, A. M., Burakoff, S. J., Eisen, H. N., and Tonegawa, S. (1985). Expression of the T-cell-specific γ gene is unnecessary in T cells recognizing class II MHC determinants. *Nature (London)* **317**, 68–70.

Heilig, J. S., and Tonegawa, S. (1986). Diversity of murine gamma genes and expression in fetal and adult T lymphocytes. *Nature (London)* **322**, 836–840.

Hercend, T., Meuer, S., Brennan, A., Edson, M. A., Acuto, O., Reinherz, E. L., Schlossman,

S. F., and Ritz, J. (1983). Identification of a clonally restricted 90 kD heterodimer on two human cloned natural killer cell lines. *J. Exp. Med.* **158,** 1547–1560.

Hershfield, M. S., Kurtzberg, J., Harden, E., Moore, J. O., Whang-Peng, J., and Haynes, B. F. (1984). Conversion of a stem cell leukemia from a T-lymphoid to a myeloid phenotype induced by the adenosine deaminase inhibitor 2'-deoxycoformycin. *Proc. Natl. Acad. Sci. U.S.A.* **81,** 253–257.

Hochstenbach, F., Parker, C., McLean, J., Gieselmann, V., Band, H., Bank, I., Chess, L., Spits, H., Strominger, J. L., and Brenner, M. B. (1988). Three forms of the human T cell receptor $\gamma\delta$: preferential use of one form in selected healthy individuals. Submitted.

Huck, S., and Lefranc, M. P. (1987). Rearrangements to the JP1, JP and JP2 segments in the human T-cell rearranging gamma gene (TRG γ) locus. *FEBS. Lett.* **224,** 291.

Huck, S., Dariavach, P., and Lefranc, M. P. (1988). Variable region genes on the human T-cell rearranging gamma (TRG) locus: V-J junction and homology with the mouse genes. *EMBO J.* **7,** 719–726.

Ikuta, K., Hattori, M., Wake, K., Kano, S., Honjo, T., Yodol, J., and Minato, N. (1986). Expression and rearrangement of the α, β, and γ chain genes of the T cell receptor in cloned murine large granular lymphocyte lines. *J. Exp. Med.* **164,** 428–442.

Imboden, J. B., and Stobo, J. D. (1985). Transmembrane signalling by the T cell antigen receptor. *J. Exp. Med.* **161,** 446–456.

Ioannides, C. G., (1987). Itoh, K., Fox, F. E., Pahwa, R., Good, R. A., and Platsoucas, C. D. (1987). Identification of a second T-cell antigen receptor in human and mouse by an anti-peptide γ-chain-specific monoclonal antibody. *Proc. Natl. Acad. Sci. U.S.A.* **84,** 4244–4248.

Iwamoto, A., Rupp, F., Ohashi, (1986). P. S., Walker, C. L., Pircher, H., Joho, R., Hengartner, H., and Mak, T. W. (1986). T cell-specific γ genes in C57BL/10 mice, *J. Exp. Med.* **163,** 1203–1212.

Jitsukawa, S., Faure, F., Lipinski, M., Triebel, F., and Hercend, T. (1987). A novel subset of human lymphocytes with a T cell receptor-γ complex. *J. Exp. Med.* **166,** 1192–1197.

Jones, N., Leiden, J., Dialynas, D., Fraser, J., Clabby, M., Kishimoto, T., Strominger, J. L., Andrews, D., Lane, W., and Woody, J. (1985). Partial primary structure of the alpha and beta chains of human tumor T-cell receptors. *Science* **227,** 311–314.

Kabat, E. A., Wu, T. T., Reid-Miller, M., Perry, H. M., and Gottesman, K. S., eds. (1987). Sequences of Proteins of Immunological Interest. 4th Edition. U.S. Department of Health and Human Services.

Kagan, J., Finan, J., Letofsky, J., Besa, E. C., Nowell, P. C., and Croce, C. M. (1987). α-chain locus of the T-cell antigen receptor is involved in the t(10;14) chromosome translocation of T-cell acute lymphocytic leukemia. *Proc. Natl. Acad. Sci. U.S.A.* **84,** 4543–4546.

Kamps, W. A., Cooper, M. D. (1984). Development of lymphocyte subpopulations identified by monoclonal antibodies in human fetus. *J. Clin. Immunol.* **4,** 36.

Kanellopoulos, J. M., Wigglesworth, N. M., Owen, M. J., and Crumpton, M. J. (1983). Biosynthesis and molecular nature of the T3 antigen of human T lymphocytes. *EMBO J.* **2,** 1807–1814.

Kappler, J., Kubo, R., Haskins, K., Hannum, C., Marrack, P., Pigeon, M., McIntyre, B., Allison, J., and Trowbridge, I. (1983). The major histocompatibility complex-restricted antigen receptor on T cells in mouse and man: Identification of constant and variable peptides. *Cell* **35,** 295–302.

Kaye, J., Porcelli, S., Tite, J., Jones, B., and Janeway, C. A., Jr. (1983). Both a monoclonal antibody and antisera specific for determinants unique to individual cloned helper T cell lines can substitute for antigen and antigen-presenting cells in the activation of T cells. *J. Exp. Med.* **158,** 836–856.

Kennaugh, A. A., Butterworth, S. V., Hollis, R., Baer, R., Rabbitts, T. H., and Taylor, A. M. R. (1986). The chromosome breakpoint at 14q32 in an ataxia telangiectasia t(14;14) T cell clone is different from the 14q32 breakpoint in Burkitts and an inv(14) T cell lymphoma. *Hum. Genet.* **73,** 254–259.

Kindred, B., and Shreffler, D. C. (1972). H-2 dependence of co-operation between T and B cells *in vivo. J. Immunol.* **109,** 940–943.

Kohler, G., and Milstein, C. (1975). Continuous cultures of fused cells secreting antibody of predefined specificity. *Nature (London)* **256,** 495–497.

Koning, F., Stingl, G., Yokoyama, W. M., Yamada, H., Maloy, W. L., Tschachler, E., Shevach, E. M., and Coligan, J. E. (1987a). Identification of a T3-associated $\gamma\delta$ T cell receptor on Thy-1+ dendritic epidermal cell lines. *Science* **236,** 834–837.

Koning, F., Maloy, W. L., Cohen, D., and Coligan, J. E. (1987b). Independent association of T cell receptor β and γ chains with CD3 in the same cell. *J. Exp. Med.* **166,** 595–600.

Korman, A. J., Marusic-Galesic, S., Spencer, D., Kruisbeek, A. M., and Raulet, D. (1988). Limited repertoire of variable region genes expressed by γ/δ T cell receptor bearing cells in the adult thymus. *J. Exp. Med.* In press.

Krangel, M. S., Band, H., Hata, S., McLean, J., and Brenner, M. B. (1987a). Structurally divergent human T cell receptor γ proteins encoded by distinct Cγ genes. *Science* **237,** 64–67.

Krangel, M. S., Bierer, B. E., Devlin, P., Clabby, M., Strominger, J. L., McLean, J., and Brenner, M. B. (1987b). T3 glycoprotein is functional although structurally distinct on human T-cell receptor gamma T lymphocytes. *Proc. Natl. Acad. Sci. U.S.A.* **84,** 3817–3821.

Kranz, D. M., Saito, H., Disteche, C. M., Swissheim, K., Pravtcheva, D., Ruddle, F. H., Eisen, H. N., and Tonegawa, S. (1985a). Chromosomal locations of the murine T-cell receptor alpha-chain gene and the T-cell gamma gene. *Science* **227,** 941–945.

Kranz, D. M., Saito, H., Heller, M., Takagaki, Y., Haas, W., Eisen, H. N., and Tonegawa, S. (1985b). Limited diversity of the rearranged T-cell γ gene. *Nature (London)* **313,** 752–755.

Kronenberg, M., Siu, G., Hood, L. E., and Shastri, N. (1986). The molecular genetics of the T-cell antigen receptor and T-cell antigen recognition. *Annu. Rev. Immunol.* **4,** 529–591.

Kung, P. C., Goldstein, G., Reinherz, E. L., and Schlossman,– S. F. (1979). Monoclonal antibodies defining distinctive human T cell surface antigens. *Science* **206,** 347–349.

Kuziel, W. A., Takashima A., Bonyhadi, M., Bergstresser, P. R., Allison, J. P., Tigelaar, R. E., and Tucker, P. W. (1987). Regulation of T-cell receptor γ-chain RNA expression in murine Thy-1+ dendritic epidermal cells. *Nature (London)* **328,** 263–266.

Lanier, L. L., and Weiss, A. (1986). Presence of Ti (WT31) negative T lymphocytes in normal blood and thymus. *Nature (London)* **324,** 268–270.

Lanier, L. L., Ruitenberg, J. J., and Phillips, J. H. (1986). Human CD3$^+$ T lymphocytes that express neither CD4 nor CD8 antigens. *J. Exp. Med.* **164,** 339–344.

Lanier, L. L., Federspiel, N. A., Ruitenberg, J. J., Phillips, J. H., Allison, J. P., Littman, D., and Weiss, A. (1987). The T cell antigen receptor complex expressed on normal peripheral blood CD4$^-$, CD8$^-$ T lymphocytes. *J. Exp. Med.* **165,** 1076–1094.

LeBeau, M. M., McKeithan, T. W., Shima, E. A., Goldman-Leikin, R. E., Chan, S. J., Bell, G. I., Rowley, J. D., and Diaz, M. O. (1986). T-cell receptor α-chain gene is split in a human T-cell leukemia cell line with at (11;14)(p15;q11). *Proc. Natl. Acad. Sci. U.S.A.* **83,** 9744–9748.

Ledbetter, J. A., Evans, R. L., Lipinski, M., Cunningham-Rundles, C., Good, R. A., and Herzenberg, L. A. (1981). Evolutionary conservation of surface molecules that distinguish T lymphocyte helper/inducer and cytotoxic/suppressor subpopulations in mouse and man. *J. Exp. Med.* **153,** 310–323.

Leeuwenberg, J. F. M., Spits, H., Tax, W. J. M., and Capel, P. J. A. (1985). Induction of nonspecific cytotoxicity by monoclonal anti-T3 antibodies. *J. Immunol.* **134,** 3770–3775.

LeFranc, M.-P., and Rabbitts, T. H. (1985). Two tandemly organized human genes encoding the T-cell γ constant-region sequences show multiple rearrangements in different T-cell types. *Nature (London)* **316,** 464–466.

LeFranc, M.-P., Forster, A., and Rabbitts, T. H. (1986a). Rearrangement of two distinct T-cell γ-chain variable-region genes in human DNA. *Nature (London)* **319,** 420–422.

LeFranc, M.-P., Forster, A., and Rabbitts, T. H. (1986b). Genetic polymorphism and exon changes of the constant regions of the human T-cell rearranging gene γ. *Proc. Natl. Acad. Sci. U.S.A.* **83,** 9596–9600.

LeFranc, M.-P., Forster, A., Baer, R., Stinson, M. A., and Rabbitts, T. H. (1986c). Diversity and rearrangement of the human T cell rearranging γ genes: Nine germ-line variable genes belonging to two subgroups. *Cell* **45,** 237–246.

Leo, O., Foo, M., Sachs, D. H., Samelson, L. E., and Bluestone, J. A. (1987). Identification of a monoclonal antibody specific for a murine T3 polypeptide. *Proc. Natl. Acad. Sci. U.S.A.* **84,** 1374–1378.

Lew, A. M., Pardoll, D. M., Maloy, W. L., Fowlkes, B. J., Kruisbeek, A., Cheng, S-F., Germain, R. N., Bluestone, J. A., Schwartz, R. H., and Coligan, J. E. (1986). Characterization of T-cell receptor gamma chain expression in a subset of murine thymocytes. *Science* **234,** 1401–1405.

Lewis, W. H., Michalopoulos, E. E., Williams, D. L., Minden, M. D., and Mak, T. W. (1985). Breakpoints in the human T-cell antigen receptor α-chain locus in two T-cell leukaemia patients with chromosomal translocations. *Nature (London)* **317,** 544–546.

Lindsten, T., Fowlkes, B. J., Samelson, L. E., Davis, M. M., and Chien, Y. (1987). Transient rearrangements of the T cell antigen receptor α locus in early thymocytes. *J. Exp. Med.* **166,** 761–765.

Littman, D. R., Newton, M., (1987). Crommie, D., Ang, S.-L., Seidman, J. G., Gettner, S. N., and Weiss, A. (1987). Characterization of an expressed CD3-associated Ti γ-chain reveals Cγ domain polymorphism. *Nature (London)* **326,** 85–88.

Lobach, D. F., and Haynes, B. F. (1987). Ontogeny of the human thymus during fetal development. *J. Clin. Immunol.* **7,** 81–97.

Loh, E. Y., Lanier, L. L., Turck, C. W., Littman, D. R., Davis, M. M., Chien, Y.-H., and Weiss, A. (1987). Identification and sequence of a fourth human T cell antigen receptor chain. *Nature (London)* **330,** 569–572.

McDonald, H. R., Bron, C., Rousseaux, M., Horvath, C., and Cerottini, J.-C. (1985). Production and characterization of monoclonal anti-Thy-1 antibodies that simulate lymphokine production by cytolytic T cell clones. *Eur. J. Immunol.* **15,** 495–501.

MacDonald, H. R., Miescher, G. C., and Howe, R. C. (1988). Anti-Thy-1 induced proliferation of immature thymocytes expressing the CD3-associated γ/δ heterodimer. *Eur. J. Immunol.* In press.

Maeda, K., Nakanishi, N., Rogers, B. L., Haser, W. G., Shitara, K., Yoshida, H., Takagaki, Y., Augustin, A. A., and Tonegawa, S. (1987). Expression of the T-cell receptor γ-chain gene products on the surface of peripheral T cells and T-cell blasts generated by allogeneic mixed lymphocyte reaction. *Proc. Natl. Acad. Sci. U.S.A.* **84,** 6536–6540.

Marrack, P., and Kappler, J. (1986). The antigen-specific major histocompatibility complex-restricted receptor on T cells. *Adv. Immunol.* **38,** 1–30.

Mathieson, B. J., and Fowlkes, B. J. (1984). Cell surface antigen expression on thymocytes: Development and phenotypic differentiation of intrathymic subsets. *Immunol. Rev.* **82,** 141–173.

Mathieson, B., Sharrow, S., Rosenberg, Y., and Hammerling, U. (1981). Ly1$^+$23$^-$ cells appear in the thymus before Ly123$^+$ cells. *Nature (London)* **289,** 179.

Mathieu-Mahul, D., Caubet, J. F., Bernheim, A., Mauchauffe, M., Palmer, E., Berger, R., and Larsen, C.-J. (1985). Molecular cloning of a DNA fragment from human chromosome 14(14q11) involved in T-cell malignancies. *EMBO J.* **4,** 3427–3433.

Mathieu-Mahul, D., Sigaux, F., Zhu, C., Bernheim, A., Mauchauffe, M., Daniel, M.-T., Berger, R., and Larsen, C.-J. (1986). A t(8;14)(q24;q11) translocation in a T-cell leukemia (L1-ALL) with c-myc and TcR-alpha chain locus rearrangements. *Int. J. Cancer* **38,** 835–840.

Matis, L. A., Cron, R., and Bluestone, J. A. (1987). Major histocompatibility complex-linked specificity of $\gamma\delta$ receptor-bearing T lymphocytes. *Nature (London)* **330,** 262–264.

McIntyre, B. W., and Allison, J. P. (1983). The mouse T cell receptor: Structural heterogeneity of molecules of normal T cells defined by xenoantiserum. *Cell* **34,** 739–746.

McKeithan, T. W., Shima, E. A., LeBeau, M. M., Minowada, J., Rowley, J. D., and Diaz, M. O. (1986). Molecular cloning of the breakpoint junction of a human chromosomal 8;14 translocation involving the T-cell receptor α-chain gene and sequences on the 3' side of MYC. *Proc. Natl. Acad. Sci. U.S.A.* **83,** 6636–6640.

Mengle-Gaw, L., Willard, H. F., Smith, C. I. E., Hammarstrom, L., Fischer, P., Sherrington, P., Lucas, G., Thompson, P. W., Baer, R., and Rabbitts, T. H. (1987). Human T-cell tumours containing chromosome 14 inversion or translocation with breakpoints proximal to immunoglobulin joining regions at 14q32. *EMBO J.* **6,** 2273–2280.

Mentzer, S. J., Barbosa, J. A., and Burakoff, S. J. (1985). T3 monoclonal antibody activation of nonspecific cytolysis: A mechanism of CTL inhibition. *J. Immunol.* **135,** 34–38.

Meuer, S. C., Fitzgerald, K. A., Hussey, R. E., Hodgdon, J. C., Schlossman, S. F., and Reinherz, E. L. (1983). Clonotypic structures involved in antigen-specific human T cell function. *J. Exp. Med.* **157,** 705–719.

Meuer, S. C., Hussey, R. E., Fabbi, M., Fox, D., Acuto, O., Fitzgerald, K. A., Hodgdon, J. C., Protentis, J. P., Schlossman, S. F., and Reinherz, E. L. (1984). An alternative pathway of T-cell activation: A functional role for the 50 kd T11 sheep erythrocyte receptor protein. *Cell* **36,** 897–906.

Michell, R. H. (1982). Inositol lipid metabolism in diving and differentiating cells. *Cell Calcium* **3,** 429–440.

Moingeon, P., Ythier, A., Goubin, G., Faure, F., Nowill, A., Delmon, L., Rainaud, M., Forestier, F., Daffos, F., Bohuon, C., and Hercend, T. (1986). A unique T-cell receptor complex expressed on human fetal lymphocytes displaying natural-killer-like activity. *Nature (London)* **323,** 638–640.

Moingeon, P., Jitsukawa, S., Faure, F., Troalen, F., Triebel, F., Graziani, M., Forestier, F., Bellet, D., Bohuon, C., and Hercend, T. (1987). A γ-chain complex forms a functional receptor on cloned human lymphocytes with natural killer-like activity. *Nature (London)* **325,** 723–726.

Moller, G. (1981). T cell clones. *Imm. Rev.* **54**.
Moller, G. (1983). T cell hybrids. *Imm. Rev.* **76**.
Moore, M. A. S., and Owen, J. J. T. (1967). Experimental studies on the development of the thymus. *J. Exp. Med.* **126**, 715–726.
Moretta, L., Pende, D., Bottino, C., Migone, N., Ciccone, E., Ferrini, S., Mingari, M. C., and Moretta, A. (1988). Human CD3$^+$4$^-$8$^-$WT31$^-$ T lymphocyte populations expressing the putative T cell receptor γ-gene product. A limiting dilution and clonal analysis. *Eur. J. Immunol.* In press.
Murre, C., Waldman, R. A., Morton, C. C., Bongiovanni, K. F., Waldmann, T. A., Shows, T. B., and Seidman, J. G. (1985). Human γ-chain genes are rearranged in leukaemic T cells and map to the short arm of chromosome 7. *Nature (London)* **316**, 549–552.
Nakanishi, N., Maeda, K., Ito, K-I., Heller, M., and Tonegawa, S. (1987). Tγ protein is expressed on murine fetal thymocytes as a disulfide-linked heterodimer. *Nature (London)* **325**, 720–723.
Nishizuka, Y. (1984). Turnover of inositol phospholipids and signal transduction. *Science* **225**, 1365.
Nixon-Fulton, J. L., Bergstresser, P. R., and Tigelaar, R. E. (1986). Thy-1$^+$ epidermal cells proliferate in response to concanavalin A and interleukin 2. *J. Immunol.* **136**, 2776–2786.
Oettgen, H. C., Kappler, J., Tax, W. J. M., and Terhorst, C. (1984). Characterization of the two heavy chains of the T3 complex on the surface of human T lymphocytes. *J. Biol. Chem.* **259**, 12039-12048.
Oettgen, H. C., Terhorst, C., Cantley, L. C., and Rosoff, P. M. (1985). Stimulation of the T3-T cell receptor complex induces a membrane-potential-sensitive calcium influx. *Cell* **40**, 583–590.
O'Flynn, K., Zanders, E. D., Lamb, J. R., Beverley, P. C. L., Wallace, D. L., Tatham, P. E. R., Tax, W. J. M., and Linch, D. C. (1985). Investigation of early T cell activation: Analysis of the effect of specific antigen, interleukin 2 and monoclonal antibodies on intracellular free calcium concentration. *Eur. J. Immunol.* **15**, 7–11.
Pantaleo, G., Ferrini, S., Zocchi, M. R., Bottino, C., Biassoni, R., Moretta, L., and Moretta, A. (1988). Analysis of signal transducing mechanisms in CD3$^+$ CD4$^-$ CD8$^-$ cells expressing the putative T cell receptor γ-gene product. *J. Immunuol.* In press.
Pardoll, D. M., Fowlkes, B. J., Bluestone, J. A., Kruisbeek, A., Maloy, W. L., Coligan, J. E., and Schwartz, R. H. (1987). Differential expression of two distinct T-cell receptors during thymocyte development. *Nature (London)* **326**, 79–81.
Pardoll, D. M., Raulet, D. H., Fowlkes, B. J., Lew, A. M., Maloy, W. L., Weston, M. A., Bluestone, J. A., Schwartz, R. H., Coligan, J. E., and Kruisbeek, A. (1988). Thymus dependent and Thymus independent developmental pathways for peripheral T-cell receptor $\gamma\delta$ bearing lymphocytes. *J. Immunol.* In press.
Pelicci, P. G., Subar, M., Weiss, A., Dalla-Favera, R., and Littman, D. R. (1987). Molecular diversity of the human T-gamma constant region genes. *Science* **237**, 1051–1055.
Pelkonen, J., Traunecker, A., and Karjalainen, K. (1987). A new mouse TCRVγ gene that shows remarkable evolutionary conservation. *EMBO J.* **6**, 1941–1944.
Pernis, B., and Axel, R. (1985). A one and a half receptor model for MHC-restricted antigen recognition by T lymphocytes. *Cell* **41**, 13–16.
Phillips, J. H., Weiss, A., Gemlo, B. T., Rayner, A. A., and Lanier, L. L. (1987). Evidence that the T cell antigen receptor may not be involved in cytotoxicity mediated by γ/δ and α/β thymic cell lines. *J. Exp. Med.* **166**, 1579–1584.

Picker, L. J., Weiss, L. M., Medeiros, L. J., Wood, G. S., and Warnke, R. A. (1987). Immunophenotypic criteria for the diagnosis of non-Hodgkin's lymphoma. *Am. J. Pathol.* **128**, 181–201.

Platsoucas, C. D., and Good, R. A. (1981). Inhibition of specific cell-mediated cytotoxicity by monoclonal antibodies to human T cell antigens. *Proc. Natl. Acad. Sci. U.S.A.* **78**, 4500–4505.

Poggi, A., Bottino, C., Zocchi, M. R., Pantaleo, G., Ciccone, E., Mingari, C., Moretta, L., and Moretta, A. (1987). CD3+WT31− peripheral T lymphocytes lack T44 (CD28), a surface molecule involved in activation of T cells bearing the α/β heterodimer. *Eur. J. Immunol.* **17**, 1065–1068.

Pont, S., Regnier-Vigouroux, A., Naquet, P., Blanc, D., Pierres, A., Marchetto, S., and Pierres, M. (1985). Analysis of the Thy-1 pathway of T cell hybridoma activation using 17 rat monoclonal antibodies reactive with distinct Thy-1 epitopes. *Eur. J. Immunol.* **15**, 1222–1228.

Quertermous, T., Murre, C., Dialynas, D., Duby, A. D., Strominger, J. L., Waldman, T. A., and Seidman, J. G. (1986a). Human T-cell γ chain genes: Organization, diversity, and rearrangement. *Science* **231**, 252–255.

Quertermous, T., Strauss, W., Murre, C., Dialynas, D. P., Strominger, J. L., and Seidman, J. G. (1986b). Human T-cell γ genes contain N segments and have marked junctional variability. *Nature (London)* **322**, 184–187.

Quertermous, T., Strauss, W. M., van Dongen, J. J. M., and Seidman, J. G. (1987). Human T cell γ chain joining regions and T cell development. *J. Immunol.* **138**, 2687–2690.

Rabbitts, T. H., LeFranc, M.-P., Stinson, M. A., Sims, J. E., Schroder, J., Steinmetz, M., Spurr, N. L., Solomon, E., and Goodfellow, P. N. (1985). The chromosomal location of T-cell receptor genes and a T cell rearranging gene: Possible correlation with specific translocations in human T cell leukaemia. *EMBO J.* **4**, 1461–1465.

Rao, A., Ko, W. W-P., Faas, S. J., and Cantor, H. (1984). Binding of antigen in the absence of histocompatibility proteins by arsonate-reactive T-cell clones. *Cell* **36**, 879–888.

Raulet, D. H., Garman, R. D., Saito, H., and Tonegawa, S. (1985). Developmental regulation of T-cell receptor gene expression. *Nature (London)* **314**, 103–107.

Reilly, E. B., Kranz, D. M., Tonegawa, S., and Eisen, H. N. (1986). A functional γ gene formed from known γ-gene segments is not necessary for antigen-specific responses of murine cytotoxic T lymphocytes. *Nature (London)* **321**, 878–880.

Reinherz, E. L., Meuer, S. C., Fitzgerald, K. A., Hussey, R. E., Hodgdon, J. C., Acuto, O., and Schlossman, S. F. (1983). Comparison of T3-associated 49- and 43-kilodalton cell surface molecules on individual human T-cell clones: Evidence for peptide variability in T-cell receptor structures. *Proc. Natl. Acad. Sci. U.S.A.* **80**, 4104–4108.

Ritz, J., Campen, T. J., Schmidt, R. E., Royer, H. D., Hercend, T., Hussey, R. E., and Reinherz, E. L. (1985). Analysis of T-cell receptor gene rearrangement and expression in human natural killer clones. *Science* **228**, 1540–1543.

Roehm, N., Herron, L., Cambier, J., DiGuisto, D., Haskins, K., Kappler, J., and Marrack, P. (1984). The major histocompatibility complex-restricted antigen receptor on T cells: Distribution on thymus and peripheral T cells. *Cell* **38**, 577–584.

Romani, N., Stingl, G., Tschachler, E., Witmer, M. D., Steinman, R. M., Shevach, E. M., and Schuler, G. (1985). The Thy-1-bearing cell of murine epidermis. *J. Exp. Med.* **161**, 1368–1383.

Rothenberg, E., and Lugo, J. P. (1985). Differentiation and cell division in the mammalian thymus. *Dev. Biol.* **112**, 1–17.

Royer, H. D., Acuto, O., Fabbi, M., Tizard, R., Ramachandran, K., Smart, J. E., and Reinherz, E. L. (1984). Genes encoding the Ti β subunit of the antigen/MHC receptor

undergo rearrangement during intrathymic ontogeny prior to surface T3-Ti expression. *Cell* **39,** 261–266.
Royer, H. D., Ramarli, D., Acuto, O., Campen, T. J., and Reinherz, E. L. (1985). Genes encoding the T-cell receptor α and β subunits are transcribed in an ordered manner during intrathymic ontogeny. *Proc. Natl. Acad. Sci. U.S.A.* **82,** 5510–5514.
Rupp, F., Frech, G., Hengartner, H., Zinkernagel, R. M., and Joho, R. (1986). No functional γ-chain transcripts detected in an alloreactive cytotoxic T-cell clone. *Nature (London)* **321,** 876–878.
Saito, H., Kranz, D. M., Takagaki, Y., Hayday, A. C., Eisen, H. N., and Tonegawa, S. (1984a). A third rearranged and expressed gene in a clone of cytotoxic T lymphocytes. *Nature (London)* **312,** 36–40.
Saito, H., Kranz, D. M., Takagaki, Y., Hayday, A. C., Eisen, H. N., and Tonegawa, S. (1984b). Complete primary structure of a heterodimeric T-cell receptor deduced from cDNA sequences. *Nature (London)* **309,** 757–762.
Samelson, L. E., Germain, R. N., and Schwartz, R. H. (1983). Monoclonal antibodies against the antigen receptor on a cloned T-cell hybrid. *Proc. Natl. Acad. Sci. U.S.A.* **80,** 6972–6976.
Sangster, R. N., Minowada, J., Suciu-Foca, N., Minden, M., and Mak, T. W. (1986). Rearrangment and expression of the α, β and γ chain T cell receptor genes in human thymic leukemia cells and functional T cells. *J. Exp. Med.* **163,** 1491–1508.
Schmidt, R. E., Bartley, G. T., Lee, S. S., Daley, J. F., Royer, H. D., Levine, H., Reinherz, E. L., Schlossman, S. F., and Ritz, J. (1986). Expression of the NKTa clonotype in a series of human natural killer clones with identical cytotoxic specificity. *J. Exp. Med.* **163,** 812–825.
Schuler, G., and Steinman, R. M. (1985). Murine epidermal Langerhans cells mature into potent immunostimulatory dendritic cells in vitro. *J. Exp. Med.* **161,** 526–546.
Scollay, R., Bartlett, P., and Shortman, K. (1984). T cell development in the adult murine thymus: Changes in the expression of the surface antigens Ly2, L3T4 and B2A2 during development from early precursor cells to emigrants. *Immunol. Rev.* **82,** 79–103.
Shearer, G. M., and Schmitt-Verhulst, A. M. (1977). Major histocompatibility complex restricted cell mediated immunity. *Adv. Immunol.* **25,** 55–91.
Shima, E. A., LeBeau, M. M., McKeithan, T. W., Minowada, J., Showe, L. C., Mak, T. W., Minden, M. D., Rowley, J. D., and Diaz, M. O. (1986). Gene encoding the α chain of the T-cell receptor is moved immediately downstream of c-myc in a chromosomal 8;14 translocation in a cell line from a human T-cell leukemia. *Proc. Natl. Acad. Sci. U.S.A.* **83,** 3439–3443.
Siliciano, R. F., Hemesath, T. J., Pratt, J. C., Dintzis, R. Z., Dintzis, H. M., Acuto, O., Shin, H. S., and Reinherz, E. L. (1986). Direct evidence for the existence of nominal antigen binding sites on T cell surface Ti α-β heterodimers of MHC-restricted T cell clones. *Cell* **47,** 161–171.
Sim, G. K., Yague, J., Nelson, J., Marrack, P., Palmer, E., Augustin, A., and Kappler, J. (1984). Primary structure of human T-cell receptor α-chain. *Nature (London)* **312,** 771–775.
Smith, L. (1987). CD4+ murine T cells develop from CD8+ precursors *in vivo*. *Nature (London)* **326,** 798–800.
Snodgrass, H. R., Dembic, Z., Steinmetz, M. and von Boehmer, H. (1985a). Expression of T-cell antigen receptor genes during fetal development in the thymus. *Nature (London)* **315,** 232–233.
Snodgrass, H. R., Kisielow, P., Kiefer, M., Steinmetz, M., and von Boehmer, H. (1985b).

Ontogeny of the T-cell antigen receptor within the thymus. *Nature (London)* **313,** 592–595.

Sowder, J. T., Chen, C. H., Ager, L. L., Chan, M. M., and Cooper, M. D. (1988). A large subpopulation of avian T cells express a homologue of the mammalian T$\gamma\delta$ receptor. *J. Exp. Med.* **167,** 315–322.

Spits, H., Borst, J., Terhorst, C., and de Vries, J. E. (1982). The role of T cell differentiation markers in antigen-specific and lectin-dependent cellular cytotoxicity mediated by T8+ and T4+ human cytotoxic T cell clones directed at class I and class II MHC antigens. *J. Immunol.* **129,** 1563–1569.

Spits, H., Borst, J., Tax, W., Capel, P. J. A., Terhorst, C., and DeVries, J. E. (1985a). Characteristics of a monoclonal antibody (WT-31) that recognizes a common epitope on the human T cell receptor for antigen. *J. Immunol.* **135,** 1922–1928.

Spits, H., Yssel, H., Leeuwenberg, J., and DeVries, J. E. (1985b). Antigen-specific cytotoxic T cell and antigen-specific proliferating T cell clones can be induced to cytolytic activity by monoclonal antibodies against T3. *Eur. J. Immunol.* **15,** 88–91.

Stingl, G., Gunter, K. C., Tschachler, E., Yamada, H., Lechler, R. I., Yokoyama, W. M., Steiner, G., Germain, R. N., and Shevach, E. M. (1987a). Thy-1+ dendritic epidermal cells belong to the T-cell lineage. *Proc. Natl. Acad. Sci. U.S.A.* **84,** 2430–2434.

Stingl, G., Koning, F., Yamada, H., Yokoyama, W. M., Tschachler, E., Bluestone, J. A., Steiner, G., Samelson, L. E., Lew, A. M., Coligan, J. E., and Shevach, E. M. (1987b). Thy-1+ dendritic epidermal cells express T3 antigen and the T-cell receptor γ chain. *Proc. Natl. Acad. Sci. U.S.A.* **84,** 4586–4590.

Strauss, W. M., Quertermous, T., and Seidman, J. G. (1987). Measuring the human T cell receptor γ-chain locus. *Science* **237,** 1217–1219.

Stutman, O. (1985). Ontogeny of T cells. *Clin. Immunol. Allergy* **5,** 191–234.

Tax, W. J. M., Willems, H. W., Reekers, P. P. M., Capel, P. J. A., and Koene, R. A. P. (1983). Polymorphism in mitogenic effect of IgG1 monoclonal antibodies against T3 antigen on human T cells. *Nature (London)* **304,** 445–447.

Tentori, L., Pardoll, D. M., Zuniga, J. C., Hu-Li, J., Paul, W. E., Bluestone, J. A., and Kruisbeek, A. M. (1988). Proliferation and production of IL-2 and BSF-1/IL-4 in early fetal thymocytes by activation through Thy-1 and CD3. *J. Immunol.* **140,** 1089–1094.

Tonegawa, S. (1983). Somatic generation of antibody diversity. *Nature (London)* **302,** 575–581.

Tonegawa, S. (1985). The molecules of the immune system. *Sci. Am.* **253,** 122–131.

Toyonaga, B., and Mak, T. W. (1987). Genes of the T-cell antigen receptor in normal and malignant T cells. *Annu. Rev. Immunol.* **5,** 585–620.

Toyonaga, B., Yoshikai, Y., Vadasz, V., Chin, B., and Mak, T. W. (1985). Organization and sequences of the diversity, joining and constant region genes of the human T cell receptor β chain. *Proc. Natl. Acad. Sci. U.S.A.* **82,** 8624–8628.

Traunecker, A., Oliveri, F., Allen, N., and Karjalainen, K. (1986). Normal T cell development is possible without "functional" γ chain genes. *EMBO J.* **5,** 1589.

Trowbridge, I. S., Lesley, J., Trotter, J., and Hyman, R. (1985). Thymocyte subpopulation enriched for progenitors with an unrearranged T-cell receptor β-chain gene. *Nature (London)* **315,** 666–669.

Tschachler, E., Schuler, G., Hutterer, J., Leibl, H., Wolff, K., and Stingl, G. (1983). Expression of Thy-1 antigen by murine epidermal cells. *J. Invest. Derm.* **81,** 282–285.

Tsien, R. Y., Pozzan, T., and Rink, T. J. (1982). T cell mitogens cause early changes in cytoplasmic free Ca++ and membrane potential in lymphocytes. *Nature (London)* **295,** 68.

Ueshima, Y., Rowley, J. D., Variakojis, D., Winter, J., and Gordon, L. (1984). Cytogenetic studies on patients with chronic T cell leukemia/lymphoma. *Blood* **63,** 1028.
Van de Griend, R. J., Tax, W. J. M., van Krimpen, B. A., Vreugdenhil, R. J., Ronteltap, C. P. M., and Bolhuis, R. L. H. (1987). Lysis of tumor cells by $CD3^+4^-8^-16^+$ T cell receptor $\alpha\beta^-$ clones, regulated via CD3 and CD16 activation sites, recombinant interleukin 2, and interferon β. *J. Immunol.* **138,** 1627–1633.
Van de Griend, R. J., Borst, J., Tax, W. J. M., and Bolhuis, R. L. H. (1988). Functional reactivity of WT31 monoclonal antibody with TCR gamma expressing $CD3^+4^-8^-$ T cells. *J. Immunol.* In press.
Van den Elsen, P., Shepley, B. A., Borst, J., Coligan, J. E., Markham, A. F., Orkin, S., and Terhorst, C. (1984). Isolation of cDNA clones encoding the 20K T3 glycoprotein of human T-cell receptor complex. *Nature (London)* **312,** 413–418.
Van Dongen, J. J. M., Quertermous T., Bartram, C. R., Gold, D. P., Wolvers-Tettero, I. L. M., Comans-Bitter, W. M., Hooijkaa, H., Adriaansen, H. J., de Klein, A., Raghavachar, A., Ganser, A., Duby, A. D., Seidman, J. G., van den Elsen, P., and Terhorst, C. (1987). T cell receptor-CD3 complex during early T cell differentiation. *J. Immunol.* **138,** 1260–1260.
Van Dongen, J. J. M., Wolvers-Tettero, I. L. M., Seidman, J. G., Ang. S.-L., van de Griend, R. J., De Vries, E. F. R., and Borst, J. (1988). Two types of gamma T cell receptors expressed by T cell acute lymphoblastic leukemias. *Eur. J. Immunol.* In press.
Van Ewijk, W., Jenkinson, E., and Owen, J. (1982). Detection of thy-1, T200, Lyt-1 and Lyt-2 bearing cells in the developing lymphoid organs of the mouse thymus *in vivo* and *in vitro*. *Eur. J. Immunol.* **12,** 262.
Van Wauwe, J. P., DeMay, J. R., and Goossens, J. G. (1980). OKT3: A monoclonal anti-human T lymphocyte antibody with potent mitogenic properties. *J. Immunol.* **124,** 2708–2712.
Weiss, A., and Stobo, J. D. (1984). Requirement for the coexpression of T3 and the T cell antigen receptor on a malignant human T cell line. *J. Exp. Med.* **160,** 1284–1299.
Weiss, A., Imboden, J., Shoback, D., and Stobo, J. (1984). Role of T3 surface molecules in human T-cell activation: T3-dependent activation results in an increase in cytoplasmic free calcium. *Proc. Natl. Acad. Sci. U.S.A.* **81,** 4169–4173.
Weiss, A., Newton, M., and Crommie, D. (1986). Expression of T3 in association with a molecule distinct from the T-cell antigen receptor heterodimer. *Proc. Natl. Acad. Sci. U.S.A.* **83,** 6998–7002.
Weiss, M. J., Daley, J. F., Hodgdon, J. C., and Reinherz, E. L. (1984). Calcium dependency of antigen specific (T3-Ti) and alternative (T11) pathways of human T cell activation. *Proc. Natl. Acad. Sci. U.S.A.* **81,** 6836–6840.
Williams, D. L., Look, A. T., Melvin, S. L., Roberson, P. K., Dahl, G., Falke, T., and Stass, S. (1984). New chromosomal translocations correlate with specific immunophenotypes of childhood acute lymphoblastic leukemia. *Cell* **36,** 101–109.
Wu, Y.-J., Tian, W.-T., Snider, M., Rittershaus, C., Rogers, P., LaManna, L., and Ip, S. H. (1988). Signal transduction of $\gamma\delta$ T cell antigen receptor with a novel anti-δ antibody. *J. Immunol.* In press.
Yanagi, Y., Yoshikai, Y., Leggett, K., Clark, S. P., Aleksander, I., and Mak, T. W. (1984). A human T cell-specific cDNA clone encodes a protein having extensive homology to immunoglobulin chains. *Nature (London)* **308,** 145–149.
Yanagi, Y., Caccia, N., Kronenberg, M., Chin, B., Roder, J., Rohel, D., Kiyohara, T., Lauzon, R., Toyonaga, B., Rosenthal, K., Dennert, G., Acha-Orbea, H., Hengartner, H., Hood, L., and Mak, T. W. (1985). Gene rearrangement in cells with natural killer

activity and expression of the β-chain of the T-cell antigen receptor. *Nature (London)* **314**, 631–633.

Yaoita, Y., Matsunami, N., Choi, C. Y., Sugiyama, H., Kishimoto, T., and Honjo, T. (1983). The D-J$_H$ complex is an intermediate to the complete immunoglobulin heavy chain V-region gene. *Nucleic Acids Res.* **11**, 7303.

Yoshikai, Y., Clark, S. P., Taylor, S., Sohn, U., Wilson, B. I., Minden, M. D., and Mak, T. W. (1985). Organization and sequences of the variable, joining and constant region genes of the human T cell receptor α-chain. *Nature (London)* **316**, 837–840.

Yoshikai, Y., Reis, M. D., and Mak, T. W. (1986). Athymic mice express a high level of functional γ-chain but greatly reduced levels of α- and β-chain T-cell receptor messages. *Nature (London)* **324**, 482–485.

Yoshikai, Y., Toyonaga, B., Koga, Y., Kimura, N., Griesser, H., and Mak, T. W. (1987). Repertoire of the human T cell gamma genes: High frequency of nonfunctional transcripts in thymus and mature T cells. *Eur. J. Immunol.* **17**, 119–126.

Zech, L., Gahrton, G., Hammarstrom, L., Juliusson, G., Mellstedt, H., Robert, K. H., and Smith, C. I. E. (1984). Inversion of chromosome 14 marks human T-cell chronic lymphocytic leukaemia. *Nature (London)* **308**, 858–860.

Zinkernagel, R. M., and Doherty, P. C. (1979). MHC-restricted cytotoxic T cells: Studies on the biological role of polymorphic major transplantation antigens determining T-cell restriction-specificity, function, and responsiveness. *Adv. Immunol.* **27**, 51–177.

Specificity of the T Cell Receptor for Antigen

STEPHEN M. HEDRICK

*Department of Biology and Cancer Center,
University of California, San Diego,
La Jolla, California 92093*

I. Introduction

Thymus-dependent T lymphocytes can be subdivided based on the cell surface expression of polypeptide chains involved in immune recognition. The predominant subset of T lymphocytes expresses the CD3 polypeptides (γ, δ, ε, and ζ) noncovalently associated with an $\alpha-\beta$ chain receptor heterodimer (Allison et al., 1982; Meuer et al., 1983; Haskins et al., 1983; Borst et al., 1983; Samelson et al., 1985a) termed the T cell receptor (TCR). This set of T cells can be once again subdivided into class II major histocompatibility complex (MHC)-specific T cells, expressing CD4 accessory molecules, and class I MHC-specific T cells, expressing CD8 accessory molecules (Swain, 1981; Spits et al., 1982; Meuer et al., 1982; Marrack et al., 1983; Greenstein et al., 1984). Virtually all of the known, antigen-specific cell-mediated effector and regulatory mechanisms of the immune system are carried out by these two subsets of T lymphocytes. Another subset of T lymphocytes expresses the CD3 polypeptide chains in association with a $\gamma-\delta$ chain receptor heterodimer, and most of these cells do not express either the CD4 or CD8 molecules (Brenner et al., 1986; Bank et al., 1986; Moingeon et al., 1986; Lanier et al., 1986). The role of this T cell subset in the immune system is not yet known, and these cells will not be discussed in this article.

The goal of this article is to understand the specificity of the TCR, and thus, the mechanisms by which T cells carry out the process of self- and nonself-discrimination. The TCR α and β chains that determine clonal specificity are encoded by variable (V), diversity (D), and joining (J) region gene elements that rearrange to form a mature, transcribable gene, and the number and diversity of these gene elements determine the repertoire potential for TCR specificity. A description of the organization of these gene elements and of the structure of TCR heterodimer has been reviewed extensively elsewhere (Kronenberg et al., 1986; Allison and Lanier, 1987; Toyonaga and Mak, 1987; Davis and Bjorkman, 1988). Understanding the specificity of recognition in the immune system is uniquely complicated because of the vast diversity of the "specific" receptors. Just as antibodies and antigens were for years circularly defined by one another, this article will focus on the structure of the determinant recognized by the TCR and the structure of the receptor itself. In addition, a recently proposed model of the TCR–antigen–MHC interaction will be discussed. Beyond a simple biochemical explanation for various observed immune phenomena, with this information, we can attempt to understand the evolutionary and organismal influences that may have contributed to this particular strategy of immune recognition.

II. Nature of the Determinant Recognized by T Cells

A. ANTIGEN CONFORMATION

In analyzing the specificity of T cells for many protein antigens and synthetic peptide antigens, the most conspicuous characteristic that emerged concerned the form of the antigen recognized. In several different assays of T cell function, the consistent conclusion was that a denatured form of antigen served to stimulate a response (Gell and Benacerraf, 1959; Senyck et al., 1971; Thompson et al., 1972; Ishizaka et al., 1975). Moreover, synthetic peptides were at least as immunogenic as the native proteins (Sela, 1969). Such information seemed to point to a distinction between the mechanisms of recognition used by T cells and surface immunoglobulin expressed as the B cell receptor. Whereas antibodies were often specific for determinants dependent on the native conformation of the protein antigen (Brown et al., 1959; Freedman 1966; Crumpton and Sonall, 1967; Gerwing and Thompson, 1968), T cell responses were not (see below). Several different observations contributed to an initial understanding of this phenomenon.

B. RECOGNITION OF MAJOR HISTOCOMPATIBILITY COMPLEX (MHC) MOLECULES

T cells recognize antigen presented by a second cell type functionally referred to as an antigen-presenting cell (APC) (Möller, 1978; Unanue, 1981). These cells can be dendritic cells (Van Voorhis et al., 1983), macrophages (Unanue, 1984), or B cells (McKean et al., 1981; Chestnut et al., 1982b) and share the property of expressing class I and II MHC molecules. The unexpected observation was that T cells recognize antigen fragments, and simultaneously, determinants on MHC molecules; this interaction was termed MHC restriction. Experiments using the adoptive transfer of carrier-primed T cells and hapten-primed B cells showed that the two cell types had to come from mice bearing the same *I* region allele of the MHC in order to cooperate in the induction of antibody synthesis (Katz et al., 1973, 1975). Experiments reconstituting athymic mice with T cells from different strains gave similar results (Kindred, 1975). The interpretation of these experiments was contentious due to the possibility of allogeneic effects that could suppress the *in vivo* response. Restriction was also shown in experiments in which guinea pig peritoneal exudate T cells from immunized animals were cultured with antigen-pulsed macrophages from syngeneic or allogeneic guinea pigs, and only the syngeneic macrophages would effectively present antigen (Rosenthal and Shevach, 1973). This experiment still held the possibility of negative influences caused by allogeneic effects, although no such postulated effects interfered with the presentation of antigen by F_1 heterzygous APCs expressing both MHC alleles. Several experiments using other assays did seem to indicate at the time that allogeneic effects (Heber-Katz and Wilson, 1975) or a lack of tolerance (von Boehmer et al., 1975) could account for an apparent MHC restriction.

The experiments that most convincingly demonstrated MHC restriction as a fundamental property of T cell recognition were carried out with cytotoxic T cells (CTLs). In experiments with virus-specific T cells (Zinkernagel and Doherty, 1974), minor histocompatibility-specific T cells (Bevan, 1975; Gordon et al., 1975), and hapten-specific T cells (Shearer et al., 1975), it was clearly demonstrated that cytotoxicity required the simultaneous recognition of antigen determinants and MHC class I determinants. The assay for CTL activity was a 4 hour ^{51}Cr release assay, and in this short-term culture in which T cells lyse appropriate target cells, the influence of allogeneic effects were shown to be inconsequential (Cerrotini and Brunner, 1974).

Since class II MHC and antigen recognition by T helper cells always required an extended culture period for proliferation or helper cell activity, other experimental systems were used to rule out allogeneic effects. If allogeneic effects were eliminated either by acute depletion of alloreactive cells or by tolerance in radiation bone marrow chimeras (Sprent and von Boehmer, 1976; Sprent, 1978), it was clearly shown that T helper cells always recognized antigen in association with allelic differences of MHC class II antigens.

Finally, the long-term growth of T cells in culture, combined with the ability to grow T cells from a single cell precursor, allowed the specificity of T cells to be examined in a clonal population. No extraneous cells would thus confuse the results. This technology was applicable to T helper cells assayed by cell proliferation or secretion of lymphokines (Watson, 1979; Kimoto and Fathman, 1980; Sredni and

Schwartz, 1980) and to CTLs assayed for proliferation or cytolytic activity (Nabholz et al., 1978; von Boehmer et al., 1979; Glasebrook and Fitch, 1980). One further technique that proved to be useful was the immortilization of T cells by fusion with the thymoma BW5147. Analogous to the production of antibody-secreting hybridomas, T cell hybridomas were produced and assayed for stimulation by the secretion of the lymphokine interleukin 2 (IL-2) (Kappler et al., 1981). All of the specificity experiments carried out with these clonal T cell lines showed clearly that the determinant recognized depended on the allelic form of MHC class I or class II molecule and the amino acid sequence of the antigen. Thus, MHC restriction was an integral part of the specificity of the TCR and there was a pleasing symmetry: T helper cells mainly recognized antigen and class II MHC antigens and CTLs mainly recognized antigen and class I MHC antigens. One further characteristic of the specificity seen for T cell clones was a cross-reactivity for allogeneic MHC molecules (Sredni and Schwartz, 1980). Frequently, a T cell clone reactive to antigen in association with a particular allelic form of MHC molecule would react with another allelic form of MHC molecule in the absence of added antigens. This allocross-reactivity usually reflected the MHC class of the original response. Class I-restricted T cells were cross-reactive with allelic variations of class I molecules, whereas class II-restricted T cells were cross-reactive with allelic variations of class II molecules. These cross-reactions appeared to indicate that the TCR had an intrinsic reactivity with MHC molecules.

C. Antigen Processing

The notion that T cells would recognize denatured or fragmented antigen as efficiently as native antigen implied that the natural ligand of T cells was a form of processed antigen, and thus, antigen processing was an obligatory step in the induction of T cells. Experiments showed that antigen presentation by macrophages was a time-dependent process that required active cellular metabolism (Ziegler and Unanue, 1981). Furthermore, agents that raise the pH of lysosomes, ammonia and chloroquine, interfere with the ability of cells to present antigen if added early, but not later in the reaction (Ziegler and Unanue, 1982; Chestnut et al., 1982a; Grey et al., 1982; Lee et al., 1982). Finally, experiments directly showed the requirements for the processing and presentation of antigen. The B cell lymphoma A20-2J was shown to serve as an antigen-presenting cell for T cell hybridomas specific for ovalbumin. If the A20-2J cells were pulsed with ovalbumin and then lightly glutaraldehyde fixed, the cells would still effectively present antigen, although they were no longer metabolically active. However, if the A20-2J cells were fixed prior to ovalbumin addition, antigen-presenting function was abolished. The interesting result was that fixed A20-2J cells would present ovalbumin to some, but not all, T cell hybridomas, if the ovalbumin had been previously cleaved into fragments by trypsin, chymotrypsin, or cyanogen bromide (Shimonkevitz et al., 1983). Denatured ovalbumin was not presented by fixed A20-2J cells. Furthermore, it was shown that fixed cells would present synthetic peptides corresponding to the antigenic ovalbumin peptide sequence, e.g., 323–339 (Shimonkevitz et al., 1984). These experiments showed that cleavage of antigens was a necessary and sufficient requirement for antigen processing as required by T cells.

III. Studies on the Specificity of T Cells for Peptide Antigens

A. APPROACHES

With the determination of MHC restriction as a fundamental property of T cell recognition, attention turned to the precise chemical structure of antigens that would stimulate a response. Two types of model antigens were investigated: synthetic amino acid polymers and naturally occurring proteins.

Synthetic polypeptide chains were made with different amino acid compositions and sizes, and these studies established many of the fundamental properties of antigenicity for antibody responses and T cell recognition. For reasons still not well understood, homopolymers of α-amino acids are very poorly immunogenic, whereas homopolymers with haptenic groups or copolymers often elicit both delayed-type hypersensitivity, a cellular response reflecting T cell activation, or antibody synthesis, a humoral response reflecting T cell and B cell activation (Sela, 1969). The lower size limit for simple antigens appeared to be about seven amino acids, since dinitrophenylhepta-L-lysine (DNP-poly-L-lys$_7$) was immunogenic in guinea pigs, whereas a dinitrophenylhexa-L-lysine (DNP-poly-L-lys$_6$) was not (Stashenko and Schlossman, 1977). In later studies, it was shown that the minimum size of other antigenic fragments varied from 8 for angiotensin II (Thomas et al., 1981) to 17 for ovalbumin (Shimonkevitz et al., 1984). Clearly, the size requirement was dependent on the chemical composition of the antigen and could reflect conformational requirements or the size of the determinant that most effectively binds to the combining site of MHC molecules, T cell receptors, or both. If the assay is the generation of antibodies, the size requirements could in addition reflect the size of the antibody-combining site.

The advantage of using naturally occurring proteins as immunogens was that a group of species variants of known sequence could be obtained and used to precisely map the amino acid positions involved in antigenicity. Once an antigenic determinant was mapped, enzymatic or biochemical cleavage of the native molecule would yield oligopeptide fragments that could even more precisely map the required structure of the antigen determinant. Finally, the techniques of solid-phase oligopeptide synthesis could be used to generate even more peptide variants. This approach has proved to be instrumental in understanding the mechanisms of T cell recognition and the specificity of the T cell receptor.

B. Specificity of T cells for Linear Sequences in Naturally Occurring Proteins

The first proteins to be used as model antigenic systems to investigate T cell specificity were insulins (Barcinski and Rosenthal, 1977), staphlococcal nuclease (Schwartz et al., 1978), cytochromes c (Coradin and Chiller, 1979; Solinger et al., 1979), myoglobin (Berzofsky et al., 1979; Infante et al., 1981), and egg lysozymes (Maizels et al. 1980). The antigenic epitopes of all of these antigens were shown to be present on oligopeptides. These experiments combined with previous work (Senyck et al., 1971) showed clearly that the molecular form of antigen recognized by T cells was different than the native structure recognized by B cells.

Cytochrome c was originally chosen as an antigen for its properties, as a well-characterized nonglycosylated protein of 103 or 104 amino acids. In studying the evolution of proteins, the sequences of over 90 cytochrome c molecules had been determined from a large variety of species from yeast to humans (Schwartz and Dayhoff, 1978). These cytochrome c species variants offered a set of protein antigens with subtle sequence variations, perfect for determining the antigen fine specificity of T cells and the constraints on the immune repertoire. In addition, the three-dimensional structure of three cytochrome c molecules had been determined (Dickerson, 1980), potentially allowing an examination of the structure of an antigenic determinant. This last notion, although reasonable at the time, was based on the incorrect assumption that the TCR could recognize native proteins. As later deduced from these and other studies, the native structure of cytochrome may not bear any relation to the structure of the antigenic peptide recognized by the TCR.

Initial studies on the immunogenicity and antigenicity of the cytochrome *c* molecules by Corradin and Chiller and by Schwartz and colleagues showed that mice responded to various different xenogeneic cytochrome *c* molecules. These immune responses were initiated by immunizations with cytochrome *c* in complete freund's adjuvant, and the resulting responses were characterized by the antigen-specific restimulation of cultured T cells from the draining lymph nodes or peritoneal cavity of immunized mice.

T cells from (C57BL/6 × DBA/2)F_1 mice, immunized with beef cytochrome *c*, could be restimulated in culture with beef and mouse, but not horse and rabbit cytochromes *c*. A comparison of the sequences indicated that the T cells recognized a determinant that included an amino acid at position 89 (Corradin and Chiller, 1979). Data consistent with this mapping showed that fragment 81–104 of beef cyctochrome *c* would stimulate a response, whereas fragment 81–104 of horse cytochrome *c*, differing at only position 89, would not. This further indicated that the determinant recognized did not require the native structure of cytochrome *c*. Surprisingly, mouse cytochrome *c* would stimulate a response in culture at the same concentrations as beef cytochrome, and the reasons for this autoreactivity were not clear. Although not extensively investigated in this antigenic response, other strains recognized different antigenic epitopes. The MHC dependence of determinant recognition was more clearly shown in the response to insulins (Barcinski and Rosenthal, 1977) and lysozymes (Katz *et al.*, 1982) and probably represents a general property of T cell specificity.

Similar experiments using pigeon cytochrome *c* as an immunogen showed that T cells specific for pigeon cytochrome *c* mainly recognized amino acid substitutions at positions 100 and 104 (Solinger *et al.*, 1979). Initial evidence for the recognition of an amino acid at position 3 in pigeon cytochrome *c* fragment 1–65 was apparently the result of a separate population of pigeon cytochrome *c*-specific T cells, and this determinant was never further characterized. Again consistent with the localization of the antigenic determinant in pigeon cytochrome *c* was the finding that the cyanogen bromide cleavage fragment 81–104 contained all of the immunogenicity and antigenicity of the native molecule. In contrast to the response to beef cytochrome *c*, the T cells from B10.A mice specific for pigeon cytochrome *c* did not respond significantly to mouse cytochrome *c*. Thus, there seemed to be a significant difference in the degree of self-tolerance that was either dependent on the immunogen or on the strain of mice. The relevance of this to the immune repertoire will be discussed in the last section (Section VI).

A characteristic of the response to pigeon cytochrome c involved a curious twist to the antigen fine specificity and this was one of the first indications for antigen specificity not involving the TCR. Immunizations with pigeon cytochrome c resulted in a population of reactive T cells that could be restimulated with cytochrome c molecules from other species by cross-reaction. Surprisingly, at least two cytochrome c molecules were more stimulatory than the pigeon cytochrome c immunogen. Cytochromes c from tobacco hornworm moth and screwworm fly stimulated the T cell population at lower concentrations of antigen added to culture than did pigeon cytochrome c. These two antigens shared the property of an alanine deletion at position 103, resulting in a protein of one less amino acid. Further experiments showed that the critical difference was the spacing of the antigenic Gln-100 and Lys-104 positions (Hansburg and Appella, 1985). Regardless of the fact that the T cell population was immunized with Gln-100–Lys-104, it virtually always responded more effectively to Gln-100–Lys-103.

Although there was a precedent for "heteroclicity" in the antibody response to the nitrophenyl hapten (Imanishi and Mäkelä, 1973), it was unexpected that the T cell population should recognize a cross-reactive antigen better than the immunogen. Theoretically, a shift in dose–response curves could be due to more efficient uptake by the antigen-presenting cells, a higher affinity association with the E^k class II MHC molecules, or a higher affinity association with the TCR. A serendipitous finding appeared to provide evidence that the antigen (or processed peptide) had an affinity for MHC that could be defined independent of the T cell population. Hansburg et al. (1981) tested the response to cytochrome c that was acetimidylated to block the free amino groups. This was done as a first step toward the coupling of cytochrome c to a carrier protein and was tested for immunogenicity as a control. The surprising finding was that the population of T cells specific for acetimidylated cytochrome c did not respond to native cytochrome c molecules, but responded to acetimidylated species variants of cytochrome c with the same hierarchy as seen in the response to native cytochrome c. Thus, acetimidylated moth and fly cytochrome c stimulated the response at lower concentrations that did the acetimidylated pigeon cytochrome c immunogen. These data suggested that the heterocyclicity of the moth cytochrome c was independent of the T cell, and further experiments showed that it was dependent on the MHC allele of the antigen-presenting cell (Heber-Katz et al., 1983).

Other experiments showed that the critical chemical modification affecting the cross-reactivity of the T cell population was the acetimidylation of the ε-amino group on Lys-99 on cytochrome c, fragment 81–104

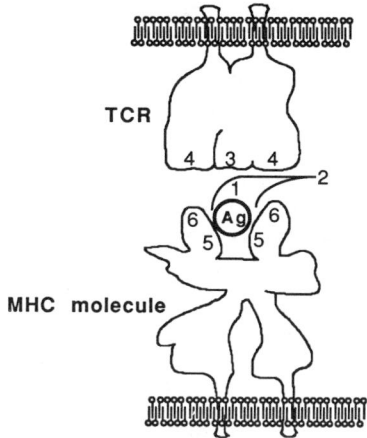

FIG. 1. Model of the TCR–antigen–MHC interaction. 1, Epitope; 2, agretope; 3, paratope; 4, restitope; 5, desetope; 6, histotope.

(Hansburg et al., 1983). A model was put forth, based on these data, that described the TCR recognition site to include Lys-99, and this was termed the epitope. The MHC-binding site included Gln-100 and Lys-103–104, and this was termed the agretope. A slightly updated version of the model is shown in Fig. 1. This model will be described in more detail in the last section (Section VI).

Another antigen-specific effect not dependent on T cells was shown by Shastri et al. (1986b). In an examination of the response to hen egg lysozyme (HEL), it was found that virtually all T cell clones from H-2^b mice specific for HEL would response to ring-necked pheasant lysozyme (REL) at concentrations of 100-fold less than that required to stimulate a response by the immunogen, HEL. The determinant recognized by these T cell clones was shown to be contained within the cyanogen bromide fragment 13–105, and it was shown that different T cell clones recognized different epitopes within this large polypeptide fragment. The interesting result was that the preference for REL over HEL was not seen when using fragments REL 13–105 and HEL 13–105, and furthermore, that the preference was dependent on active antigen processing in the APC population. These results indicated that the antigen dose–response curves were influenced by amino acids outside the recognized epitope and suggested that there was a specificity for the antigen contained within the process of antigen processing and presentation.

C. MHC-Restricted Antigen Recognition

In broad terms, the simultaneous recognition of antigen and MHC molecules could be viewed in terms of *altered self*, in which the TCR recognized MHC molecules modified by bound antigen (Zinkernagel and Doherty, 1977; Matzinger, 1981), or *dual recognition*, in which the TCR had separate binding sites for the two determinants (Langman, 1978; Cohn and Epstein, 1978). The model (see above), based on the ternary interaction between the cytochrome c, MHC molecules, and the TCR, was an elaboration of the altered self model. A prediction, based on this model, was that the antigen determinant recognized by the TCR would depend on the allelic form of the MHC molecule to which it was bound, i.e., the same antigen binding to different MHC molecules would display different determinants. This prediction was born out in experiments using T cell clones that recognized cytochrome c in association with two different MHC alleles. In several examples, the specificity of a T cell clone for cross-reactive cytochromes c was different depending on which MHC allele was simultaneously recognized (Heber-Katz *et al.*, 1982; Hedrick *et al.*, 1982; Matis *et al.*, 1983). This was also shown for MHC class I-restricted cells, in which the specificity for minor antigens recognized by a T cell clone was dependent on the MHC allele that was simultaneously recognized (Hünig and Bevan, 1982). As such, antigen and either class I or class II MHC molecules were shown not to be independently recognized.

Ashwell and Schwartz (1986b) provided independent evidence supporting the model of TCR, antigen, and MHC molecule interactions using another approach. The antigen dose–response curve of a T cell clone reflects several characteristics of the biological response. It reflects the efficiency with which antigen is taken up and processed by APCs, the association of antigen with MHC molecules, and the association of the TCR with the antigen–MHC ligand. With respect to the latter interaction, the T cell response is proportional to the concentration of free (unbound by the TCR) antigen–MHC ligand. If normalized antigen dose–response curves of a T cell clone are analyzed at increasing concentrations of T cells, eventually a point will be reached where there are sufficient T cells that specifically bind antigen–MHC to significantly decrease the concentration of free ligand (Hedrick *et al.*, 1985; Ashwell *et al.*, 1986). At this point, the antigen dose–response curves will be shifted to higher concentrations of antigen proportional to the number of T cells in the reaction. This was termed the transition point, and this point reflects the relative affinity of the TCR for its ligand. The lower the transition point, the higher the affinity of the TCR ↔ antigen–MHC interaction. Using this transition point analysis, one can distinguish between antigen dose–response effects that result from changes in TCR ↔ ligand affinity from changes that result from antigen uptake, antigen processing, or antigen ↔ class II MHC affinity.

When moth cytochrome c peptides were substituted at either positions 99 (Lys → Arg) or 103 (Lys → Arg), the antigen dose–response curves were shifted to much higher concentrations of antigen. The qualitative difference in the antigenicity of these peptides was seen in the transition point analysis. Whereas the position 99 substitution dramatically changed the transition point, the position 103 substitution showed a barely detectable change. These results affirmed the model whereby position 99 was part of the epitope binding to the TCR, and position 103 was part of the agratope binding to the class II MHC molecule (Ashwell et al., 1986). In another set of experiments, a cytochrome peptide analog was tested for antigenicity in association with B10.A (E^k) APCs and B10.A(5R) (E^b) APCs (Ashwell and Schwartz, 1986). The dose–response curve in association with the B10.A(5R) APCs was shifted to 12-fold higher concentrations relative to that in the presence of B10.A APCs. The interesting point was that the transition point in these two situations was nonetheless, identical. This result indicated that the dose–response difference was due to an interaction that was MHC specific, but independent of the TCR ↔ antigen–MHC interaction. The conclusion was that the antigen ↔ E^k was of higher affinity than the antigen ↔ E^b interaction.

D. Conformation of the Antigenic Fragment Recognized in Association with Class II MHC Molecules

Berzofsky and co-workers (1979) have investigated the antigenic properties of sperm whale myoglobin in T cell clones that recognize this antigen in association with E^d (Berkower et al., 1986). These workers found that an 11-amino acid peptide, 136–146, was stimulatory for T cell clones and that substitutions at 4 of these amino acids, Lys-133, Glu-136, Lys-140, and Lys-145, had large effects on antigenicity. Modeling studies of the peptide suggested a possible explanation for the periodicity of these four, important, hydrophilic amino acids. In its native α-helical conformation, the structure of the peptide would line up all four of these charged amino acids on one side of the peptide. At the same time, the hydrophobic amino acids, Leu-135, Phe-138, Ile-142, and Tyr-146, would be found on the opposite side of the helical structure.

The contribution of antigen conformation was also investigated in the response to pigeon cytochrome c. Although amino acids 99, 100, and

103 were shown to be important in forming the antigenic determinant, short peptides such as 97–103 would not stimulate a response at concentrations measured. The shortest peptide that stimulated a detectable response was 94–103, and 88–103 was much more effective. Thus, antigenic potency involved perhaps more than the simple TCR and MHC contact amino acids. Modeling of the secondary structure of the stimulatory and nonstimulatory cytochrome c peptides suggested that an α-helical conformation was necessary for antigenicity and that 88–98 might stabilize this conformation (Schwartz *et al.*, 1985). Carbone *et al.* (1987) provided more evidence for this by producing peptides containing the α-helical constraining amino acid, α-aminoisobutyric acid (Aα). Experiments showed that peptide $(A\alpha)_3$-95–104 stimulated a response, whereas $(A\alpha)_4$-96–104 did not. This indicated that the minimum determinant was 95–103(4), but that this determinant needed to be in an α-helical conformation.

To understand the general properties of antigenic fragments, De Lisi and Berzofsky (1985) analyzed 12 antigenic peptides from 6 proteins for amphipathic propensities, similar to that of the antigenic myoglobin peptide, and found that 10 of 12 sites were found to display a periodicity of hydrophobicity of $100 \pm 20°$, conducive to stabilization of an amphipathic α-helix. These authors found this correlation to be highly significant and suggested that conformation, in particular an amphipathic α-helix, is a common property of antigenic peptide fragments from native proteins. The authors speculated that the hydrophobic face would contact the MHC molecule, and the hydrophilic face would constitute the contact residues comprising the determinant recognized by the T cell receptor. As discussed by Livingstone and Fathman (1987), this hypothesis can be used to predict likely regions within proteins that would give rise to antigenic fragments. Within the sperm whale myoglobin protein, there are three such regions: 136–146 as discussed above, 106–118 also shown to be antigenic for T cells (Berkower *et al.*, 1986; Cease *et al.*, 1986), and a third falling within the region 69–81. In fact, Livingstone and Fathman (1987) independently identified an epitope within the nine amino acid sequence 70–78, obviously consistent with the predictions of the model.

The property of amphipathicity was not predicted to be essential for a TCR epitope, but to simply increase the probability that a peptide would interact with MHC molecules and stimulate a T cell response. A second model has been put forth by Rothbard and Taylor (1988) based on an empirical analysis of 28 antigenic sites known at the time (Rothbard *et al.*, 1986). Each amino acid was assigned to be hydrophobic, charged, polar, or glycine–proline. From this analysis, the authors found a pattern of a charged amino acid or glycine, followed by two hydrophobic amino acids, and used this pattern to successfully predict one of the antigenic peptides from the influenza nucleoprotein. Each of these models for antigenic structure made no distinction between class I and class II MHC-restricted antigens, and perhaps this and other details may help in finally describing the motifs for antigenic peptides.

E. Nature of the Determinant Recognized in Association with MHC Class I Molecules

The striking conclusions concerning the specificity of the TCR for class II MHC molecules and antigenic peptide fragments had further implications that were perhaps even more surprising. The TCR-combining site structurally comprises different combinations of α and β chain segments, and these segments appear to be interchangable in class I and class II-specific T cells (Rupp et al., 1985). As such the overall structure of the TCR-combining site is similar, whether it is specific for class I or class II MHC molecules in association with antigen. Thus, the structure of the class I–antigen and class II–antigen determinant may also assume the same basic overall conformation. This belies an intuitive notion of the determinant recognized by class I MHC-specific CTLs.

CTLs can be readily induced to respond to foreign antigens that are endogenously synthesized. The classic example is the CTL response to virus infection, although a response can be induced to endogenously encoded allelic molecules that comprise minor histocompatibility antigens. These observations led to the idea that CTLs recognize membrane antigens in association with MHC molecules, i.e., virus infections induced CTLs specific for viral glycoproteins expressed on the cell surface, and minor histocompatibility antigens were allelic forms of non-MHC membrane-bound molecules. Analysis of the specificity of CTLs, carried out at approximately the same time as that on class II-specific antigens, showed these ideas to be too simplistic.

Townsend and colleagues showed that a portion of the CTL population specific for influenza-infected cells were specific for the nucleoprotein, not the cell surface hemagglutinin (Townsend and Skehel, 1984). Since the nucleoprotein was not detected on the plasma membrane and its sequence contained no obvious characteristics of an integral membrane protein, there must be a another mechanism for transporting a least an antigenic portion of the molecule to the cell surface. It was shown that a clone or a population of CTL from H-2^b mice would lyse L cells cotransfected with D^b and the 498-amino acid nucleoprotein or truncated proteins (1–386) or (1–2, 328–498)

(Townsend et al., 1985). Thus, the determinant appeared to be located in the overlapping region 328–386, and transport to the cell surface for recognition by CTLs did not seem to require that the determinant be specifically associated with either the N-terminus or the C-terminus of the nucleoprotein. It was further shown that the CTLs would lyse target cells that were incubated with synthetic peptides corresponding to part of the nucleoprotein sequence, and the most efficiently recognized peptide corresponded to the 366–379 sequence (Townsend et al., 1986b).

These experiments clearly established that CTLs do not necessarily recognize membrane-associated molecules. Considering that these data parallel that of class II MHC recognition, the explanation put forth was that a degraded form of endogenously synthesized proteins is transported to the cell surface, and it is this antigenic fragment that constitutes the determinant for TCR recognition. Carrying the analogy to class II MHC–antigen recognition one step further, the degraded peptide may be transported to the cell surface bound to a MHC class I molecule.

A tangential, but important, question concerned the association between intracellular antigens (e.g., viruses) and class I MHC recognition and extracellular antigens (e.g., soluble proteins or bacteria) and class II MHC recognition. An explanation was put forth by Germain (1986) and elegantly shown in experiments by Morrison et al. (1986). The presentation of influenza hemagglutinin was examined separately for class I and class II-restricted T cells. Experiments showed that the antigen only became associated class I MHC molecules if the hemagglutinin molecule was endogenously synthesized by the APCs, whereas the antigen became associated with class II molecules if the cells were capable of endocytosis and proteolytic processing. Thus, the restriction was at least, in part, a result of compartmentalized cellular processing mechanisms.

IV. Formation of the Antigen–MHC Determinant Recognized by the TCR: Stable Formation in the Absence of T Cells

A. Immune Response Gene Control

Genetic control of the immune response, as investigated extensively using synthetic polymers as antigens, maps to the MHC region (Paul and Benacerraf, 1977). It was also shown to be a universal phenomenon of the immune responses to well-defined protein antigens. The consistent result was that genetic control of the response always mapped to the same gene controlling the MHC restriction of the response (Longo et al., 1980). For example, the responses to several antigens, such as GLφ (Dorf and Benacerraf, 1975) and pigeon cytochrome c (Solinger et al., 1979), were shown to be highly dependent on the class II MHC allele mapping to the I-A and I-E subregions, the same genes encoding the MHC molecules recognized by the respective antigen-specific T cells. This, we now know, reflects the genetics of the class II MHC region in the mouse, in which a nonpolymorphic α chain is encoded in the I-E subregion and is separated by recombination from its paired poly morphic β chain, which is encoded in the I-A subregion. Thus, E^k molecules are expressed if the strain has an I-A^k allele and an I-E allele that allows for the productive expression of an E-encoded α chain (I-$E^{d,k,p,r,u}$). In strains bearing I-$E^{b,s,m,q}$, no α chain is expressed, and thus, no E molecule is present at the cell surface. Most responses to soluble synthetic or protein antigens map to the I-A region, and the I-E-encoded allele has no effect on the response. These responses are restricted to the A molecules (Berzofsky, 1980). For antigens that are endogenously synthesized, such as viruses and minor histocompatibility antigens, the response phenotype maps to the MHC class I antigens consistent with their restriction specificity (Zinkernagel, 1978).

These observations led to the speculation that an Ir gene phenotype was the result of differential binding of oligopeptide antigens to MHC molecules. This model of Ir gene control followed from the determinant selection model for MHC restriction (Barcinski and Rosenthal, 1977; Benacerraf, 1978). In this model, it was first proposed that an apparent MHC recognition was imposed by the way in which antigenic peptides bound to MHC molecules and not by the direct recognition of MHC molecules themselves. Different peptides from a processed protein could be bound by different MHC molecules, and even the same peptides could be bound in different conformations. Strains expressing a MHC allele to which an antigen could effectively bind would respond to that antigen, otherwise no response was possible. An alternative effect of MHC molecules on the response phenotype could be due to selective differences in the T cell repertoire caused by MHC-dependent selection mechanisms. For example, the T cell population is presumably deleted of any cells that recognize self-protein products in association with self-MHC molecules. If this self-antigen–MHC determinant should cross-react with a foreign antigen in the context of that same MHC determinant, then the strain would be unresponsive to the foreign antigen (Schwartz, 1978). In this light, it is curious that beef cytochrome c can elicit a response that is cross-reactive with mouse cytochrome c, since this population of T cells would be expected to be deleted. These models were tested directly in further experiments defining the structure of antigenic peptides and their interactions with MHC molecules and the TCR.

B. ANTIGENIC COMPETITION

The model describing the ternary interaction between the TCR, MHC molecules, and antigen peptide fragments presupposes that there exists an affinity that can be defined for the binding of antigen and MHC molecules, and this affinity is different for different antigen–MHC combinations. A prediction of this model is that the binding of antigen to MHC molecules is saturable, and thus, unrelated antigens could compete for the same MHC-binding site. Based on these assumptions, Werdelin (1982) investigated the competition between the amino acid copolymer GL (glutamic acid and lysine) and DNP–PLL (dinitrophenol–poly-L-lysine). In these experiments using guinea pigs, Werdelin showed that preincubation of APCs with GL would significantly inhibit the ability of the APC population to present DNP–PLL to DNP–PLL-specific T cells; however, the inhibition was not reciprocal, and preincubation with either GL or DNP–PLL would not inhibit the presentation of ovalbumin to ovalbumin-specific T cells. The "binding" of GL to the APCs appeared to be very stable, since washing the APCs and incubation for 2 hours in the absence of GL did not restore their ability to present DNP–PLL. Other structurally related amino acid copolymers such as poly-L-lysine would show this inhibition, and it was shown not to be simply due to charge, since neither poly-D-lysine nor poly-L-arginine would inhibit the response.

This type of antigen competition was also shown for the T cell responses in mice by Rock and Benacerraf (1983), and these experiments showed even more specificity. The response of a T cell hybridoma specific for the amino acid ter-polymer GAT (poly-L-glutamic acid, poly-L-alanine, and poly-L-tyrosine) could be inhibited by the addition of the copolymer poly(GT) (glutamic acid and tyrosine), but not the addition of poly(GA) (glutamic acid and alanine) or poly(GL) (glutamic acid and lysine). poly(GT) would not inhibit the presentation of ovalbumin, KLH (keyhole limpet hemocyanin), or TNP (trinitrophenol) to the appropriate T cell hybridomas. All of these responses were specific for antigen in association with A^d, and thus, if the inhibition by poly(GT) was due to competition for binding sites on the A^d-encoded MHC molecule, the authors proposed that there must be specific binding sites for each antigen or antigen fragment. The antigen competition in these experiments seemed to indicate that the binding of antigen was in dynamic equilibrium. The APCs that were prepulsed with poly(GT) were more effective at presenting the GAT antigen than those that were simultaneously incubated with poly(GT) and GAT. In addition, the presentation by APCs prepulsed with GAT was still inhibited by the inclusion of poly(GT) in culture. These data will be considered below in light of the measured kinetic rates of association and dissociation.

Antigen competition was clearly shown in the response to the λ phage repressor protein (λ-r). Guillet et al. (1986) investigated three groups of T cell hybridomas that recognized λ-r 12–26 in association with A^d. One type of hybridoma, which responded to λ-r 12–24, could be inhibited by the addition of the nonstimulatory peptide λ-r 15–26. To test whether unrelated antigens could similarly compete for binding to A^d, 14 overlapping peptides, derived from the sequence of the antigen *Staphylococcus aureus* nuclease, were screened for the ability to inhibit a λ-r-specific response. One of the peptides, 61–80 (Nase I), was a potent inhibitor of all of the T cells specific for λ-r 12–26. In fact, Nase I comprises the immunodominant determinant of nuclease as recognized in association with A^d, a finding consistent with the notion that this peptide binds strongly to A^d.

These investigators examined the sequences of λ-r 12–24 and Nase I and found that nuclease 66–78 was remarkably similar to λ-r 12–24. All, but two of the amino acids, were conservative substitutions. When the nuclease peptide was synthesized incorporating the two amino acids from λ-r, the resulting peptide was stimulatory for λ-r-specific T cells. The peptide from ovalbumin that was recognized in association with A^d did not show any obvious sequence homology with either λ-r or Nase I, and yet, ovalbumin 323–339 competed effectively with the presentation of λ-r to all λ-r-specific T cells tested. Surprisingly, synthesis of an ovalbumin peptide with the two epitopic amino acids from λ-r appropriately placed resulted in a peptide that could stimulate at least one set of the T cell hybridomas specific for λ-r.

These experiments appeared to show that apparently unrelated peptides compete for the binding to a class II MHC molecule and that, once bound, present a epitope for T cells that can be mimicked, to an extent, by two amino acids. The competition was seen only for particular stimulating antigens and T cell hybridomas, and the authors speculated that this could be due to an influence of T cells on the stabilization of the antigen–MHC binding. In contrast to the experiments of Werdelin (1982) and Rock and Benacerraf (1983), there appeared to be no antigen specificity to the competition. However, Werdelin examined a heterogeneous population of T cells specific for

DNP–PLL, and thus, the competition could have been limited to antigens that bound to MHC molecules with exceptionally high affinity. Rock and Benacerraf (1983) examined clones of T cells specific for GAT, and no attempt was made to find a suboptimal stimulatory antigen, as shown to be required by Guillet et al. (1986).

Guillet et al. (1986) stated that the structure of the λ-r 12–26 was unlikely to form an amphipathic α-helix, as predicted by DeLisi and Berzofsky (1985); however, they attempted to expand on their homology data by aligning various A^d-restricted antigens and E^k-restricted antigens. Surprisingly, several pairs of antigenic peptides restricted to the same MHC molecule showed significant amino acid homology, although no attempt was made to deduce "universal" MHC-binding motifs.

The structural requirements for antigen competition was further exemplified in studies by Allen et al. (1987). These workers showed that T cell clones reactive with HEL and A^k recognized a 10-amino acid antigenic fragment, 52–61. The amino acids involved in antigenicity were determined by alanine substitutions at each position in synthetic oligopeptides. Three substitutions, Gly-54, Ile-55, and Asn-59, did not affect the antigenicity and were considered to be spacer positions; Ser-60 was also considered to be a spacer residue, in that it could be deleted without affecting antigenicity. The other six substitutions that abolished the antigenicity were considered to be important as TCR–MHC contact residues or important in the required, overall conformation. These six amino acids were then further catagorized as being part of the MHC-binding agretope, if the peptides substituted at these positions would not compete for the antigenic stimulation of the T cell clone by the native 52–61 sequence, and as part of the TCR epitope, if they would compete. The results of this analysis, combined with a molecular modeling of the 10-amino acid fragment, indicated that the MHC contact amino acids, Ala-52, Ile-58, and Arg-61, lined up on one side of an α-helix and that the TCR contact amino acids, Tyr-53, Leu-56, and Gln-57, lined up on the other. Thus, once again, the antigenic fragment may have the conformation of an α-helix, but in contrast to the results and implications of the work of DeLisi and Berzofsky, there was no evidence for an amphipathic α-helix. In fact, the two faces showed no particular tendency to either hydrophobicity or hydrophilicity, each consisted of a combination of hydrophobic amino acids, and to either polar or charged amino acids.

These functional experiments showed convincingly that the determinant recognized by the TCR on MHC class II-specific T helper cells consists of a short oligopeptide bound to the class II MHC molecule. The competition experiments showed that, at least for some oligopeptides, a

MHC molecule can bind only one fragment at a time, but each MHC molecule is, nonetheless, capable of binding a diverse group of peptide fragments.

C. Binding of Antigen and MHC Molecules in Solution

The functional experiments described above, especially those showing competition between unrelated antigens, indicated that antigen molecules bind to MHC molecules independently of the TCR. Physical evidence for this process was provided by Babbitt et al. (1985), who showed, by equilibrium dialysis, that purified A^k molecules, solubilized in detergent, could bind to antigenic peptides of HEL. The equilibrium constant for the interaction was calculated to be on the order of 10^{-6} M. Moreover, the HEL peptide would not bind detectably to A^d. Strikingly, this specificity corresponded with the Ir gene phenotype of mice bearing A^k and A^d; A^k-bearing mice are responders to insulin, whereas A^d-bearing mice are nonresponders. As an aside, it was fortuitous that the sensitivity of binding, as measured by equilibrium dialysis in detergents, corresponded closely with the sensitivity of the immune system for responsiveness to antigen.

It was also shown that certain nonstimulatory peptides would, nonetheless, inhibit the response to stimulatory peptides and that this characteristic corresponded well with the ability of these nonstimulatory peptides to bind to A^k. Thus, binding appeared to be required for antigenicity, and functional competition resulted from a competition for A^k binding.

Further experiments by Buus et al. (1986a) showed that the same type of correlations held for the antigen of chicken ovalbumin. The antigenic peptide 323–339 was shown to bind A^d, but not E^d, E^k, or A^k, exactly correlating with the specificity of T cells. Mice of the H-2^d MHC haplotype, immunized with ovalbumin, respond to this immunodominant peptide in association with A^d and not E^d, and H-2^k mice do not respond to this antigenic determinant.

A remarkable feature of this peptide–MHC interaction is the rate constants for the binding reaction. Buus et al. (1986b) showed that the on-rate for the reaction was slow, perhaps 100,000-fold slower that than of antigen–immunoglobulin (Ig) binding, but this slow on-rate was compensated by a slow off-rate, making for a very stable interaction. In fact, the binding was sufficiently stable to allow the separation of bound and free antigen by G-50 chromatography, a feature that may be important in the analysis of the structure of MHC molecules by crystallography and X-ray diffraction. In light of this stable binding, it is interesting to consider the competition studies of Rock and Benacerraf (1983). As described above, in these functional studies, it appeared as if the competing molecule could be washed off in a matter of hours and was most effective if added simultaneously with the stimulating molecule. These results could indicate that the rate constants for antigen–MHC binding are quite different under physiological conditions or that different antigens exhibit different binding kinetics. For example, highly charged, repeating peptides might bind differently to MHC molecules than oligopeptides from native sequences.

The caveats to the experiments that showed binding in solution were, first, that the binding reaction was carried out under conditions in which the MHC molecules presumably formed mixed micelles with detergent, and second, that the binding studies were carried out using only two MHC molecules, and thus, correspondence between *Ir* gene phenotype and binding could have been a coincidence.

In an elegant confirmation of the proposed correlation between antigen–MHC interactions and immune-response phenotype, Buus *et al.* (1987) extended these experiments to 12 different antigenic peptides and 4 different class II molecules. The binding was examined directly, as described above, and by an inhibition assay in which the 12 unlabeled peptides were used to inhibit the binding of a labeled peptide known to bind to a particular MHC molecule. The first result was that peptides that would bind to a particular MHC molecule would inhibit the binding of any other peptide to that MHC molecule. Thus, all peptides compete for the same MHC-binding site, or MHC molecules can bind only one peptide at a time. The second result was that binding (or competition) virtually always correlated with the known MHC specificity of the immune response. There were three cases in which this correlation failed. First, the response to HEL 74–86 is preferentially restricted to A^k, although the peptide bound weakly to A^k, E^k, A^d, and E^d with no particular preference. Second, the response to HEL 81–96 is restricted to E^k and Shastri *et al.* (1986a) have shown that there is no response to HEL 85–96 restricted to A^k, even in B10.A(4R) mice lacking the E^k molecule. In contrast, HEL 81–96 binds to A^k and effectively competes with labeled HEL 46–61 for binding to A^k. Third, and perhaps most striking, λ-r protein 12–26 bound significantly to A^d, but much stronger to E^d, and yet, the T cell response has been shown to be restricted to A^d and not E^d.

The correlation between binding and *Ir* gene phenotype was strong and established the point, but the lack of correlation in the response to λ-r 12–26 was shown to be the exception that proved the rule. Guillet *et al.* (1987), in analyzing the response to λ-r protein, found that the sequence of 12–26 bears a detectable homology to the polymorphic region of the $E^d \beta$ chain. They reasoned that the lack of an E^d-restricted response might be due to a clonal deletion of those T cells able to recognize E^d, and therefore, those T cells capable of recognizing the mimicking λ-r 12–26 bound to E^d. To test this notion, the 12–26 peptide sequence was modified to eliminate the E^d homology, but retained the ability to bind E^d. When H-2^d mice were immunized with the new peptide, the response was preferentially restricted to E^d over A^d. These experiments together showed that a response to antigen requires that the antigen bind to a MHC molecule, and this is usually sufficient to elicit a T cell response. The known remaining exceptions, namely, the responses to HEL 74–86 and HEL 81–96, may be based on tolerance or other, as yet unknown, influences on the immune repertoire.

From these observations, we can deduce that the specificity of the TCR is extensive enough to distinguish the difference between the set of self-peptides bound to the self-encoded MHC molecules and almost any new peptide that can bind to a self-encoded MHC molecule. The exceptions constitute the holes in the repertoire, and these holes would appear to be limited and due to selection influences, such as the maintenance of tolerance (or MHC-specific positive selection in the thymus), rather than intrinsic limitations on the diversity potential of the TCR. This does not mean that the diversity potential of the TCR is limitless, but at least the germ-line TCR potential is capable of recognizing a vast number of variations of the determinants created by peptides bound to MHC molecules.

D. Three-Dimensional Structure of a MHC Molecule

The experiments described in the preceding sections provided convincing evidence both functionally and biochemically that the ligand of the TCR is a peptide bound to either class I or class II MHC molecules. It is not presently known whether the determinant recognized by the TCR comprises the amino acids from the MHC molecule specifically modified in conformation by the binding of a peptide (altered self) (Zinkernagel and Doherty, 1977), the amino acids from the peptide in a conformation dictated by the MHC molecule to which it is bound (peptiditic self) (Kourilsky et al., 1987), or both. Although there is no reason to believe that the TCR would selectively recognize only a part of the antigen–MHC complex, without a description of the three-dimensional structure of the interacting molecules, it is difficult to distinguish between the possibilities. At least, the first part of such a structural determination was swift in arriving.

Bjorkman and colleagues (Bjorkman et al., 1987a) succeeded in producing crystals of papain-cleaved MHC class I molecules and determined the atomic coordinates of the α-carbon backbone and 80% of the side chains to a resolution of 3.5 Å. The structure of this human HLA-A2 molecule was simply startling.

The MHC class I molecule comprises a heavy chain encoded within the MHC complex, noncovalently associated with β_2-microglobulin (β2m) (Hood et al., 1983). The three-dimensional structure of the molecule consists of four globular domains. The two membrane proximal domains, encoded by the α3 exon of the heavy chain, and the entire β2m gene. These domains each assume a conformation similar to a prototypical immunoglobulin domain, although they associate with one another in an orientation not previously observed in other examples of the Ig superfamily (Bjorkman et al., 1987a).

Of more importance are the membrane distal domains that are encoded by the α1 and α2 heavy chains exons. The structure of these two domains consists of a platform of β-pleated sheets topped by two antiparallel α-helices. The net result is an elongated cleft, the sides formed by the α-helices, and the bottom formed by the β sheet structure. Such a structure would seem to be able to accommodate an oligopeptide, and interestingly, there was considerable electron density within the cleft that could not be attributed to the primary structure of the HLA-A2 sequence. Since the electron density in the cleft was virtually the same as the that of HLA polypeptide, it was speculated that almost all of the HLA molecules had bound another small molecule. Furthermore, since the density in this region was diffuse, it was further speculated that it was the result of a number of different molecules. This speculation was entirely consistent with the principles of processed peptides binding to MHC molecules; as shown for class II MHC molecules, the binding of a peptide to MHC molecule is sufficiently stable to allow peptides to copurify with the MHC molecules. The functional data of Townsend and others, combined with the structural data of Bjorkman et al., indicated that MHC class I molecules most likely bind peptides in a similar manner to that shown for MHC class II molecules. Surprisingly, no reports have thus far emerged showing that class I MHC molecules can bind to antigenic peptides. This is perhaps even more surprising, considering the fact that class I molecules are relatively easy to purify in the absence of detergents by papain cleavage. It remains to be seen if this is a matter of interest or if there is an intrinsic property of class I molecules that prevents soluble peptides from binding.

Whether the known specificity of a TCR is for an antigen bound to a MHC molecule or, as in an allogeneic response, to an apparent MHC allelic difference, the TCR recognizes almost every MHC polymorphism. For instance, a CTL clone that will recognize an influenza virus peptide bound to K^d will not recognize influenza-infected cells bearing any of the other several dozen class I alleles. Part of this MHC restriction could now be explained by the fact that the influenza peptide may bind only to a limited number of MHC class I molecules, but of those that it will bind, the TCR recognizes each of the allelic differences. These differences have been shown to be localized within the $\alpha 1$ and $\alpha 2$ domains of the MHC class I molecule. In actuality, the recognition of MHC differences could result from the direct interaction with the polymorphic MHC amino acids or from an indirect effect of MHC binding to the antigenic peptides. Since the peptides can presumably assume a number of different conformations, different conformations may bind to different MHC molecules. These differences might then be recognized by the TCR.

In a report directly following the description of the HLA-A2 structure, Bjorkman et al. (1987b) analyzed the structure of the class I molecule with respect to the known human and mouse sequence polymorphisms. These allelic differences were mainly located on the antiparallel α-helices, either on the sides of the cleft or on the top of the helices, distal to the cell membrane. It was speculated that the polymorphisms has two effects. Those on the sides of the cleft should affect the peptide–MHC interaction, and those at the top of the α-helices would affect TCR recognition. Thus, the species polymorphisms in MHC molecules are localized to specific regions of the membrane distal domains and, therefore, not simply the result of random genetic drift. Presumably, these changes have been evolutionarily selected from random mutations in the population. These polymorphisms thus may be influenced by two characteristics of the immune system: the need for MHC molecules to bind peptides and the need for this complex to be recognized by T cells. From these observations, one could deduce that the specificity of the TCR is for a complex determinant that includes amino acids from the antigen peptide and amino acids from the MHC molecule itself.

V. TCR Combining Site

A. Selection for T Cells Specific for Antigens Bound to MHC Molecules

From the information discussed in the previous sections, it appears that the specificity of the TCR is for peptides bound to MHC molecules, and the determinant may or may not include the structure of the MHC molecule itself. Nonetheless, the determinant recognized by the TCR is chemically uniform; in some cases, one antigen can be mimicked by as few as two amino acids substituted into another, unrelated MHC-binding peptide (Guillet et al., 1986). In order to understand the specificity of the TCR, it is important to first consider the potential mechanisms that operate to select for antigen- and MHC-reactive T cells.

One possible explanation for the absolute requirement for MHC restriction in antigen-specific responses is that activation of a T cell requires that its receptor interacts with MHC molecule. At least for activation of cultured T cell clones or mature T cells from lymph nodes, this appears not to be the case. Some T cell clones can be activated with soluble antibodies specific for the TCR (Kaye et al., 1983), and almost all others can be activated if the TCR-specific antibodies are coupled to Sepharose beads or bound to an APC (Marrack and Kappler, 1986). In other experiments, T cell clones can be activated by hybrid antibodies that have one specificity for the TCR and another for a cell surface molecule expressed on the surface of APCs (Staerz et al., 1986). This activation was shown to be independent of MHC recognition. Antibody-mediated activation is not limited to cultured T cell clones, since lymph node T cells can be activated by TCR-specific antibodies again coupled to Sepharose beads (Crispe et al., 1985). Thus, the MHC specificity of the TCR cannot be explained by a selective immune induction requiring MHC recognition. To phrase the question in practical terms, many viruses encode a membrane-bound protein that is expressed on the surface of virally infected cells; the obvious enigma is that there never exits T cells expressing a receptor with specificity for this intact cell surface protein. In every instance, the T cell has specificity for catabolic products of this or other intracellular viral proteins bound to MHC molecules.

Another possible explanation for MHC-restricted recognition is that T cells bearing MHC-specific receptors could be selectively expanded during development in the thymus, and there is a substantial amount of evidence to indicate that this type of organismal selection may occur (Sprent and Webb, 1987). An implication of these data is that T cells are expanded if they react with MHC molecules or MHC molecules that have bound self-encoded peptides. This simply places the enigma back to the level of thymus selection, as opposed to activation in the periphery, i.e., there is an activation signal rendered when the TCR interacts with a MHC molecule in the thymus, but not when it interacts with other cell surface molecules. However, another way of looking at the data would be to take the simplistic view that an interaction between the TCR and any cell surface molecule expressed on epithelial

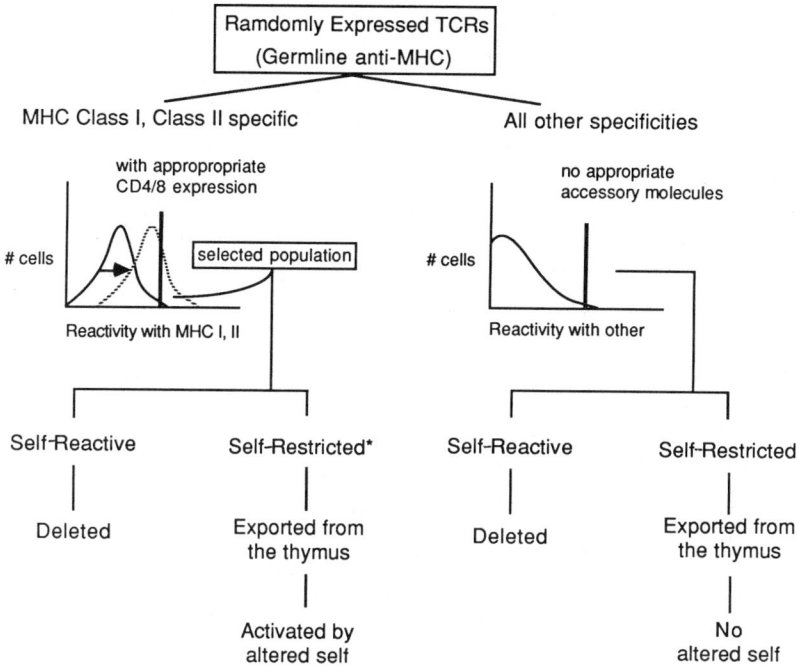

FIG. 2. Thymus selection. (*) Note reactive with self-encoded molecules unless modified by foreign antigens.

cells in the thymus could lead to the expansion of T cells and several characteristics of the immune system select for MHC-reactive T cells.

In Fig. 2, a schematic diagram representing several of the steps of T cell selection in the thymus is shown. T cells first undergo rearrangement and expression of their receptors during residence in the thymus (Raulet et al., 1985; Born et al., 1985; Samelson et al., 1985b; Trowbridge et al., 1985), and thus, selection based on the specificity of the receptor first begins there. Initially, the population expresses an assortment of TCRs formed from random combinations of α and β chain V–D–J regions. Evidence indicates that the conformation of TCR-combining sites in general is complementary to the structure of MHC class I and class II molecules, since random α and β chain combinations often have MHC reactivity (Blackman et al., 1986; Engel and Hedrick, 1988). Thus, the initial population of TCRs will have a higher average affinity for MHC molecules than for any other randomly chosen cell surface molecule. This is represented in Fig. 2 as a profile of affinities on the abscissa plotted against the number cells bearing TCRs of a given affinity on the ordinate. On the left is a hypothetical profile of cells with affinities for MHC molecules, and on the right is a similar profile of cells with affinities for any other (or all other) molecules expressed on the surface of thymic epithelial cells.

Along with the TCR, each mature T cell either expresses a CD4 or a CD8 accessory molecule (Sprent and Webb, 1987), and the net effect of this expression is to increase the avidity of the interaction if the TCR and the accessory molecule have the same specificity (Marrack et al., 1983; Greenstein et al., 1984). Thus, cells expressing a TCR specific for class II molecules will have a higher avidity for epithelial cells if the T cells also express CD4 molecules. Likewise, cells expressing a TCR specific for class I molecules will have a higher avidity for epithelial cells if they also express a CD8 molecule. The implication is that the TCR and the accessory molecule must cooperatively interact to achieve a productive increase in avidity. This is supported by recent evidence from von Boehmer and Steinmetz and their colleagues, indicating that T cells expressing a H-Y-specific, class I-restricted transgenic TCR are preferentially selected for the coexpression of CD8 molecules (H. von Boehmer, personal communication). The net effect of this avidity increase by the appropriate expression of accessory molecules can be viewed in the left-hand diagram in Fig. 2 as a shift in the reactivity profile to higher average affinities. Since presumably no accessory molecules exist that are specific for other cell surface molecules, no such shift is seen for TCRs with non-MHC specificities (right-hand diagram in Fig. 2).

If the selection of the population depends on the above-mentioned simplistic assumption, then an interaction between the expressed TCR and any epithelial cell surface molecule could potentially lead to thymocyte expansion. The criteria for expansion must be similar to many such types of receptor-mediated cellular activation, i.e., the probability of activation is proportional to the number of receptors occupied. This probability is, in turn, proportional to the affinity of the interaction and the concentration of the interacting receptors and ligands. Although the probability of activation is a continuum, it is easier to think of a defined cutoff, whereby all cells that can interact with epithelial cells above a certain avidity will be expanded, and those below this avidity will either be excluded or simply not expanded. In case of T cells specific for MHC moelcules, it is reasonable to picture a significant percentage that would be selected, whereas the percentage of T cells specific for other molecules might be relatively lower.

There are known to be two types of selections that occur in the thymus based on receptor specificity. Not assuming a known order, one selection is the deletion of high-affinity self-reactive T cells in order to avoid autoreactivity within the mature, secondary immune system. Kappler and Marrack and their colleagues showed that cells reactive with I-E-encoded class II molecules are deleted within the thymus in strains bearing I-E-encoded molecules, whereas these cells persist in the mature T cell population in strains that do not express I-E-encoded molecules (Kappler *et al.*, 1987). A second selection is for cells that can recognize the allelic form of self-MHC molecules when bound by foreign antigenic peptides. In MHC homozygous mice (MHC^a) irradiated and repopulated with MHC heterzygous bone marrow cells ($MHC^{a/b}$), the T cell population preferentially recognizes foreign antigens in association with the MHC allele expressed by the radioresistant thymus epithelium (MHC^a) (Bevan, 1977; Zinkernagel, 1978; Sprent, 1978). Thus, despite the fact that the APCs express both MHC^a and MHC^b, the T cell population reacts with antigen associated with MHC^a. The simplest conclusion that can be drawn from these experiments is that there is a selection within the thymus for T cells that are not overtly self-reactive to the point of causing autoreactivity, yet are self-MHC restricted prior to encountering foreign antigen. This concept is one of the most difficult to understand in the study of the immune response. Interpretations other than positive selection in the thymus such as the presence of suppression (Matzinger, 1981) have not been supported by experiments (Zinkernagel, 1978; Fink and Bevan, 1981). However, it is important to note that the concept of positive selection originates entirely from experiments with F1 → parent radiation chimeras, and there is sufficient complexity in this model system to warrant a further consideration of other explanations.

A final selection that occurs prior to experimental analysis of T cell specificities is that T cells are induced and expanded in the secondary lymphoid organs as a result of specific immunizations. This requires that an antigen or a fragment of antigen be capable of binding to the molecule to which the T cells are restricted in order to generate an altered self. In the case in which T cells are specific for MHC class I and class II molecules, this apparently occurs quite readily. MHC molecules bind to a large diversity of oligopeptides as described above, and in addition, there may be specific cellular compartmentalization that encourages antigen–MHC interactions (Germain, 1986). In contrast, no known ability to stably bind foreign antigens exists for non-MHC molecules, and so even if there exist T cells restricted to other cellular membrane proteins, these T cells would not normally be induced and expanded. In summary, T cells specific for foreign antigens are always

MHC restricted, but this property may result from a combination of at least four different influences: evolutionary selection for TCRs that are complementary to the MHC structure, expansion of thymic T cells based on a TCR-mediated interaction with thymic epithelial cells, coexpression of accessory molecules that increase the avidity of an interaction with MHC molecules, and the properties of MHC molecules that allow them to bind antigens.

B. STRUCTURE OF THE TCR

The TCR was first identified using monoclonal antibodies specific for clonotypic molecules, that is, molecules bearing determinants unique to a particular T cell clone. By immunoprecipitation and gel electrophoresis under reducing and nonreducing conditions, the determinants were shown to be present on an 85,000-MW disulfide-bonded heterodimer (Allison *et al.*, 1982; Meuer *et al.*, 1982; Haskins *et al.*, 1983; Bigler *et al.*, 1983; Kaye *et al.*, 1983; Samelson *et al.*, 1983). Two-dimensional electrophoresis identified a more acidic polypeptide chain, designated the α chain, and a more basic polypeptide chain, designated the β chain (Acuto *et al.*, 1983).

The complete primary structure of the β chain was determined from the cDNA sequences reported for the human and the mouse homologs. In the case of the human β chain clones, a cDNA library made from T cell mRNA was differentially screened with labeled cDNA from a T cell lymphoma and a B cell lymphoma. Clones that screened positive with the T cell, but not the B cell, cDNA were sequenced, and one was found to encode a protein with significant homology to immunoglobulin molecules (Yanagi *et al.*, 1984). The mouse β chain was cloned by making a subtracted cDNA probe using membrane-associated mRNA from a T cell hybridoma and by hybridizing the labeled cDNA with excess mRNA from a B cell lymphoma. The probe, enriched for T cell-specific membrane or secreted sequences, was used to screen a cDNA library made from T cell-specific cDNA (T cell minus B cell). The clones were further screened for T cell-specific clones that hybridized with genes that rearranged in T cells, and one of the clones showed this characteristic (Hedrick *et al.*, 1984a). Sequencing of this and other clones isolated by cross-hybridization showed that the clones encoded a protein with sequence homology with immunoglobulin genes including the presence of variable, joining, and constant regions (Hedrick *et al.*, 1984b). This latter technique was also used to isolate the α chain (Saito *et al.*, 1984; Chien *et al.*, 1984).

From the amino acid sequence, the overall structure of the α–β chain heterodimer is predicted to be similar to an immunoglobulin Fab fragment with both chains anchored in the membrane. The membrane proximal constant domains are homologous to immunoglobulin constant regions in terms of size, sequence conservations, and the presence of cysteines spaced to give an intrachain disulfide bond. Immunoglobulin domains share a folding scheme described as the Ig fold (Lesk and Chothia, 1980), and Novotny et al. (1986) and Beale and Coadwell (1986) have aligned TCR sequences with typical Ig sequences in order to determine whether there is a significant conservation of amino acids at strategic positions. The amino acids that are most highly conserved within the Ig family are those positions important for the structure of the domain framework and those important for the interdomain packing. The comparison studies indicate that there is a high probability that TCR chains conform to the Ig fold.

The constant region domains consist of two antiparallel β sheets comprising seven β strands, and each of the β strands have conserved amino acids at positions where side chains point inside the sheet. The TCR α and β chain sequences can easily be aligned to correspond to the seven β strands, and many of the positions known to be important for the structure of Ig molecules are conserved in TCR α and β chain constant region domains. The C_α domain is somewhat unusual among Ig-like domains in that it is very compact in terms of the length of the β strands and the lengths of the loops connecting the strands. Whereas typical Ig constant domains range between 55 and 60 amino acids separating the two cysteines forming the intrachain disulfide bond, the α chain has only 50 amino acids. The β chain has 61 amino acids separating the cysteines in mice and 65 amino acids in humans. Nonetheless, secondary structure studies are completely consistent with the constant regions of the α and β chains conforming to the Ig fold.

The variable region sequences of the TCR α and β chains are similar enough to Ig variable regions that the identities can be seen by inspection. At least 7–12 positions, important for the framework structure of Ig V regions, and 5–6 positions, important for V_H–V_L interdomain interaction, are conserved in TCR sequences (Novotny et al., 1986; Beale and Coadwell, 1986; Kronenberg et al., 1986). In a further analysis, Chothia (1988) has analyzed 37 positions, whose residues are involved in the structure of the domain β sheets, conformation of conserved loops, and packing of the V_L–V_H domains. With few exceptions, the TCR chains have appropriately conserved amino acids at each of these positions; the exceptions are not predicted to cause significant changes in the domain structure. The conclusion is that the structure of the TCR is likely to be quite similar to that of Ig molecules, and thus, the complementarity determining regions (CDRs), forming the combining site of the TCR, may have the same spacial orientation as that of the typical immunoglobulin.

One interesting difference between TCR sequences and Ig sequences is in the lengths of the putative CDRs. Whereas these regions can vary widely in length for Ig molecules (Kabat, et al. 1987). TCR chains show a significantly smaller variation, especially in the third CDRs (G. Siu et al., 1988). The effect of especially long CDRs in immunoglobulins is to form a deep combining site or to shield shorter CDRs from interacting with ligands bound to the combining site. TCRs would thus be predicted to have a more uniform combining site shape, and each of the six CDRs would be predicted to participate in its formation. The recently solved structure of an Ig molecule, D1.3, specific for the lysozyme protein, has a particularly flat combining site structure to accommodate a face of the native lysozyme protein (Amit et al., 1986), and it is interesting to speculate that such a structure would be a good model for TCR-MHC interactions.

With these data, Davis and Bjorkman (1988) have proposed a model for the interaction between the TCR and antigen bound to MHC molecules. The combining site of D1.3 looks roughly as if the first and second CDRs of the heavy and light chains define the outer boundary and as if the third CDRs lie in the middle. If this structure is simplistically docked onto the MHC class I structure in one orientation, the first and second CDRs would contact the antiparallel α-helices of the MHC molecule and the third CDRs would contact the bound peptide. These authors make the interesting speculation that the diversity of the third CDRs in TCR chains is important for the peptide specificity, and the first and second CDRs might be more important in interactions with the MHC molecule itself. Furthermore, since the sequence of the third CDR is determined by processes involved in gene rearrangements, then the need for peptide specificity was the original evolutionary pressure that selected for rearranging genes in the immune system.

In a preliminary confirmation of this model, Hedrick et al. (1988) have shown that there is a strict conservation of an amino acid positioned in the middle of the third CDR in the β chain of TCRs specific for pigeon cytochrome c. Furthermore, site-directed mutagenesis of this residue changes the specificity for cytochrome peptides without affecting the MHC specificity for $E^{k,s,b}$ (Engel and Hedrick, 1988). Further, such experiments will be needed to confirm that such an interaction is a general property of TCR-MHC interactions.

VI. Strategies of Immune Recognition

There is great difficulty in reconstructing the stepwise selection of the immune system through evolution. However, there are now several concepts of immune recognition that are now more clearly understood, and these concepts point to at least some of the fundamental strategies developed by vertebrates in order to accomplish the complex process of self-versus nonself-discrimination.

An implication of the work described in this review is that immune recognition by T lymphocytes is fundamentally different than that of B lymphocytes. B lymphocytes recognize antigen determinants via surface immunoglobulin, and the specificity of monoclonal antibodies shows that B cells can recognize a wide diversity of chemical determinants. Thus, self-versus nonself-discrimination for B cells means that the immune system must delete (or suppress) B cells that functionally bind any self-determinant, while retaining those that recognize the rest of the universe of foreign antigens. Since binding by antibodies can range over many orders of magnitude, one of the key elements of this process is the concept of functional recognition. What is the minimum antibody affinity that constitutes a deleterious self-reactivity, and what is the minimum affinity that is still functionally effective in dispatching a foreign antigen? T lymphocytes recognize antigen determinants via the TCR, and the specificity is for a processed form of antigen bound to MHC molecules present on a second cell type. As a result, self-versus nonself-discrimination for T cells is entirely different from that of B cells. T cells must be deleted if they have a specificity for any of the processed self-peptides that bind to self-MHC molecules, and the population must retain the ability to recognize a variety of foreign peptides bound to self-MHC molecules. The TCR repertoire must discriminate between a relatively homogeneous group of determinants, a group that is but a small fraction of the biological determinants recognized by B cells. Functional recognition by T cells almost certainly spans a different range than functional recognition by soluble antibodies or even antibodies in surface receptor form. Although there is no estimate of the actual affinity of interaction between the TCR and its antigen–MHC ligand, the relative sensitivity of T cell clones from animals immunized with any particular antigenic fragment varies only from one to two orders of magnitude. This probably reflects the constraints on the cell–cell interactions necessary for T cell induction, e.g., a constant number of receptors per T cell, a relatively constant number of MHC molecules per APC, and a predetermined affinity of interaction between a particular antigen fragment and MHC molecules. In addition, it reflects the lack of affinity-selected somatic mutations in TCRs. The important point is that there is a limited diversity of determinants that are recognized by T cells and a limited set of conditions under which T cells become activated.

An attractive hypothesis for the origin of immune recognition is that it arose from a more primitive form of recognition in invertebrates evolved to encourage heterozygosity and to discourage fusion between separate organisms (Valembois, 1974; Lubbock, 1980; Scofield et al., 1982). Such a recognition would involve a receptor of limited diversity and a self-encoded determinant that marked the difference between self and other members of the species. The fact that the immune system may have usurped this recognition system is, perhaps, just a reflection of the opportunistic nature of evolution. However, the interesting facet that we can observe directly is that one part of the immune system seems to have retained this property of MHC recognition, whereas another directly evolved system seems to have developed toward a much more diverse recognition system. MHC recognition is, therefore, fundamentally important for the control or effectiveness of T cell responses, whereas MHC recognition must be relatively unimportant for B cell responses and antibody secretion. Hence, the two types of immune recognition have evolved under the influence of different selective pressures.

One difference between T and B cell recognition is that T cells are effective in dealing with intracellular parasites and in initiating several types of immune responses via cell–cell interactions, whereas antibodies are more effective in dealing with extracellular pathogens. A less-obvious difference may be in the requirements for the two recognition systems to maintain self-tolerance. The T cell population responds very poorly to self-antigen determinants, and yet, clearly the mature repertoire of T cells includes cells that are potentially self-reactive. For example, it is easy to stimulate an anti-self-MHC response in tissue culture (automixed lymphocyte reaction) (Howe, 1973; von Boehmer and Adams, 1973; Ponzio et al., 1975; Glimcher et al., 1981), and yet, *in vivo* conditions do not promote any manifestations of this response. Another example is that mentioned earlier; certain strains of mice respond to immunizations with beef cytochrome c by making a strong response specific for an autologous cytochrome c determinant. Although the experiment was not carried out for long periods of time, a prediction is that this autoreactivity would not be associated with a noticeable pathology. Cytochrome c might never gain access to the pathway of class II MHC presentation, and thus, the animal is neither tolerant to it nor susceptible to effects of cytochrome c-specific T cells. In general, T cell tolerance to self-components may require deletion of a limited diversity of cells, whereas in principle, B cell tolerance requires the deletion of all of the B cells reactive with the enormous variation of self-encoded proteins, carbohydrates, nucleic acids, etc. Since T cells must help to initiate virtually all specific immune responses, a speculation is that this level of tolerance would suffice to prevent virtually all self-reactive antibody responses, as well as self-reactive T cell responses. The specificity of the TCR may thus be constrained by function to be directed at cell surface determinants, but in addition, the specificity may be focused on the recognition of MHC molecules in order to facilitate the maintenance of self-tolerance.

Acknowledgments

This work was supported by USPHS grants AI21372 and GM35880, by NSF DCB-8452023, and by Research Career Development Award AI00662.

References

Acuto, O., Hussey, R. E., Fitzgerald, K. A., Protentis, J. P., Meuer, S. C., Schlossman, S. F., and Reingerz, E. L. (1983). The human T cell receptor: Appearance in ontogeny and biochemical relationship of alpha and beta subunits on IL-2 dependent clones and T cell tumors. *Cell* **34,** 717.

Allen, P. M., Matsueda, G. R., Evans, R. J., Dunbar, J. B., Jr., Marshall, G. R., and Unanue, E. R. (1987). Identification of the T-cell and Ia contact residues of a T-cell antigenic epitope. *Nature (London)* **327,** 713.

Allison, J. P., and Lanier, L. L, (1987). Structure, function, and serology of the T-cell antigen receptor complex. *Annu. Rev. Immunol.* **5,** 503.

Allison, J. P., McIntyre, B. W., and Bloch, D. (1982). Tumor-specific antigen of murine T lymphoma defined with monoclonal antibody. *J. Immunol.* **129,** 2293.

Amit, A. G., Mariuzza, R. A., Phillips, E. V., and Pojak, R. (1986). Three-dimensional structure of an antigen–antibody complex at 2.8 Å resolution. *Science* **233**.

Ashwell, J. D., and Schwartz, R. H. (1986). T-cell recognition of antigen and the Ia molecule as a ternary complex. *Nature (London)* **320,** 176.

Ashwell, J. D., Fox, B. S., and Schwartz, R. H. (1986). Functional analysis of the interaction of the antigen-specific T cell receptor with its ligands. *J. Immunol.* **136,** 757.

Bank, I., DePinho, R. A., Brenner, M., Cassimeris, J., Alt, F. W., and Chess, L. (1986). A functional T3 molecule associated with a novel heterodimer on the surface of immature human thymocytes. *Nature (London)* **322,** 179.

Babbitt, B. P., Allen, P. M., Matsueda, G., Haber, E., and Unanue, E. R. (1985). Binding of immunogenic peptides to Ia histocompatibility molecules. *Nature (London)* **317,** 359.

Barcinski, M. A., and Rosenthal, A. S. (1977). Immune response gene control of determinant selection. I. Intramolecular mapping of the immunogenic sites on insulin recognized by guinea pig T and B cells. *J. Exp. Med.* **145,** 726.

Beale, D., and Coadwell, J. (1986). Unusual features of the T-cell receptor C domains are revealed by structural comparisons with other members of the immunoglobulin superfamily. *Biochem. Physiol.* **85,** 205.

Benacerraf, B. (1978). A hypothesis to relate the specificity of T lymphocytes and the activity of I region-specific Ir genes in macrophages and B lymphocytes. *J. Immunol.* **120,** 1809.

Berkower, I., Buckenmeyer, G. K., and Berzofsky, J. A. (1986). Molecular mapping of a histocompatibility-restricted immunodominant T cell epitope with synthetic and natural peptides: Implications for T cell antigenic structure. *J. Immunol.* **136,** 2498.

Berzofsky, J. A. (1980). Immune response genes in the regulation of mammalian immunity. *In* "Biological Regulation and Development" (R. F. Goldberger, ed.), Vol. 2, p. 467. Plenum, New York.

Berzofsky, J. A., Richman, L. K., and Killion, D. J. (1979). Distinct H-2 linked Ir genes control both antibody and T cell responses to different determinants on the same antigen, myoglobin. *Proc. Natl. Acad. Sci. U.S.A.* **76,** 4046.

Bevan, M. J. (1975). Interaction antigens detected by cytotoxic T cells with the major histocompatibility complex as modifier. *Nature (London)* **256,** 419.

Bevan, M. J. (1977). In a radiation chimaera, host H-2 antigens determine immune responsiveness of donor cytoxic cells. *Nature (London)* **269,** 417.

Bigler, R. D., Fisher, D. E., Wang, C. Y., Tinnoykan, E. A., and Kunkel, H. G. (1983). Idiotype like molecules on cells of a human T cell leukemia. *J. Exp. Med.* **158,** 1000.

Bjorkman, P. J., Saper, M. A., Samraoui, B., Bennett, W. S., Strominger, J. L., and Wiley, D. C. (1987a). Structure of the human class I histocompatibility antigen, HLA-A2. *Nature (London)* **329,** 506.

Bjorkman, P. J., Saper, M. A., Samraoui, B., Bennett, W. S., Strominger, J. L., and Wiley, D. C. (1987b). The foreign antigen binding site and T cell recognition regions of class I histocompatibility antigens. *Nature (London)* **329,** 512.

Blackman, M., Yague, J., Kubo, R., Gay, D., Coleclough, C., Palmer, E., Kappler, J., and Marrack, P. (1986). The T cell repertoire may be biased in favor of MHC recognition. *Cell* **47,** 349.

Born, W., Yague, J., Palmer, E., Kappler, J., and Marrack, P. (1985). Rearrangement of T-cell receptor β-chain genes during T-cell development. *Proc. Natl. Acad. Sci. U.S.A.* **82,** 2925.

Borst, J., Alexander, S., Elder, J., and Terhorst, C. (1983). The T3 complex on human T lymphocytes involves four structurally distinct glycoproteins. *J. Biol. Chem.* **258,** 5135.

Brenner, M. B., Mclean, J., Dialynas, D. P., Strominger, J. L., Smith, J. A., Owen, F. L., Seidman, J. G., Ip, S., Rosen, F., and Krangel, M. S. (1986). Identification of a putative second T-cell receptor. *Nature (London)* **322,** 145.

Brown, R. K., Delaney, R., Levine, L., and van Vunakis, H. (1959). Studies on the antigenic structure of ribonuclease. *J. Biol. Chem.* **234,** 2043.

Buus, S., Colon, S., Smith, C., Freed, J. H., Miles, C., and Grey, H. M. (1986a). Interaction between a "processed" ovalbumin peptide and Ia molecules. *Proc. Natl. Acad. Sci. U.S.A.* **83,** 4509.

Buus, S., Alessandro, S., Colon, S. M., Jenis, D. M., and Grey, H. M. (1986b). Isolation and characterization of antigen–Ia complexes involved in T cell recognition. *Cell* **47,** 1071.

Buus, S., Sette, A., Colon, S. M., Miles, C., and Grey, H. M. (1987). The relation between major histocompatibility complex (MHC) restriction and the capacity of Ia to bind immunogenic peptides. *Science* **235,** 1353.

Carbone, F. R., Fox, B. S., Schwartz, R. H., and Paterson, Y. (1987). The use of hydrophobic, α-helix defined peptides in delineating the T cell determinant for pigeon cytochrome C. *J. Immunol.* **138,** 1838.

Cease, K. B., Berkower, I., York-Jolley, J., and Berzofsky, J. A. (1986). T cell clones specific for an amphipathic alpha helical region of sperm whale myoglobin show differing fine specificities for synthetic peptides: A multiview/single structure interpretation of immunodominance. *J. Exp. Med.* **164**, 1779.

Cerottini, J. C., and Brunner, K. T. (1974). Cell-mediated cytotoxicity allograft rejection and tumor immunity. *Adv. Immunol.* **19**, 67.

Chestnut, R. W., Colon, S., and Grey, H. M. (1982a). Requirements for the processing of antigen by antigen-presenting B cells. I. Functional comparison of B cell tumors and macrophages. *J. Immunol.* **129**, 2382.

Chestnut, R. W., Colon, S., and Grey, H. M. (1982b). Antigen presentation by normal B cells, B cell tumors, and macrophages: Functional and biochemical comparison. *J. Immunol.* **128**, 1764.

Chien, Y., Becker, D., Lindsten, T., Okamura, M., Cohen, D., and Davis, M. (1984). A third type of murine T-cell receptor gene. *Nature (London)* **312**, 31.

Chothia, C. (1988). A structural model for the T-cell receptor. In preparation.

Cohn, M., and Epstein, R. (1978). T-cell inhibition of humoral responsiveness. II. Theory on the role of restrictive recognition in immune regulation. *Cell. Immunol.* **39**, 125.

Corradin, G., and Chiller, J. M. (1979). Lymphocytes specificity to protein antigens. II. Fine specificity of T-cell activation with cytochrome c and derived peptides as antigenic probes. *J. Exp. Med.* **149**, 436.

Crispe, I. N., Bevan, M. J., and Stearz, U. D. (1985). Selective activation of Ly+2+ precursor T cells by ligation of antigen receptors. *Nature (London)* **317**, 627.

Crumpton, M. J., and Sonall, P. A., Jr. (1967). Conformation of immunologically active fragments of sperm whale myoglobin in aqueous solution. *J. Mol. Biol.* **26**, 143.

Davis, M., and Bjorkman, M. (1988). T cell receptor genes and T cell recognition. In preparation.

DeLisi, C., and Berzofsky, J. A. (1985). T-cell antigenic sites tend to be amphipathic structures. *Proc. Natl. Acad. Sci. U.S.A.* **82**, 7048.

Dickerson, R. E. (1980). Cytochrome c and the evolution of energy metabolism. *Sci. Am.* **242**, 136.

Dorf, M. E., and Benacerraf, B. (1975). Complementation of H-2 linked Ir genes in the mouse. *Proc. Natl. Acad. Sci. U.S.A.* **72**, 3671.

Engel, I., and Hedrick, S. M. (1988). Peptide specificity of the T cell receptor is influenced by the third complementary determining region. *Cell.* In press.

Fink, P. J., and Bevan, M. J. (1981). Studies on the mechanism of the self restriction of T cell responses in radiation chimeras. *J. Immunol.* **127**, 694.

Freedman, M. H., and Sela, M. (1966). Recovery of antigenic activity upon reoxidation of completely reduced polyalanylated rabbit immunoglobulin G. *J. Biol. Chem.* **241**, 2383.

Gell, P. G., and Benacerraf, B. (1959). Studies on hypersensitivity. II. Delayed hypersensitivity to denatured proteins in guinea pigs. *Immunology* **2**, 64.

Germain, R. N. (1986). The ins and outs of antigen processing and presentation. *Nature (London)* **322**, 687.

Gerwing, J., and Thompson, K. (1968). Studies on antigenic properties of eggwhite lysozyme. I. Isolation and characterization of a tryptic peptide from reduced and alkylated lysozyme exhibiting haptenic activity. *Biochemistry* **7**, 3888.

Glasebrook, A. L., and Fitch, F. W. (1980). Alloreactive cloned T cell lines. I. Interactions between cloned amplifier and cytolytic T cell lines. *J. Exp. Med.* **151**, 876.

Glimcher, L. H., Longo, D. L., Green, I., and Schwartz, R. H. (1981). Murine syngeneic mixed lymphocyte response. *J. Exp. Med.* **154**, 1652.

Gordon, R. D., Simpson, E., and Samelson, L. E. (1975). In vitro cell-mediated immune responses to the male specific (H-Y) antigen in mice. *J. Exp. Med.* **142,** 1108.

Greenstein, J. L., Kappler, J., Marrack, P., and Burakoff, S. J. (1984). The role of L3T4 in recognition of Ia by a cytotoxic, H-2D^d-specific T cell hybridoma. *J. Exp. Med.* **159,** 1213.

Grey, H. M., Colon, S., and Chestnut, R. S. (1982). Requirements for the processing of antigen by antigen-presenting B cells. II. Biochemical comparison of the fate of antigen in B cell tumors and macrophages. *J. Immunol.* **129,** 2389.

Guillet, J. G., Lai, M. Z., Briner, T. J., Smith, J. A., and Gefter, M. L. (1986). Interaction of peptide antigens and class II major histocompatibility complex antigens. *Nature (London)* **324,** 20.

Guillet, J. G., Lai, M. Z., Briner, T. J., Buus, S., Sette, A., Grey, H. M., Smith, J. A., and Gefter, M. L. (1987). Immunological self, nonself discrimination. *Science* **235,** 865.

Hansburg, D., and Appella, E. (1985). The sites of antigen–T cell and antigen–MHC interactions overlap. *J. Immunol.* **135,** 3712.

Hansburg, D., Hannum, C., Inman, K., Appella, E., Margoliash, E., and Schwartz, R. H. (1981). Parallel cross-reactivity patterns of 2 sets of antigenically distinct cytochrome c peptides: Possible evidence for a presentational model of Ir gene function. *J. Immunol.* **127,** 1844.

Hansburg, D., Farirwell, T., Schwartz, H., and Appella, E. (1983). The T lymphocyte response to cytochrome c. VIV. Distinguishable sites on a peptide antigen which affect antigenic strength and memory. *J. Immunol.* **131,** 319.

Haskins, K., Kubo, R., White, J., Pigeon, M., Kappler, J., and Marrack, P. (1983). The major histocompatibility complex-restricted antigen receptor on T cells. I. Isolation with a monoclonal antibody. *J. Exp. Med.* **157,** 1149.

Heber-Katz, E., and Wilson, D. B. (1975). Collaboration of allogeneic T and B lymphocytes in the primary antibody response to sheep erythrocytes *in vitro. J. Exp. Med.* **142,** 928.

Heber-Katz, E., Schwartz, R. H., Matis, L. A., Hannum, C., Fairwell, T., Appella, E., and Hansburg, D. (1982). Contribution of antigen-presenting cell major histocompatibility complex gene products to the specificity of antigen-induced T cell activation. *J. Exp. Med.* **155,** 1086.

Heber-Katz, E., Hansburg, D., and Schwartz, H. (1983). The Ia molecule of the antigen-presenting cell plays a critical role in immune response gene regulation of T cell activation. *J. Mol. Cell. Immunol.* **1,** 3.

Hedrick, S. M., Matis, L. A., Hecht, T. T., Samelson, L. E., Longo, D. E., Heber-Katz, E., and Schwartz, R. H. (1982). The fine specificity of antigen and Ia determinant recognition by T cell hybridoma clones specific for pigeon cytochrome c. *Cell* **30,** 141.

Hedrick, S. M., Cohen, D. I., Nielsen, E. A., and Davis, M. M. (1984a). Isolation of a cDNA clone encoding T cell-specific membrane-associated proteins. *Nature (London)* **308,** 149.

Hedrick, S. M., Nielsen, E. A., Kavaler, J., Cohen, D. I., and Davis, M. M. (1984b). Sequence relationships between putative T-cell receptor polypeptides and immunoglobulins. *Nature (London)* **308,** 153.

Hedrick, S. M., Ashwell, J. D., and Matis, L. A. (1985). T cell activations determined by receptor occupancy: Implications for selection of the repertoire and activation in immune responses. *In* "Recognition and Regulation in Cell-Mediated Immunity" (J. Marbrook and J. D. Watson, eds.), p. 149. Dekker, New York.

Hedrick, S. M., Engel, I., McElligott, D. L., Fink, P. J., Hsu, M. L., Hansburg, D., and Matis, L. A. (1988). Selection of amino acid sequences in the beta chain of the T cell antigen receptor. *Science* **239,** 1541.

Hood, L., Steinmetz, M., and Malissen, B. (1983). Genes of the major histocompatibility complex of the mouse. *Annu. Rev. Immunol.* **1,** 529.

Howe, M. L. (1973). Isogenic lymphocyte interaction: Responsiveness of murine thymocytes to self antigens. *J. Immunol.* **100,** 1090.

Hünig, T. R., and Bevan, M. J. (1982). Antigen recognition by cloned cytotoxic T lymphocytes follows rules predicted by the altered-self-hypothesis. *J. Exp. Med.* **155,** 111.

Imanishi, T., and Mäkelä, O. (1973). Strain differences in the fine specificity of mouse antihapten antibodies. *Eur. J. Immunol.* **3,** 323.

Infante, A. J., Atassi, M. Z., and Fathman, C. G. (1981). T cell clones reactive with sperm whale myoglobin. Isolation of clones with specificity for individual determinants on myoglobin. *J. Exp. Med.* **154,** 1342.

Ishizaka, K., Okudaira, H., and King, T. (1975). Immunogenic properties of modified antigen E. II. Ability of urea-denatured antigen and E-polypeptide chain to prime T cells specific for antigen E. *J. Immunol.* **114,** 110.

Kabat, E. A., Wu, T. T., Reid-Miller, M., Perry, H. M., and Gottesman, K. S. (1987). Sequences of proteins of immunological interests, 4th ed. U.S. Dept. of Health and Human Services. NIH Bethesda, MD.

Kappler, J. W., Skidmore, B., White, J., and Marrack, P. (1981). Antigen-inducible, H-2 restricted, interleukin-2-producing T-cell hybidoma: Lack of independent antigen and H-2 recognition. *J. Exp. Med.* **153,** 1198.

Kappler, J. W., Roehm, N., and Marrack, P. (1987). T cell tolerance by clonal elimination in the thymus. *Cell* **49,** 273.

Katz, D. H., Hamaoka, T., and Benacerraf, B. (1973). Cell interactions between histoincompatible T and B lymphocytes. Failure of physiologic cooperative interactions between T and B lymphocytes from allogeneic donor strains in humoral response to hapten–protein conjugates. *J. Exp. Med.* **137,** 1405.

Katz, D. H., Graves, M., Dorf, M. E., Dimuzio, H., and Benacerraf, B. (1975). Cell interactions between histoincompatible T and B lymphocytes. VII. Cooperative responses between lymphocytes are controlled by genes in the I region of the H-2 complex. *J. Exp. Med.* **141,** 263.

Katz, M. E., Maizels, R. M., Wicker, L., Miller, A., and Sercarz, E. E. (1982). Immunological focusing by the mouse major histocompatibility complex: Mouse stains confronted with distantly related lysozymes confine their attention to very few epitopes. *Eur. J. Immunol.* **12,** 535.

Kaye, J., Procelli, S., Tite, J., Jones, B., and Janeway, C. A., Jr. (1983). Both a monoclonal antibody and antisera specific for determinants unique to individual cloned helper T cell lines can substitute for antigen and antigen-presenting cells in the activation of T cells. *J. Exp. Med.* **158,** 836.

Kimoto, M., and Fathman, C. G. (1980). Antigen reactive T cell clones. II. Transcomplementing hybrid I-A region gene products function effectively in antigen presentation. *J. Exp. Med.* **152,** 759.

Kindred, B. (1975). Can tolerant alogeneic cells restore nude mice? *Cell Immunol* **20,** 241.

Kourilsky, P. Chaouat, G., Rabourdin-Combe, C., and Claverie, J.-M. (1987). Working principles in the immune system implied by the "peptic self" model. *Proc. Natl. Acad. Sci. U.S.A.* **84,** 3400.

Kronenberg, M., Siu, G., Hood, L. E., and Shastri, N. (1986). The molecular genetics of the T-cell antigen receptor and T-cell antigen recognition. *Annu. Rev. Immunol.* **4,** 529.

Langman, R. E. (1978). Cell-mediated immunity and the major histocompatibility complex. *Rev. Physiol. Biochem. Pharmacol.* **81,** 1.

Lanier, L., Ruitenberg, J. J., and Phillips, J. H. (1986). Human CD3+ T lymphocytes that express neither CD4 nor CD8 antigens. *J. Exp. Med.* **164,** 339.

Lee, K. C., Wong, M., and Spitzer, D. (1982). Chloroquine as a probe of antigen processing by accessory cells. *Transplanation* **34,** 150.

Lesk, A. M., and Chothia, C. (1980). How different amino acid sequences determine similar protein structures: The structure and evolutionary dynamics of the globins. *J. Mol. Biol.* **136,** 225.

Livingstone, A. M., and Fathman, C. G. (1987). The structure of T-cell epitopes. *Annu. Rev. Immunol.* **5,** 477.

Longo, D. L., and Schwartz, R. H. (1980). T-cell specificity for H-2 and Ir gene phenotype correlates with the phenotype of thymic antigen-presenting cells. *Nature (London)* **287,** 44.

Lubbock, R. (1980). Clone-specific cellular recognition in a sea anemone. *Proc. Natl. Acad. Sci. U.S.A.* **77,** 6667.

Maizels, R. M., Clarke, J. A., Harvey, M. A., Miller, A., and Sercarz, E. E. (1980). Epitope specificity of the T-cell proliferative response to lysozyme: Proliferative T-cells react predominantly to different determinants from those recognized by B-cells. *Eur. J. Immunol.* **10,** 509.

Marrack, P., and Kappler, J. (1986). The antigen-specific, major histocompatibility complex-restricted receptor on T cells. *Adv. Immunol.* **38,** 1.

Marrack, P., Endress, R., Schimonkevitz, R., Zlotnik, A., Dialynas, D., Fitch, F., and Kappler, J. (1983). The major histocompatibility complex-restricted antigen receptor on T cells. II. Role of the L3T4 product. *J. Exp. Med.* **158,** 1077.

Matis, L. A., Longo, D. L., Hedrick, S. M. Hannum, C., Margoliash, E., and Schwartz, R. H. (1983). Clonal analysis of the major histocompatibility complex restriction in the T cell proliferative response to cytochrome c. *J. Immunol.* **130,** 1527.

Matzinger, P. (1981). A one receptor view of T-cell behavior. *Nature (London)* **292,** 497.

McKean, D. J., Infante, A. J., Nilson, A., Kimoto, M., Fathman, C. G., Walker, E., and Warner, N. L. (1981). Major histocompatibility complex-restricted antigen presentation to antigen-reactive T cells by B lymphocyte tumor cells. *J. Exp. Med.* **154,** 1419.

Meuer, S. C., Schlossman, S. F., and Reinherz, E. (1982). Clonal analysis of human cytotoxic T lymphocytes: T4+ and T8+ effector T cells recognize products of different major histocompatibility complex regions. *Proc. Natl. Acad. Sci. U.S.A.* **79,** 4395.

Meuer, S. C., Fitzgerald, K., Hussey, R., Hodgdon, J., Schlossman, S., and Reinherz, E. (1983). Clonotypic structures involved in antigen-specific human T cell function. Relationship to the T3 molecular complex. *J. Exp. Med.* **157,** 705.

Möller, G. (1978). Role of macrophages in the immune response. *Immunol. Rev.* **40,** 3.

Moingeon, P., Ythier, A., Goubin, G., Faure, F., Nowill, A., Delmon, L., Rainaud, M., Forrestier, F., Daffos, F., Bohuon, C., and Hercend, T. (1986). A unique T-cell receptor complex expressed on human fetal lymphocytes displaying natural killer-like activity. *Nature (London)* **323,** 638.

Morrison, L. A., Lukacher, A. E., Braciale, V. L., Fan, D. P., and Braciale, T. J. (1986). Differences in antigen presentation to MHC Class I- and Class II-restricted influenza virus-specific cytolytic T lymphocyte clones. *J. Exp. Med.* **163,** 903.

Nabholz, M., Engers, H. D., Colavo, D., and North, M. (1978). Cloned T cell lines with specific cytolytic activity. *Curr. Top. Microbiol. Immunol.* **81,** 176.

Novotny, J., Tonegawa, S., Saito, H., Kranz, D. M., and Eisen, H. N. (1986). Secondary,

tertiary, and quaternary structure of T-cell-specific immunoglobulin-like polypeptide chains. *Proc. Natl. Acad. Sci. U.S.A.* **83**, 742.
Paul, W. E., and Benacerraf, B. (1977). Functional specificity of thymus-dependent lymphocytes. *Science* **195**, 1293.
Ponzio, N. M., Fink, J. H., and Battisto, J. R. (1975). Adult murine lymph node cells respond blastogenically to a new differentiation antigen on isologous and autologous B lymphocytes. *J. Immunol.* **114**, 971.
Raulet, D. H., Garman, R. D., Saito, H., Tonegawa, S. (1985). Developmental regulation of T-cell receptor gene expression. *Nature (London)* **314**, 103.
Rock, K. L., and Benacerraf, B. (1983). Inhibition of antigen-specific T lymphocyte activation by structurally related Ir gene-controlled polymers. Evidence of specific competition for accessory cell antigen presentation. *J. Exp. Med.* **157**, 1618.
Rosenthal, A. S., and Shevach, E. M. (1973). Function of macrophages in antigen recognition by guinea pig T lymphocytes. I. Requirement for histocompatible macrophages and lymphocytes. *J. Exp. Med.* **138**, 1194.
Rothbard, J. B., and Taylor, W. R. (1988). A sequence pattern common to T cell epitopes. *EMBO* **7**, 93.
Rupp, F., Acha-Orbea, H., Hengartner, H., Zinkernagel, R., and Joho, R. (1985). Identical V beta T-cell receptor genes used in alloreactive cytotoxic and antigen plus I-A specific helper T cells. *Nature (London)* **315**, 425.
Saito, H., Kranz, D., Takagaki, Y., Hayday, A., Eisen, H., and Tonegawa, S. (1984). A third rearranged and expressed gene in a clone of cytotoxic T lymphocytes. *Nature (London)* **312**, 36.
Samelson, L. E., Germain, R. N., and Schwartz, R. H. (1983). Monoclonal antibodies against the antigen receptor on a cloned T cell hybrid. *Proc. Natl. Acad. Sci. U.S.A.* **80**, 6972.
Samelson, L. E., Harford, J. B., and Klausner, R. D. (1985a). Identification of the components of the murine T cell antigen receptor complex. *Cell* **43**, 223.
Samelson, L. E., Lindsten, T., Fowlkes, B. J., van den Elsen, P., Terhorst, C., Davis, M. M., Germain, R. M., and Schwartz, R. H. (1985b). Expression of genes of the T-cell antigen receptor complex in precursor thymocytes. *Nature (London)* **315**, 765.
Schwartz, R. H. (1978). A clonal deletion model for Ir gene control of the immune response. *Scand. J. Immunol.* **7**, 3.
Schwartz, R. H., Berzofsky, J. A., Horton, C. L., Schechter, A. N., Sachs, D. H. (1978). Genetic control of the T-lymphocyte proliferative response to staphylococcal nuclease: Evidence for multiple MHC-linked Ir gene control. *J. Immunol.* **120**, 1741.
Schwartz, R. H., Fox, B. S., Fraga, E., Chen, C., and Singh, B. (1985). The T lymphocyte response to cytochrome c. V. Determination of the minimal peptide size required for simulation of T cell clones and assessment of the contribution of each residue beyond this size to antigenic potency. *J. Immunol.* **135**, 2598.
Schwartz, R. M., and Dayhoff, M. O. (1978). Cytochromes. *In* "Atlas of Protein Sequence and Structure," Vol. 5, Suppl. 3. National Biomedical Research Foundation, Washington, D.C.
Scofield, V. L., Schlumpberger, J. M., West, L. A., and Weissman, I. L. (1982). Protochordate allorecognition is controlled by a MHC-like gene system. *Nature (London)* **295**, 499.
Sela, M. (1969). Antigenicity: Some molecular aspects. *Science* **166**, 1365.
Senyck, G., Williams, E. B., Nitecki, D., and Goodman, J. W. (1971). The functional

dissociation of an antigen molecule: Specificity of humoral and cellular immune responses to glucagon. *J. Exp. Med.* **133**, 1295.

Shastri, N., Gammon, G., Miller, A., and Sercarz, E. E. (1986a). Ia molecule-associated selectivity in T cell recognition of a 23-amino-acid peptide of lysozyme. *J. Exp. Med.* **164**, 882.

Shastri, N., Gammon, G., Miller, A., and Sercarz, E. E. (1986b). Amino acid residues distinct from the determinant region can profoundly affect activation of T cell clones by related antigens. *J. Immunol.* **136**, 371.

Shearer, G. M., Rhen, T. G., and Garbarino, C. A. (1975). Cell mediated lympholysis of trinitrophenyl-modified autologous lymphocytes. Effector cell specificity to modified cell surface components controlled by the H-2K and H-2D serological regions of the murine major histocompatibility complex. *J. Exp. Med.* **141**, 1348.

Shimonkevitz, R, Colon, S., Kappler, J. W., Marrack, P., and Grey, H. M. (1984). Antigen recognition by H-2-restricted T cells. II. A tryptic ovalbumin peptide that substitutes for processed antigen. *J. Immunol.* **133**, 2067.

Shimonkevitz, R. Kappler, J. W., Marrack, P., and Grey, H. M. (1983). Antigen recognition by H-2-restricted T cells. I. Cell-free antigen processing. *J. Exp. Med.* **158**, 303.

Siu, G., Miller, J. E., Hsu, M-I., Klotz, J. L., Halrick, S. M., and Kronenberg, M. (1988). Diversity of gamma gene junctional sequences: Implications from comparisons with other gene families. In preparation.

Solinger, A. M., Ultee, M. E., Margoliash, E., and Schwartz, R. H. (1979). T-lymphocyte response to cytochrome c. I. Demonstration of a T-cell heteroclitic proliferative response and identification of a topographic antigenic determinant on pigeon cytochrome c whose immune recognition requires two complementing major histocompatibility complex-linked immune response genes. *J. Exp. Med.* **150**, 830.

Spits, H., Borst, J., Terhorst, C., and de vries, J. E. (1982). The role of T cell differentiation markers in antigen-specific and lectin-dependent cellular cytotoxicity mediated by T8+ and T4+ human cytotoxic T cell clones directed at class I and class II MHC antigens. *J. Immunol.* **129**, 1563.

Sprent, J. (1978). Restricted helper fuction of F1 → parent bone marrow chimeras controlled by K-end of H-2 complex. *J. Exp. Med.* **147**, 1838.

Sprent, J., and von Boehmer, H. (1976). Helper function of T cells depleted of alloantigen-reactive lymphocytes by filtration through irradiated F_1 hybrid recipients. I. Failure to collaborate with allogeneic B cells in a secondary response to sheep erythrocytes measured in vivo. *J. Exp. Med.* **144**, 617.

Sprent, J., and Webb, S. R. (1987). Function and specificity of T cell subsets in the mouse. *Adv. Immunol.* **41**, 39.

Sredni, B., and Schwartz, R. H. (1980). Alloreactivity of an antigen-specific T-cell clone. *Nature (London)* **287**, 855.

Staerz, U. D., Kanangawa, O., and Bevan, M. J. (1985). Hybrid antibodies can target sites for attack by T cells. *Nature (London)* **314**, 628.

Stashenko, P. P., and Schlossman, S. (1977). Antigen recognition: The specificity of an isolated T lymphocyte population. *J. Immunol.* **118**, 544.

Swain, S. (1981). Significance of Lyt phenotypes: Lyt 2 antibodies block activities of T cells that recognize class I major histocompatibility complex antigens regardless of their function. *Proc. Natl. Acad. Sci. U.S.A.* **78**, 7101.

Thomas, D. W., Hsieh, K. H., Schauster, J. L., and Wilner, G. D. (1981). Fine specificity of genetic regulation of guinea pig T lymphocyte responses to angiotensin II and related peptides. *J. Exp. Med.* **153**, 583.

Thompson, K., Harris, M., Benjamini, E., Mitchell, G., and Noble, M. (1972). Cellular and humoral immunity: A distinction in antigenic recognition. *Nature (London) New Biol.* **238**, 20.

Townsend, A. R. M., and Skehel, J. J. (1984). The influenza A virus nucleoprotein gene controls the induction of both subtype specific and crossreactive cytotoxic T cells. *J. Exp. Med.* **160**, 552.

Townsend, A. R. M., Gotch, M., and Davey, J. (1985). Cytotoxic T cells recognize fragments of the influenza nucleoprotein. *Cell* **42**, 457.

Townsend, A. R. M., Bastin, J., Gould, K., and Brownlee, G. G. (1986a). Cytotoxic T lymphocytes recognize influenza haemagglutinin that lacks a signal sequence. *Nature (London)* **324**, 575.

Townsend, A. R. M., Rothbard, J., Gotch, F. M., Bahadur, G., Wraith, D., and McMichael, A. J. (1986b). The epitopes of influenza nucleoprotein recognized by cytotoxic T lymphocytes can be defined with short synthetic peptides. *Cell* **44**, 959.

Toyonaga, B., and Mak, T. W. (1987). Genes of the T-cell antigen receptor in normal and malignant T cells. *Annu. Rev. Immunol.* **5**, 585.

Trowbridge, I. S., Lesley, J., Trotter, J., and Hyman, R. (1985). Thymocyte subpopulation enriched for progenitors with an unrearranged T-cell receptor β-chain gene. *Nature (London)* **315**, 666.

Unanue, E. R. (1981). The regulatory role of macrophages in antigenic stimulation. Part Two: Symbiotic relationship between lymphocytes and macrophages. *Immunol. Rev.* **31**, 1.

Unanue, E. R. (1984). Antigen-presenting function of the macrophage. *Annu. Rev. Immunol.* **2**, 395.

Valembois, P. (1974). Cellular aspects of graft rejection in earthworm and some other metazoa. *Contemp. Top. Immunobiol.* **4**, 121.

Van Voorhis, W. E., Valinsky, J., Hoffman, E., Luban, J., Hair, L. S., and Steinman, R. M. (1983). The relative efficacy of human monocytes and dendritic cells as accessory cells for T cell replication. *J. Exp. Med.* **158**, 174.

von Boehmer, H., and Adams, P. B. (1973). Syngeneic mixed lymphocyte reaction between thymocytes and peripheral lymphoid cells in mice: Strain specificity and nature of the target cell. *J. Immunol.* **110**, 376.

von Boehmer, H., Hudson, L., and Sprent, L. (1975). Collaboration of histoincompatible T and B lymphocytes using cells from tetraparental bone marrow chimeras. *J. Exp. Med.* **142**, 989.

von Boehmer, H., Hengartner, H., Nabholz, M., Lernhardt, W., Schreier, M. H., and Haas, W. (1979). Fine specificity of a continuously growing killer cell clone specific for H-Y antigen. *Eur. J. Immunol.* **9**, 592.

Watson, J. (1979). Continuous proliferation of murine antigen-specific helper T lymphocytes in culture. *J. Exp. Med.* **150**, 1510.

Werdelin, O. (1982). Chemically related antigens compete for presentation by accessory cells to T cell. *J. Immunol.* **129**, 1883.

Yanagi, Y., Yoshikai, Y., Leggett, K., Clark, S. P., Aleksander, I., and Mak, T. W. (1984). A human T cell-specific cDNA clone encodes a protein having extensive homology to immunoglobulin chains. *Nature (London)* **308**, 145.

Ziegler, H. K., and Unanue, E. R. (1981). Identification of a macrophage antigen processing event required for I-region-restricted antigen presentation to T lymphocytes. *J. Immunol.* **127**, 1869.

Ziegler, H. K., and Unanue, E. R. (1982). Decrease in macrophage antigen catabolism by ammonia and chloroquine is associated with inhibition of antigen presentation to T cells. *Proc. Natl. Acad. Sci. U.S.A.* **79,** 175.

Zinkernagel, R. M. (1978). Thymus and lymphohemopoietic cells: Their role in T cell maturation in selection of T cells' H-2-restriction-specificity and in H-2 linked Ir gene control. *Immunol. Rev.* **42,** 224.

Zinkernagel, R. M., and Althage, A. (1979). Search for suppression of T cells specific for the second nonhost H-2 haplotype in F1 → P irradiation bone marrow chimeras. *J. Immunol.* **122,** 1742.

Zinkernagel, R. M., and Doherty, P. C. (1974). Immunological surveillance against altered self components by sensitized T lymphocytes in lymphocytic choriomeningitis. *Nature (London)* **251,** 547.

Zinkernagel, R. M., and Doherty, P. C. (1977). Major transplantation antigens virus and specificity of surveillance T cells. The "altered self" hypothesis. *Contemp. Top. Mol. Immunol.* **7,** 179.

Transcriptional Controlling Elements in the Immunoglobulin and T Cell Receptor Loci

KATHRYN CALAME*,† AND SUZANNE EATON‡

*The Department of Biological Chemistry, UCLA School of Medicine,
University of California, Los Angeles,
Los Angeles, California 90024 and
†The Molecular Biology Institute and ‡The Department of Microbiology,
University of California, Los Angeles,
Los Angeles, California 90024

I. Introduction

A unique and critical function of the immune system is its ability to recognize an extremely diverse set of foreign molecular structures with exquisite specificity and high affinity. The proteins which fulfill this function for the humoral response are immunoglobulins (Igs) and for the cellular response are T cell receptors (TCRs). Since both Igs and TCRs carry out antigen recognition, it is not surprising that the structure of Ig and TCR proteins and the structure, organization, and rearrangement of the genes which encode them are quite similar (Honjo, 1983; Kronenberg et al., 1986). Given these functional and structural similarities, it is reasonable to ask if the regulated expression of Ig and TCR genes is also similar.

General similarities are evident in that expression of both Ig and TCR genes is tissue specific: Igs are expressed only in B cells and TCRs are expressed only in T cells. Furthermore, identical gene rearrangements occur and may be expected to play a role in the expression of both Ig and TCR genes. However, since Ig molecules are secreted and recognize soluble antigens, while TCR molecules remain on the T cell surface in close association with other membrane proteins and recognize cellular antigens, many more Ig molecules are synthesized by terminally differentiated B cells than TCR molecules are synthesized by mature T cells. These observations suggest that we may also expect to find some significant differences between the regulatory events controlling Ig and TCR gene expression.

There are several reasons why it is important to study the regulated expression of Ig and TCR genes. First, we can learn general principles governing tissue-specific and developmentally regulated gene expression in eukaryotes. Second, it is reasonable to hope that elucidation of important regulatory mechanisms for these key genes may aid future understanding of the precise molecular defects responsible for various autoimmune or immune deficiency diseases. And finally, unique regulatory features of Ig and TCR loci, which are intrinsic to gene rearrangement and the generation of diversity, may provide insight into similar regulatory mechanisms in other complex systems, which may generate information by combinatorial mechanisms like those used in the immune system.

Gene regulation occurs at many levels. In the following sections, we present overviews of the various ways in which Ig and TCR genes are regulated during development. However, the detailed portions of the review will be limited to transcriptional regulation and will focus on cis-acting DNA elements and trans-acting cellular proteins, which are required to regulate transcription of Ig and TCR genes. Since more work has been done on Ig gene regulation, the bulk of the review will deal with Ig genes. Helpful related reviews on Ig (Honjo, 1983; Yancopoulos and Alt, 1986; Alt et al., 1987) and TCR (Davis, 1985; Kronenberg et al., 1986; Chou et al., 1987a,b; Chien et al., 1987a) gene structure and rearrangement, general gene rearrangement (Borst and Greaves, 1987), general transcriptional regulatory elements. (Serfling et al., 1985; Wasylyk, 1986; Voss et al., 1986; Hatzopoulos et al., 1987), transcription factors (Short, 1987), Ig regulatory elements (Calame, 1985; Matthias et al., 1987), and variable region assembly (Yancopoulos and Alt, 1986) have appeared recently.

II. Immunoglobulin Gene Expression

IMMUNOGLOBULIN GENE EXPRESSION IS HIGHLY REGULATED DURING B CELL DEVELOPMENT

A general outline of immunoglobulin gene organization, rearrangement, and expression is shown in Fig. 1. In all tissues other than B cells, the heavy, κ and λ loci are transcriptionally silent with the single exception of C_μ-only (sterile) transcripts from the heavy chain region, which are sometimes detected in T cells and myeloid cells (Kemp et al., 1980; Mather and Perry, 1981; Lennon and Perry, 1985). The earliest detectable cells, which are destined to become B cells, termed pre–pre B cells in this review, produce low steady-state levels of V_H-only transcripts (Yancopoulos and Alt, 1985). These are probably similar or the same as Ly1$^+$ Pro-B cells described by Palacios et al. (1987). The only V_H

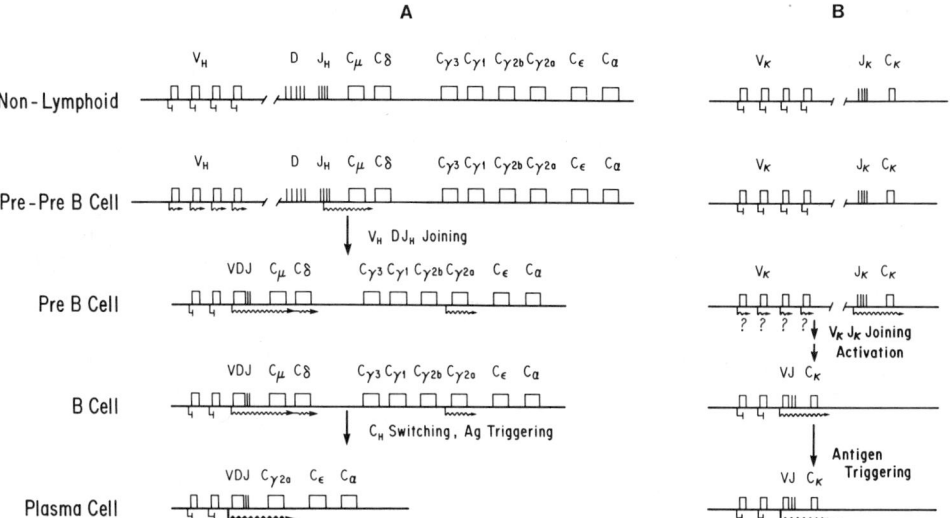

FIG. 1. Summary of (A) heavy chain and (B) κ chain gene rearrangement and expression during B cell development. Raised boxes indicate gene segments, wavy lines represent transcription; distances are not precisely to scale. λ genes are not shown here, but if neither κ rearrangement led to a functional κ gene, λ rearrangement would occur prior to IgM expression on the surface of the immature B cell.

family from which V-only transcripts have been detected is the J558 family; however, it is likely that most or all V_H genes are transcribed in pre–pre B cells, but only J558 transcripts are detectable, since many hundreds of J558 V_H gene segments are present (Livant et al., 1986). "Sterile" or C_μ-only transcripts initiating within the D or J regions are also found in early B cells (Lennon and Perry, 1985; Reth and Alt, 1984); these transcripts are probably IgH enhancer dependent (Grosschedl and Baltimore, 1985). Alt and colleagues (Blackwell et al., 1986) suggest that (1) altered chromatin structure, resulting from this low-level transcription, allows the recombinase enzymes access to transcribed gene segments so that V–D–J joining can occur and (2) specific transcription of Ig or TCR gene segments targets rearrangement of the correct genes in B and T cells, respectively. In support of these ideas, they have shown that when exogenous recombination substrates are introduced into pre-B cells, rearrangement increases with increased transcription through the substrate (Blackwell et al., 1986). They have also demonstrated an inherent lack of recombinase specificity by showing that exogenously introduced TCR genes can be rearranged in B cells. Thus, early in B cell development, V_H-only and C_μ-only transcription appears to be critical for targeting heavy chain V–D–J joining. It is not known if there is any differential transcription of particular V_H genes or V_H families at unique stages of early B cell development, which might target rearrangement and functional expression of particular antigen–recognition specificities.

After V–D–J joining, the rearranged V_H promoter is juxtaposed approximately 2 kb 5' of the IgH transcriptional enhancer element which activates the promoter (Banerji et al., 1983; Gillies et al., 1983; Mercola et al., 1983; Neuberger, 1983). Ultimately V_H promoters become entirely enhancer dependent, because no transcription of V_H genes distant from the enhancer is detectable in mature B cells (Yancopoulos and Alt, 1985; Wang and Calame, 1985). It is likely that enhancer dependence occurs relatively soon after V–D–J joining, but this question has not been addressed experimentally.

Intact μ chains, probably associated with ω polypeptides (Pillai and Baltimore, 1987), provide a positive signal for light chain V–J joining in which κ rearrangement precedes λ gene rearrangement (Alt et al. 1987). By analogy to heavy chains, one expects that these rearrangement events must be preceded by V_L- and C_L-only transcription. V_κ-only transcription has not been reported, but J_κ- and C_κ-only transcripts, initiating about 8 kb 5' of unrearranged C_κ gene segments, have been well documented and shown to occur in pre-B cells prior to V_κ–J_κ joining (Van Ness et al., 1981; Perry et al., 1981). If neither κ chromosome joins to give a functional κ gene, λ rearrangement occurs. Consistent with the model of transcription preceding DNA rearrangement, it has been reported that germ-line V_λ genes are transcribed in κ- producing cells (Picard and Schaffner, 1984b). There is also some evidence for C_λ-only transcripts, although these have not been carefully studied (Perry et al., 1980; Hollis et al., 1982).

Rearranged V_κ promoters are activated by a κ transcriptional enhancer located between J_κ and C_κ (Picard and Schaffner, 1984a); there is disagreement about the existence of λ enhancers (Picard and Schaffner, 1984a; Spandidos and Anderson, 1984). Rearrangement alone is not sufficient for activation of light chain expression, since the pre-B cell line 70Z/3 contains a functionally rearranged κ gene, which is not transcribed until the cells are stimulated with lipopolysaccharide (LPS) (Maki et al., 1980; Parslow and Granner, 1982). Pre-B cells from mice harboring κ transgenes also confirm this fact (Storb et al., 1985). κ transcription in 70Z/3 can be induced with cycloheximide (Wall et al., 1986), suggesting that new protein synthesis is not required for transcriptional activation. Complete Ig tetramers of two μ heavy chains and two light chains signal V-D-J joining to cease. The IgM molecule is displayed on the surface of the B cell, where it functions as an antigen receptor. Differential termination and splicing of heavy chain transcripts may also lead to expression of IgD on the surface of unstimulated B cells.

The latter stages of B cell development into terminally differentiated plasma cells occur in response to complex signals from antigen, T helper cells, and T cell factors (Melchers and Andersson, 1986; Hamaoka and Ono, 1986; Cambier and Ransom, 1987). Terminal differentiation involves (1) differential transcription termination, polyadenylation, and splicing to give predominantly secreted forms of heavy chain mRNA (reviewed in Calame, 1985 and Guise et al., 1988); (2) for cells expressing isotypes other than IgM and IgD, a second DNA rearrangement called C_H switching (Calame, 1985); (3) for secretion of IgM and IgA, expression of J chains (Koshland, 1985); and (4) a large increase in steady-state levels of Ig mRNA and protein (Schibler et al., 1978; Kelley and Perry, 1986).

Recent work (F. Alt, personal communication) shows that the C_H switching rearrangement is preceded by transcription through unrearranged C_H gene segments. Transcription initiates at multiple cryptic promoters 5' of unrearranged C_H gene segments and is responsive to signals from LPS and IL-4 (interleukin 4). This suggests that transcription may target rearrangement to particular isotypes in individual B cell clones in a manner similar to that by which transcription of the germ-line IgH locus promotes V-D-J joining in very early B cells.

The large increase in steady-state levels of Ig mRNAs in terminally differentiated plasma cells appears to involve both transcriptional (Schibler et al., 1978; Mather et al., 1984; Yuan and Tucker 1984; Kelley and Perry, 1986; Chen-Bettecken et al., 1987) and posttranscriptional events (Mather et al., 1984; Kelley and Perry, 1986; Gerster et al., 1986). There is disagreement as to the relative contribution of transcriptional and posttranscriptional regulation (Gerster et al., 1986; Chen-Bettecken et al., 1987; Yuan and Tucker, 1984; Kelley and Perry, 1986). When normal splenic B cells are stimulated with LPS, the increased levels of μ and κ mRNA can be accounted for completely by increases in transcription (Yuan and Tucker, 1984; Chen-Bettecken et al., 1987), but when cell lines, particularly myelomas are studied, a posttranscriptional component is observed (Kelley and Perry, 1986).

The two DNA elements, promoters and enhancers, which are important for Ig transcriptional regulation, both display a strong B cell preference (Banerji et al., 1983; Gillies et al., 1983; Queen et al., 1986; Grosschedl and Baltimore, 1985; Mason et al., 1985; Foster et al., 1985).

When tested in transgenic mice, κ transcription is more stringently B cell specific than heavy chain gene transcription, which also occurs in some T cells (reviewed in Storb, 1987). Heavy chain promoters and enhancers display synergy; that is, while IgH promoters can be activated by heterologous enhancers and IgH enhancers can activate heterologous promoters, they show greater activity in combination with one another (Garcia et al., 1986). It is not clear whether the enhancer or the promoter is responsible for observed B cell stage-specific differences in transcription rate, since both the IgH enhancer (Gerster et al., 1986) and a V_H promoter (Grosschedl and Baltimore, 1985) have been shown to have similar activities in B cell lines at different developmental stages. Ig promoters display enhancer-independent activity early in B cell development, enhancer-dependent activity later, and may also be induced by growth and maturation signals late in B cell development.

There have been reports that other elements may be important in regulating IgH transcription. Grosschedl and Baltimore (1985) noted an effect of "intragenic" sequences, but did not determine whether their effect was transcriptional or posttranscriptional. Gregor et al. (1986) showed that IgA transcription decreased when regions 3' to C_α were removed, but the putative activator has not been further characterized. A binding site for the protein NF1 is found 3' of the IgH enhancer, but its functional importance, if any, has not been demonstrated (Hennighausen et al., 1985). Finally, it is important to note that when B cells are fused with fibroblasts in culture, Ig transcription is shut off, suggesting that one or more trans-acting negative regulators of Ig transcription are present in non-B cells (Junker and Pedersen, 1985).

III. Immunoglobulin Transcriptional Enhancer Elements

A. The Mechanism by Which Transcriptional Enhancer Elements Act Is Not Known

Transcriptional enhancers, which were first discovered in animal viruses (Gruss et al., 1981), are defined functionally as cis-acting DNA sequences, which activate transcription–initiation (Treisman and Maniatis, 1985; Weber and Schaffner, 1985) in an orientation- and distance-independent manner. Delayed competition experiments (Wang and Calame, 1986) suggest that enhancers act by facilitating the formation of stable transcription–initiation complexes (Brown, 1984; Mattaj et al., 1985).

The mechanism by which enhancers activate transcription is not known, although several have been considered (for a review, see Ptashne, 1986). Early models (Moreau et al., 1981) suggested that enhancers were a loading site for transcription factors, which bound at the enhancer and then "threaded" along the DNA in either direction until they came to a transcription–initiation site. A variation on this idea is that factors initially bound at the enhancer reach transcription–initiation sites by some sort of limited diffusion. Alternatively, it has been suggested (Brent and Ptashne, 1985) that proteins bound at the enhancer and at the transcription start site are able to interact with one another because the DNA between them "loops out." Others have suggested that enhancers alter the topology of DNA in their vicinity, although Plon and Wang (1986) have shown that enhancers can act even when they are topologically separated from their target promoter. A variation of this model is that enhancers alter the chromatin structure in their vicinity, and there is recent evidence that this mechanism holds for some, but not all, yeast activator elements (Chen et al., 1987). Finally, it may be that enhancer activity involves targeting DNA regions to particular sites in the nuclear matrix (Cockerill and Garrard, 1986; Cockerill et al., 1987).

Since the mechanism of enhancer action is not clear, it is quite possible that different elements, operating by different molecular mechanisms, may all be functionally labeled "enhancers." To make matters more complicated, other transcription-activating elements, which are found closer to the start site of transcription and are usually considered part of the "promoter," often display partial distance or orientation independence. Some of these elements may act through the same mechanism as elements now called enhancers. Thus, the terms enhancer and promoter are not based on a precise mechanistic understanding of transcription activation and may be inaccurately applied in some cases. However, keeping the above limitations in mind, we will continue to use the terms in this review for convenience and consistency with past writing on the subject.

B. Ig Enhancers Are Important for Activation of Transcription upon V–D–J Joining

Consideration of several aspects of heavy chain transcription suggested the possible existence of a transcriptional enhancer between J_H and C_μ gene segments. Mather and Perry (1981) demonstrated that unrearranged V_κ genes were transcriptionally silent in plasmacytomas, even when the homologous V_κ gene was rearranged and highly transcribed. This ruled out a soluble activator acting on particular V promoters and implicated the rearrangement event in activation.

Clarke et al. (1982) subsequently showed that no alteration occurred 5' of V_H genes after V–D–J joining, implicating sequences 3' to the V genes.

Soon after, several laboratories (Banerji et al., 1983; Gillies et al., 1983; Mercola et al., 1983; Neuberger, 1983) discovered a transcriptional enhancer element in the heavy chain locus. The human heavy chain locus contains a similar enhancer at the comparable location (Rabbitts et al., 1983); however, only mouse elements will be discussed in this review. The murine enhancer activity is entirely contained on a 992-bp Xba restriction fragment located 542 bp 3' of J_{H4} and about 1.5 kb 5' of S_μ (Gillies et al., 1983). No enhancers have been found near germ-line V_H genes (Mercola et al., 1985). Thus, the location of the IgH enhancer 3' of the J_H cluster and 5' of S_μ explains the observations that transcriptional activation is coupled to the V–D–J joining rearrangement and persists after C_H switching. Enhancer activity in the κ locus is similarly located between J_κ and C_κ gene segments (Picard and Schaffner, 1984a); however, there are conflicting reports concerning enhancers in the λ locus (Picard and Schaffner, 1984a; Spandidos and Anderson, 1984).

These Ig enhancers have typical enhancer characteristics: (1) activation of transcription–initiation, (2) activation in cis, (3) activation independent of distance and orientation, and (4) activation of heterologous promoters. The IgH enhancer was the first eukaryotic cellular enhancer to be discovered, and it has continued to provide a paradigm for study of these fascinating elements.

C. The IgH Enhancer Appears to Establish Transcription–Initiation in a B Cell-Specific Manner

An important feature of IgH enhancer, noted in the initial work (Gillies et al., 1983; Banerji et al., 1983), was the B cell preference of the enhancer. The IgH enhancer functions at a low level in fibroblasts, but is much more active in B cells (Wasylyk and Wasylyk, 1986). It appears to be equally active in B cells representing all developmental stages, suggesting that it is tissue specific, but not developmentally regulated (Gerster et al., 1986). Several transgenic mouse studies have further underscored the tissue specificity of the IgH enhancer by conferring B cell-specific expression on heterologous genes under control of the IgH enhancer (Gerlinger et al., 1986; Adams et al., 1985; Grosschedl et al., 1984; Reik et al., 1987; reviewed in Storb, 1987). These studies have also revealed that expression of genes controlled by the IgH enhancer occurs in some T cells as well as in B cells (Grosschedl et al., 1984), and when low levels of expression are detected by malignant transformation, expression is found in lung and other tissues (Suda et al., 1987).

The IgH enhancer in its normal chromosomal location can activate V_H promoters located as far away as 17.5 kb; in this context, it can also activate tandem V_H promoters (Wang and Calame, 1985). These findings emphasize the strength of this activating element and suggest, but do not prove, that activation may occur by a mechanism involving DNA looping, rather than binding and processive movement of a factor (see Ptashne, 1986, for review of enhancer mechanisms).

Several laboratories have reported B cell lines in which the IgH enhancer has been deleted, but which, nevertheless, express normal levels of heavy chain (Wabl and Burrows, 1984; Klein et al., 1984; Eckhart and Birshtein, 1985; Aguilera et al., 1985). Originally, these findings led to the suggestion that there was no normal requirement for an enhancer for IgH gene expression and that the observed "enhancer requirement" was an artifact of transfection (Wabl and Burrows, 1984). However, this possibility has been ruled out (assuming no technical artifacts associated with expression of a human gene in mice) by the lack of expression of a human $\gamma1$ heavy chain gene lacking an enhancer in transgenic mice (Yamamura et al., 1984). Additional studies with the IgH enhancer-deleted lines have shown that, when the enhancer-deleted IgH genes are cloned and transfected into B cells, they are not transcribed, unless an enhancer is added (Zaller and Eckhardt, 1985; Klein et al., 1985). These findings have led to the suggestion that the IgH enhancer may be required to establish active transcription in the heavy chain locus, but that it is not required to maintain it. Consistent with this idea, the SV40 enhancer has been shown to establish active transcription, which is stable to subsequent challenge by competing enhancer sequences (Wang and Calame, 1986), presumably by the formation of stable initiation complexes (Brown, 1984).

D. *In Vivo* Protein-Binding Sites Have Been Observed within the IgH Enhancer

In vivo competition experiments (Mercola et al., 1985) demonstrated that binding of cellular factors to the IgH enhancer DNA was required for enhancer function and further showed that at least one of these enhancer factors was B cell specific. Activation of a rearranged heavy chain gene stably transfected into a fibroblast cell line by the injection of B cell nuclear proteins via erythrocyte ghosts directly demonstrated that B cells contain necessary positive factors for IgH transcription, which are probably targeted to the enhancer (Maeda et al., 1986). Finally, *in vivo* dimethyl sulfate (DMS) protection experiments (Ephrussi et al., 1985; Church et al., 1985) distinguished four regions, termed μE1–4, within the enhancer where B cell-specific protections

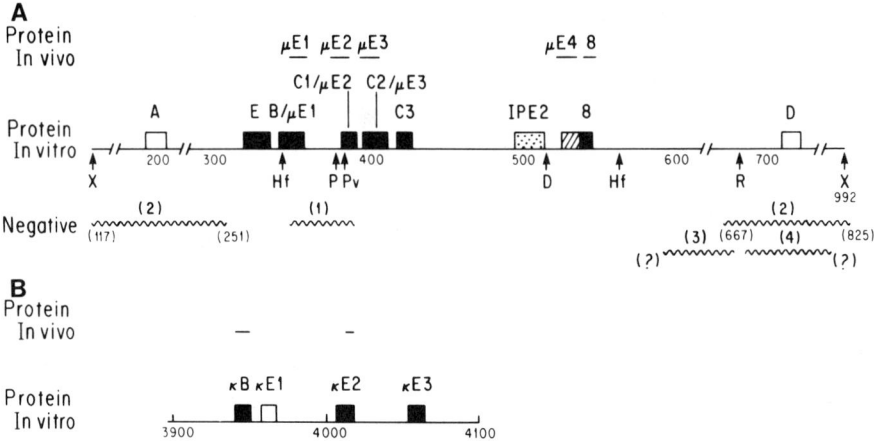

FIG. 2. Protein-binding sites in the (A) heavy and (B) κ chain enhancers. Solid boxes indicate sites in which protein binding *in vitro* has been observed and which have been shown to be important for enhancer function by mutation and transfection. Hatched boxes indicate sites which have been shown to be important for function, but in which protein binding has not been detected *in vitro*. Dotted boxes indicate sites in which protein binding *in vitro* has been observed, but which have not been tested for function. Open boxes indicate sites in which protein binding has been observed *in vitro*, but which have been shown to be unimportant for function. Numbering in A is from Ephrussi *et al.* (1985) and in B is from Max *et al.* (1981). Arrows in A indicate restriction enzyme sites: Hf, *Hin*F; P, *Pst*; Pv, *Pvu*II; D, *Dde*I; R, *Eco*RI, and X, *Xba*I. Wavy lines indicate approximate locations of regions found to have negative effects on IgH enhancer activity in fibroblasts. Numbers in parentheses (1)–(4) refer to references: (1) Kiledjian *et al.* (1988), (2) Imler *et al.* (1987), (3) Lenardo and Baltimore (1986) and Augereau and Chambon (1986), and (4) Scholer and Gruss (1986).

were observed, suggesting B cell-specific protein binding (shown in Fig. 2). These *in vivo* studies have provided a framework for subsequent protein-binding studies carried out *in vitro*.

E. MULTIPLE PROTEIN-BINDING SITES IN THE IgH ENHANCER HAVE BEEN MAPPED *in Vitro*

Several laboratories have used *in vitro* methods, such as gel shift assays (Fried and Crothers, 1981), exonuclease III assays (Wu, 1985), methylation interference (Gilman *et al.*, 1986), and footprinting assays, to map protein-binding sites within the IgH enhancer. These results are summarized in Fig. 2 and enumerated briefly below, from 5' to 3' using the numbering system of Ephrussi *et al.* (1985). A site labeled A (see Fig. 2A) at approximately 190 bp was partially mapped by exonuclease

III studies (Peterson et al., 1986). A site labeled E (see Fig. 2A) at 320–337 bp has been carefully mapped by footprinting and methylation interference studies (Peterson and Calame, 1987). A site labeled B or μE1 (see Fig. 2A) at 345–360 bp has been carefully mapped by footprinting and methylation interference (Weinberger et al., 1986; Peterson and Calame, 1987); this corresponds to μE1 identified by in vivo DMS protection (Ephrussi et al., 1985). Site C1 (see Fig. 2A) at 390 bp has been partially mapped by exonuclease III (Peterson et al., 1986); this probably corresponds to μE2 identified by in vivo DMS protection. A site labeled C2 or μE3 (see Fig. 2A) at 397–425 bp has been carefully mapped by methylation interference and footprinting (Sen and Baltimore, 1986a; Shlokat et al., 1986; Peterson and Calame, 1987); it corresponds to in vivo protection site μE3. Three regions at 433–440, 458–465, and 471–478 bp having homology to the core sequence in the SV40 enhancer (Weiher et al., 1983) have been identified in the IgH enhancer (Gillies et al., 1983); protein binding at the 5' site has been partially mapped by exonuclease III and termed site C3 (see Fig. 2A) (Peterson et al., 1986). Schlokat et al. (1986) have used methylation interference and footprinting to map a site IPE2 (see Fig. 2A) at 497–517 bp. No protein binding in vitro has been demonstrated at in vivo DMS protection site μE4. Many groups have demonstrated and mapped binding to the octamer region at 541–548 bp (Sen and Baltimore, 1986a; Peterson et al., 1986; Schlokat et al., 1986; Gerster et al., 1987). This interesting sequence occurs in all Ig promoters (Parslow et al., 1984) and many non-B cell-specific promoters as well as the IgH enhancer. Finally, site D (see Fig. 2A) has been partially mapped by exonuclease III near 720 bp (Peterson et al., 1986). See Fig. 2 and Table I for a summary of these data.

F. Multiple Sites in the IgH Enhancer Are Required for Function

Protein binding either in vivo or in vitro must be correlated with function. Most functional studies have utilized site-specific mutation or deletion combined with subsequent functional analysis of the altered enhancers by transfection. Alteration or deletion of individual protein-binding sites does not completely destroy IgH enhancer activity, suggesting that different sites may be functionally redundant. However, decreases in enhancer function have been used to demonstrate the functional importance of the following sites: octamer (Lenardo et al., 1987; Kiledjian et al., 1988; Gerster et al., 1987), E (Tsao et al., 1988), B/μE1 (Lenardo et al., 1987; Tsao et al., 1988), C1/μE2 (Kiledjian et al.,

TABLE I
Binding Site Sequences[a]

Name	Sequence	Location
E	TGAATTGAGCAATGTTGA	IgHE
	TGCATTTTGTAATAATAA	IgHP
B/μE1	GAGTCAAGATGGCCGA	IgH
C2/μE3	CAGGTCATGTGGCAAGGCTA	IgH
8	ATTTGCAT	IgHE, IgHP
		IgLP, others
NF-κB	GGGGACTTTCCCGA	IgκE
7	CTAATGA	IgHP

[a] Sequences are included only for those binding sites which have been mapped by footprinting and shown to be functionally important. Refer to text for references. IgHE, IgH enhancer; IgHP, IgH promoter; IgLP, Ig light chain promoter; IgκE, Ig κ enhancer.

1988; Tsao et al., 1988), C2/μE3 (Kiledjian et al., 1988; Tsao et al., 1988), and μE4 (Kiledjian et al., 1988). Kadesch et al. (1986) found no effect on enhancer activity in B cells when one, two, or three SV40 core homologies were deleted; however, Tsao et al. (1988) noted a slight decrease upon deletion of site C3. Wasylyk showed that IgH enhancer activity in fibroblasts was more dependent upon the SV40 core homologies than was the activity in B cells (Wasylyk and Wasylyk, 1986). Site IPE2 has not been tested by mutation and transfection. Deletion of sites D or D plus A showed no decrease in activity (Tsao et al., 1988), suggesting these sites are not involved in enhancer function as assayed by transfection. Nuclear matrix attachment sites and topoisomerase II-binding site homologies have been located near both sites A and D (Cockerill et al., 1987), and proteins binding to these sites may be involved in organization or anchoring of chromatin. When a region containing μE4 and the octamer is multimerized, B cell-specific enhancer activity is observed (Gerster et al., 1987; Kiledjian et al., 1988); multimers of a region containing sites B/μE1, C1/μE2, C2/μE3, and the SV40 core elements also shows high activity (Kiledjian et al., 1988) (see Fig. 2).

IgH enhancer-dependent transcription *in vitro* has been reported by two groups (Scholer and Gruss, 1985; Augereau and Chambon, 1986), while a third (Sen and Baltimore, 1987) reported B cell-specific IgH enhancer-dependent inhibition of transcription *in vitro*. The enhancer activation observed *in vitro* is small (usually 5–10× activation, compared with more than 10× that amount *in vivo*), difficult to reproduce, and exquisitely dependent on distance and placement of the enhancer relative to the target promoter. However, such systems may ultimately be very useful in elucidating molecular mechanisms of enhancer function.

G. IgH Enhancer-Binding Proteins Are Being Identified and Purified

Multiple proteins appear to bind to the octamer sequence and these are discussed in later sections. Until the genes encoding them are cloned or specific antibodies are raised, it is not possible to be certain how many octamer-binding proteins exist, but it is reasonable to assume that there is a family of octamer-binding proteins.

Distinct proteins have been shown to bind to sites E (Peterson and Calame, 1987), B/μE1 (Weinberger et al., 1986; Peterson and Calame, 1987), and C2/μE3 (Sen and Baltimore, 1986a; Peterson and Calame, 1987). μEBP-E, a 45-kDa protein binding to site E has been purified to homogeneity (Peterson et al., 1988). Protein binding to site C2/μE3 has been highly purified and protein binding to site B/μE1 has been partially purified (C. Peterson and K. Calame, unpublished). Proteins binding to other enhancer sites have not been highly purified to date.

Most, and possibly all, of the IgH enhancer-binding proteins also bind to regulatory elements in other genes. Octamer-binding proteins bind octamers in Ig promoters, other promoters, and other enhancers as detailed below. Purified μEBP-E binds strongly in the IgH V1 promoter region and the κ enhancer and binds more weakly to a second site in the IgH enhancer at 468–475 bp, the polyoma enhancer and the herpes virus thymidine kinase (tk) promoter (Peterson et al., 1988). Protein from crude extracts binding at sites C1/μE2 and C3 (SV40 core sequence) can be competed with SV40 enhancer DNA (Peterson et al., 1986). Purified μEBP-C2 binds to the polyoma enhancer and the C_κ enhancer.

One striking finding is that, with the exception of octamer-binding protein, all other IgH enhancer-binding proteins have a ubiquitous tissue distribution (Sen and Baltimore, 1986a; Peterson et al., 1986). Perhaps the octamer-binding protein is sufficient to confer B cell specificity on the enhancer; this question is considered in more detail in subsequent sections. Alternatively, it may be that transcriptional activity, but not binding, is B cell specific. A lack of correlation between transcriptional activity and DNA binding has been demonstrated for *Drosophila* heat-shock factor (Parker and Topol, 1984; Shuey and Parker, 1986). However, B cell-specific protein contacts were demonstrated by the *in vivo* DMS protection studies (Ephrussi et al., 1985;

Church et al., 1985). Perhaps differences between *in vitro* and *in vivo* conditions may allow binding *in vitro* of factors which are present in amounts or nuclear compartments incompatible with binding in *vivo*. It will be important to resolve this apparent paradox, as we learn more about enhancer-binding proteins and their mechanism of action.

H. Negative Regions May Exist within the IgH Enhancer

Fibroblasts appear to have trans-acting negative factors, which shut off Ig transcription, since somatic cell hybrids between fibroblasts and B cells lose Ig expression (Junker and Pedersen, 1985) and since transfected Ig genes can be activated in fibroblasts after treatment with cycloheximide (Ishihara et al., 1984). However, it is interesting to note that T cells apparently do not have a similar trans-acting negative factor for Ig transcription, since Ig transcription persists in T cell × B cell hybrids (Kozbor et al., 1987). It is possible that the IgH enhancer is a target for one or more of these negative regulators. This notion, based on the negative regulation observed for the interferon-β enhancer (Goodbourn et al., 1986), assumes that, in addition to the presence of positive enhancer proteins in B cells, there would be negative IgH enhancer proteins in non-B cells.

Several groups have reported evidence for regions which, when mutated, increase IgH enhancer function in fibroblasts and, thus, appear to be binding sites for negative factors. Based on transfection data, Kadesch et al. (1986) first suggested a negative element between the *Pst* and *Pvu*2 sites at 378 and 382 bp; subsequently, this group (Kiledjian et al., 1988) reported the negative region to be between 350 and 380 bp. These regions correspond to sites B/μE1 and C1/μE2. Imler et al. (1987) also used transfection and provided evidence of two negative elements at 117–251 and 667–825 bp; a third site at 472–564 bp appeared to inhibit the SV40 enhancer. They found the region of 378–492 bp was positive. Although the 3' and 5' elements described by Imler et al. contain protein-binding sites A and D, deletion of these sites does not increase enhancer activity in fibroblasts (Tsao et al., 1988), suggesting that sites A and D are not components of the negative elements identified by Imler et al. A central region (472–564 bp), which contains μE4 and the octamer, appears to be required for activity in B cells, but not in fibroblasts (Kiledjian et al., 1988). Lenardo et al. (1987) find evidence of a negative element 5' of the *Eco*RI site at 686 bp, although no protein binding has been reported here. Using competition for *in vitro* transcription, two groups found evidence of negative regions in the enhancer; Augereau and Chambon (1986) found a region just 5' of the EcoRI site at 686 bp, while Scholer and Gruss (1985) found that a region 3' of the EcoRI site appeared to have a negative effect in B cell extracts. No protein binding in these regions has been reported.

It is difficult to compare these studies critically, because in all cases, different constructs, different target promoters, different spacing of enhancer and promoter, and different cells have been used. Therefore, while the possibility of negative elements within the IgH enhancer, which would be partially responsible for the B cell preference of the enhancer, is intriguing, it is clear that our current data on placement of the reported negative sites are confusing and often inconsistent. Identification and characterization of fibroblast-specific proteins, which bind to putative negative elements, should help resolve the issue.

I. Ig LIGHT CHAIN ENHANCER ELEMENTS FUNCTION LIKE THE HEAVY CHAIN ELEMENT

Soon after the heavy chain enhancer was discovered, Picard and Schaffner (1984a) demonstrated similar enhancer activity between murine J_κ and C_κ; further studies delimited the κ enhancer to a 200-bp AluI fragment (Queen and Stafford, 1984). Picard and Schaffner (1984a) also searched comparable regions 5' of murine C_λ gene segments, but found no enhancing activity. Subsequently, Spandidos and Anderson (1984) reported an enhancer in the human λ locus, but this enhancer has received little additional study.

The κ enhancer is similar in many ways to the heavy chain enhancer, although in some assays, it appears to be weaker than the heavy chain enhancer (Picard and Schaffner, 1984a; Garcia et al., 1986). The κ enhancer is B cell specific, and although the κ promoter can utilize a variety of enhancers, the κ enhancer seems to prefer its homologous promoter (Queen and Stafford, 1984; Queen et al., 1986; Garcia et al., 1986). It is also capable of activating tandem κ promoters (Atchison and Perry, 1986). Interestingly, a nuclear matrix association region is also located near the κ enhancer (Cockerill and Garrard, 1986), suggesting that matrix attachment might be involved in enhancer function, although there is no direct evidence for this idea. Finally, Atchison and Perry (1987) demonstrated a situation for κ enhancers, which is functionally similar to the IgH enhancer-deleted lines. They showed that plasmacytoma S107 lacks appropriate factors to activate κ enhancers, which are transfected into the cells, although the endogenous κ gene is expressed at normal levels. This suggests that the κ enhancer, like the IgH enhancer, may act to establish transcription, but is not required to maintain it. All of these functional features (i.e., placement, activation of tandem promoters, association with matrix association regions, and role in establishing transcription) suggest that the heavy chain and κ enhancers probably work by quite similar mechanisms.

J. Multiple Protein-Binding Sites Occur in the κ Enhancer

The overall lack of sequence homology between IgH and Igκ enhancers is rather striking considering their functional similarities. The κ enhancer does not contain an octamer, but does contain three regions, termed κE1–3 (see Fig. 2B), with homology to the μE1–4 regions noted in the IgH enhancer by Ephrussi et al. (Sen and Baltimore, 1986a). Alteration of all three sites reduces, but does not abolish, κ enhancer activity (Lenardo et al., 1987). Protein binding in vitro was demonstrated at κE2 and κE3 (Sen and Baltimore, 1986a; Hromas et al., 1988). The proteins binding to these sites have a ubiquitous tissue distribution, and the protein, which binds to κE3, may be the same as the protein which binds to the C2/μE3 site in the heavy chain enhancer.

In addition, protein binding at another site in the κ enhancer, having the sequence GGGGACTTTCC, was shown to be B cell-specific, and this protein was termed NF-κB. Nuclear extracts from uninduced 70Z/3 cells, which do not express their rearranged κ gene, do not contain NF-κB activity; however, NF-κB activity is present after LPS or cycloheximide treatment (Sen and Baltimore, 1986b; Hromas et al., 1988). It is also present in λ-producing cells (Hromas et al., 1988). In vivo DMS protection studies on the human (Gimble and Max, 1987) and murine (Hromas et al., 1988) κ enhancers have shown B cell-specific protection at the NF-κB site, although there is little evidence for in vivo protection at κE1–3. Mutation and transfection studies (Lenardo et al., 1987) have demonstrated that NF-κB is required for constitutive and LPS-inducible activity of the κ enhancer; alteration of the NF-κB-binding site reduced activity to near enhancer-minus levels. Like IgH enhancer factors, NF-κB binds to other regulatory elements, including the SV40 promoter (Davidson et al., 1986; Sen and Baltimore, 1986a), class I major histocompatibility complex (MHC) promoters (Baldwin and Sharp, 1988), and the human immunodeficiency virus (HIV) long-terminal repeat (Nabel and Baltimore, 1987).

Activation of NF-κB-binding activity is particularly interesting and significant. This activity normally occurs only in B cells expressing light chains (Sen and Baltimore, 1986a). Since NF-κB can be induced in 70Z/3 cells by LPS and cycloheximide (Sen and Baltimore, 1986b), conditions which have previously been shown to induce κ transcription in these cells (Parslow and Granner, 1982; Wall et al., 1986), it seems likely that NF-κB-binding activity correlates with κ transcriptional activity. NF-κB activation in 70Z/3 occurs at a posttranslational step, since it is insensitive to inhibition of protein synthesis (Sen and Baltimore, 1986b). The nature of this posttranslational modification is not known, but it may be phosphorylation, since phorbol esters, which are known to activate protein kinase C, activate NF-κB (Sen and Baltimore, 1986b). Finally, it has been shown that the phorbol ester 12-O-tetradecanoylphorbol-13-acetate (TPA) will induce NF-κB in cells unrelated to B cells such as HeLa (Sen and Baltimore, 1986b). Thus, it appears that this key enhancer factor may be present in all cells, but is posttranslationally modified to become active only in B cells of a particular developmental stage. Furthermore, this is the first example of an immunoglobulin transcription factor which is regulated in response to signal transduction. SV40 enhancer factors, AP1 and AP2, have also been shown to be modulated in response to TPA (Angel et al., 1987; Imagawa et al., 1987). Thus, this may well be a general mechanism for altering gene expression in response to signals received at the cell surface.

Recently, Singh et al. (1988) have developed a new technique for cloning cDNAs-encoding transcription factors by screening cDNA expression libraries with DNA probes containing the binding site for the desired factor. They have used the method to clone a cDNA for NF-κB. Isolation of cDNA and genomic clones encoding the transcription factors opens the door to an exciting new level of inquiry into transcriptional regulation. The structure and regulation of the factor genes can be studied, and expression vectors containing the factor cDNAs can be used to obtain chemical amounts of purified factors for detailed biochemical analysis.

IV. Immunoglobulin Promoter Elements

A. Immunoglobulin Promoters Are Tissue and Stage Specific

Many studies have shown that the tissue specificity of both heavy and light chain gene expression is due to the preferential activity of promoters as well as enhancers. This has been demonstrated by testing the promoters in combination with a variety of tissue nonspecific enhancers for their activities in B cells and non-B cells. In cases in which the difference in transcriptional activity has been quantitated (Grosschedl and Baltimore, 1985; Picard and Schaffner, 1985), it seems that the preference is not absolute, since low levels of transcription are observed in some non-B cell lines.

B. Sequence Conservation among Heavy Chain V Gene Promoters

Several investigators have searched for regions of conservation between the promoters of different heavy chain V genes as a first step toward identifying functionally important sequences. In 1984, two groups (Parslow et al., 1984; Falkner and Zachau, 1984) identified an octanucleotide, ATGCAAAT, which occurred 5' of all heavy chain V genes and, in the inverted orientation, 5' of all light chain V genes. The same sequence is present upstream of genes encoding H2B, U1, U2, and U6 snRNAs, MHC class II α chain, herpes viral tk, and in several bacterial operons, which respond to cAMP (Falkner et al., 1986; Carbon et al., 1987; Sive et al., 1986; Parslow et al., 1987). In 1987, our group identified two additional sequences, which were common to all heavy chain V genes examined (Eaton and Calame, 1987). One of these, the heptamer CTCATGA, lies between 2 and 22 bp 5' to the octanucleotide. Ballard and Bothwell (1986) observed conservation in this region for the J558 V gene family, and Siu et al. (1987) have also noted that sequences 5' to the octamer are conserved. The second region of conservation is an area of purine–pyrimidine asymmetry no more than 46 bp upstream of the heptamer. Each of these conserved elements has been implicated in the control of V_H promoter activity in plasmacytomas (Mason et al., 1985; Ballard and Bothwell, 1986; Grosschedl and Baltimore, 1985; Eaton and Calame, 1987).

C. The Octamer Is Essential for V_H Promoter Activity

The conserved octanucleotide has been shown to be an essential part of several heavy chain V gene promoters (Mason et al., 1985; Ballard and Bothwell, 1986; Eaton and Calame, 1987). Mason et al. constructed a series of progressive 3' deletions of a Q52 heavy chain promoter fused to the human β-globin TATA box, cap site, and coding region. The V_H used in their studies contains two copies of the conserved octanucleotide, one at -48 bp and one at -249 bp from the V_H cap site. Promoter activity was determined by stable transfection into the plasmacytoma J558L, followed by S1 analysis of resulting RNAs. Deletions, which remove 3' sequences including the octanucleotide at -48, are significantly reduced in activity. Ballard and Bothwell (1986) also demonstrated the importance of the octanucleotide for the V_H 186.2 promoter. They generated mutants of this promoter region in an intact, rearranged heavy chain gene and introduced the constructs by stable transfection into J558L. A progressive 5' deletion, which removed the octanucleotide, eliminated correct transcription. Several mutants that contained single or double point mutations in the octamer were also inactive. We have shown the octamer to be important for the transcriptional activity of the V_λ heavy chain promoter; internal deletion of the octamer reduces promoter activity to <2% of wild type (Eaton and Calame, 1987).

Other investigators have shown that the octamer is also a functional part of the nonimmunoglobulin promoters in which it occurs. It is required for the normal expression of genes encoding H2B (Sive et al., 1986), U1, U2, and U6 snRNAs (Murphy et al., 1987; Mattaj et al., 1985; Carbon et al., 1987), and herpes viral tk (Parslow et al., 1987).

Since the octamer occurs in the opposite orientation in light chain promoters, one might expect a priori that its function would be independent of orientation. This is in fact the case. Heavy chain promoters in which the octamer has been inverted retain significant activity, albeit 50% reduced (Ballard and Bothwell, 1986; Eaton and Calame, 1987). This small reduction may suggest a partial requirement for interaction with surrounding DNA sequence elements or their transacting factors. With this in mind, it is interesting to note that, in the tk promoter, inversion of the octamer sequence eliminates its activity (Parslow et al., 1987).

Several studies have suggested that the octamer is responsible for the B cell-specific expression of immunoglobulin genes (Grosschedl and Baltimore, 1985; Mason et al., 1985; Dreyfus et al., 1987; Wirth et al., 1987). Wirth et al. (1987) constructed a promoter de novo consisting of a TATA box and an oligonucleotide, which contained the octamer. In the presence of either a viral or the heavy chain enhancer, this promoter was preferentially active in B cells. In related experiments, Dreyfus et al. (1987) showed that insertion of the octamer into the renin promoter (not normally active in B cells) produced a hybrid promoter that had B cell-specific activity in the presence of an enhancer. Its ability to confer B cell-specific expression under some conditions seems difficult to reconcile with its role in many promoters and enhancers whose activities are not restricted to a particular cell type. This interesting paradox will be discussed later in this review.

D. Sequences Upstream of the Octamer Are Required for Heavy Chain Gene Expression

Although the octamer is necessary for heavy chain gene expression, it does not appear to be sufficient for normal levels of transcription. Our laboratory has identified three sequence elements lying 5' to the octamer, which have a 2- to 5-fold effect on transcriptional activity of the heavy chain promoter (Eaton and Calame, 1987). We have shown that mutant promoters lacking the conserved heptamer are 17% as active as wild type. Ballard and Bothwell (1986) have also noted that progressive 5' deletions, which remove the heptamer, are significantly less active than the intact promoter. The conserved stretch of purine–pyrimidine asymmetry increases activity 2-fold. Also, an element between 251 and 125 bp from the cap site increases promoter activity by a factor of two. The DNA sequence in this latter region bears several significant homologies to the heavy chain enhancer. The specific function of each of these sequences has not been elucidated; however, they provide additional candidates for mediators of promoter–enhancer interaction, induction by lymphokines, enhancer-independent expression, and tissue specificity.

E. Multiple Proteins Interact with the Conserved Octanucleotide

Gel retardation, methylation interference, and DNase I protection experiments have shown that the octamer binds factors present in nuclear extracts. Depending on the system used, investigators have identified from one to four different binding activities (Mocikat et al., 1986; Hromas et al., 1988; Staudt et al., 1986; Landolfi et al., 1986; Rosales et al., 1987), some of which are restricted in their tissue distribution, and some of which are present in all cell types examined. An octamer factor, which is present (or active) only in lymphoid cells, might generate the tissue-restricted pattern of expression, which is observed for immunoglobulin genes, while more generally distributed octamer factors might activate genes encoding tk, snRNAs, and H2B. However restricted their activities may be *in vivo*, both ubiquitous and tissue-specific octamer-binding factors interact with immunoglobulin genes as well as nonimmunoglobulin genes *in vitro* (Sive and Roeder, 1986; Bohmann et al., 1987; Wirth et al., 1987). The abilities of the tissue-specific and general factors to activate transcription from different sets of promoters may depend on the variable organization of other elements within these promoters. With this in mind, it is interesting to note that several other nuclear factors interact with heavy chain V gene promoters (Peterson et al., 1988; S. Eaton and K. Calame, unpublished; Scheidereit et al., 1987). These factors are discussed in more detail in a subsequent section.

Both the lymphoid-specific octamer factor and a general octamer factor have recently been purified to homogeneity (Scheidereit et al., 1987; Fletcher et al., 1987; Wang et al., 1987). Each is functional in an *in vitro* transcription assay; the general factor stimulates H2B transcription, and the lymphoid factor stimulates κ transcription. Thus far, it has not been directly determined whether they do so exclusively. This will be an important issue to resolve in order to choose between models for the generation of tissue specificity.

F. A Nuclear Factor Interacts with the Conserved Heptamer

The conserved heptamer lies between 2 and 22 bp 5' to the octamer in different heavy chain V genes. Landolfi *et al.* (1987) have observed DNase protection of the heptamer, along with the octamer, in a gel retardation complex. Generation of this complex is dependent on inclusion of heptamer sequences in the probe and is present in lymphoid as well as nonlymphoid extracts. They speculate that this might reflect a different octamer-binding protein that interacts with DNA over a larger region or that, alternatively, the complex might consist of two separate proteins. If this were so, it would seem likely that the heptamer- and octamer-binding proteins interact with each other, since gel retardations are performed under conditions of probe excess. We have observed protection of the heptamer in partially purified nuclear extracts of plasmacytoma cells (S. Eaton and K. Calame, unpublished). The heptamer-binding site is illustrated in Fig. 3. The positioning of the heptamer close to the octamer suggests that it may play a role in the binding, stabilization, or activation of the tissue-specific octamer factor. Alternatively, its binding may be incompatible with the binding of a ubiquitous octamer factor. This is especially relevant to the consideration of the tissue-specific activity of heavy chain promoters.

G. Additional Tissue-Specific Interactions Occur Far Upstream of the Octamer

Our functional analysis of the V_1 heavy chain promoter demonstrated a positive transcription element located between -251 and -125 bp from the cap site (Eaton and Calame, 1987). Recent protein-binding data

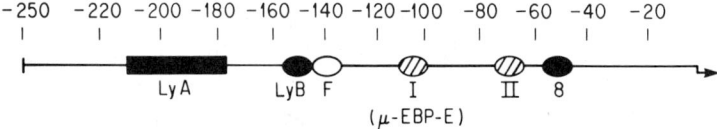

FIG. 3. Protein-binding sites of nuclear factors to the V_1 heavy chain promoter. Numbering indicates distance from the transcriptional start site. ■, present in B cells, but not in fibroblasts; ○, present in fibroblasts, but not in B cells; ⊘, present in fibroblasts and in B cells.

show a complicated pattern of protection, which has both B cell- and fibroblast-specific aspects (S. Eaton and K. Calame, unpublished). B cell nuclear extracts of all developmental stages protect a large region from -175 to -215. Contained within this protection is a strong homology to the IgH enhancer D site (see Section III,G). This factor has been termed LyA. Another B cell-specific binding site is centered at -150 (termed LyB) and partially overlaps a fibroblast-specific site (termed F). Methylation interference data indicate that these factors interact with overlapping sets of purines. Factors LyA, LyB, and F, are illustrated in Fig. 3.

Tissue-specific protection in this region strongly suggests that it is involved in restricting promoter activity to B cells. Proteins bound in this region may either activate heavy chain promoters in B cells or limit their activity in non-B cells. A previous study of a different heavy chain promoter, however, demonstrated that only 154 bp 5' to the cap site was necessary for preferential activity in B cells (Grosschedl and Baltimore, 1985). Furthermore, the conserved octanucleotide has been implicated in tissue-specific expression (Wirth *et al.*, 1987; Dreyfus *et al.*, 1987). The function of this upstream region may be redundant with that of the octamer in conferring tissue specificity on heavy chain promoters. Further redundancy exists in that both heavy chain promoters and the heavy chain enhancer independently restrict expression to B cells. Perhaps multiple tissue-specific elements are necessary to achieve the rigorous transcriptional control observed *in vivo*.

H. Immunoglobulin Heavy Chain Promoters and Enhancers Bind Two Common Factors

The heavy chain promoter and enhancer both interact with octamer-binding protein. We have recently demonstrated that the heavy chain promoter also binds to purified μEBP-E, a heavy chain enhancer-binding protein (Peterson *et al.*, 1988). Deletions of μEBP-E in the enhancer reduce activity to 36% of wild type (Tsao *et al.*, 1988). In the promoter, μEBP-E binds immediately 5' of the conserved stretch of purine–pyrimidine asymmetry (shown in Fig. 3). Although it lies in a region that progressive 5' deletion has shown to be necessary for full promoter activity, its function has not been specifically determined in the promoter. The fact that two factors bind to both the IgH promoter and enhancer is intriguing. These factors may play a role in promoter–enhancer interaction, providing the basis for the observed synergy between the two elements. Possible mechanisms of promoter–enhancer interaction will be discussed in a subsequent section. Alternatively, the presence of enhancer factors in the promoter might be involved in its enhancer-independent expression in pre-B cells. This will be discussed later in this review.

I. Sequence Conservation among Light Chain Promoters

The control of heavy and light chain gene expression is similar, although not identical. The expression of both genes is specific to B cells. At late stages of B cell development, light and heavy chains are synthesized in roughly equivalent amounts, resulting in the proper stoichiometry for formation of an intact immunoglobulin molecule (Schibler et al., 1978). At earlier stages, however, heavy chain transcription is active, while the rearranged κ gene is silent. Also, rearrangement of heavy chain genes precedes and is required for rearrangement of light chain genes (Yancopoulos and Alt, 1986; Reth and Alt, 1984). Given these facts, one might expect that heavy and light chain genes share some transcriptional regulatory elements and have others that are unique.

Both κ and λ light chain promoters are similar to heavy chain promoters in that they contain an octamer (Parslow et al., 1984; Falkner and Zachau, 1984). In addition to the octamer, a pentadecanucleotide is conserved upstream of the octamer in all light chain V genes, but is not seen in heavy chain V genes (Falkner and Zachau, 1984). The pentadecanucleotide is located approximately 100 bp from the cap site and also exists, along with the octamer, in the J chain promoter (Matsuuchi et al., 1986).

J. The Octamer Is Required for κ Promoter Activity

The octamer has been shown to be required for κ promoter function in plasmacytomas. Progressive 5' deletions, which interfere with this element, abolish promoter activity in transfection assays (Falkner and Zachau, 1984; Bergman et al., 1984; Queen et al., 1986). Although the pentadecanucleotide is conserved, these studies have not elucidated its function. Deleting the sequence has no effect on κ promoter function in plasmacytoma cells (Bergman et al., 1984; Queen et al., 1986). It is possible that the pentadecanucleotide affects transcription at some other stage of B cell development. There is evidence for different transcriptional states for κ promoters; it has recently been shown that, when transfected into normal B cells, the κ 41 gene utilizes a different start site than when transfected into plasmacytomas (Bich-Thuy and

Queen, 1988). Alternatively, the pentadecanucleotide may play a role in some nontranscriptional process.

K. MULTIPLE PROTEINS INTERACT WITH κ V GENE PROMOTERS

Several investigators have observed nuclear factors that interact with the conserved octanucleotide in κ promoters (Mocikat et al., 1986; Hromas et al., 1988; Staudt et al., 1986; Scheidereit et al., 1987). Recent studies have also described an activity in partially purified nuclear extracts, which footprints a sequence immediately 5' to the octanucleotide (not the pentadecanucleotide) in the T1 V_κ promoter (Scheidereit et al., 1987). Transfection studies using progressive 5' deletion mutants of this promoter (Falkner and Zachau, 1984) show that this sequence lies in a 70-bp region necessary for transcriptional activity. Other experiments utilizing a different κ promoter indicate that the sequence can be altered without greatly affecting expression (Bergman et al., 1984). The function of this protein is, as yet, unclear; however, its close apposition to octamer-binding protein is suggestive of some sort of interaction between them. Scheidereit et al. (1987) have also identified two additional binding activities by gel retardation, which are less well described. It is beginning to appear that, like the heavy chain promoter, the light chain promoter is more complex than was first imagined.

V. Possible Mechanisms Mediating Transcriptional Regulation of Immunoglobulin Genes

We cannot fully understand how enhancers and promoters work together to form stable transcription–initiation complexes until we understand the nature and interactions of all factors that bind to each element and the structure of the initiation complex itself. However, it is helpful to consider testable models, which explain the known features of Ig transcriptional regulation. Enhancers probably facilitate the formation of stable transcription complexes (Wang and Calame, 1986) and appear to be required for establishment, rather than maintenance, of transcription (Wang and Calame, 1985; Atchinson and Perry, 1986). Both Ig promoters and enhancers have a strong B cell preference. The octamer sequence appears to play an important role in determining B cell specificity of both promoters and the heavy chain enhancer, although other elements may also be involved. There is differential transcriptional regulation of IgH and IgL genes at different developmental stages; V_H promoters are enhancer independent and enhancer dependent at different developmental stages. Finally, while capable of interacting with heterologous elements, heavy chain enhancers and promoters display synergy, i.e., they are more active together than with heterologous elements (Garcia et al., 1986). In the following section, we suggest several models which are consistent with these facts.

A. TISSUE AND STAGE SPECIFICITY

The complexity of different factors, which have been shown to bind to Ig transcriptional promoters and enhancers, appears enormous; similar complexity has been found for other elements, which have been studied (Nomiyama et al., 1987; Xiao et al., 1987). However, most and possibly all of these factors have been shown to bind to more than one promoter or enhancer element, and many have a wide tissue distribution. Thus, there is little evidence of "gene-specific" factors.

These data are consistent with an emerging view that the specific regulatory characteristics of a particular enhancer or promoter element may be determined by the unique nature, number, and physical arrangement of binding sites for relatively common factors (Voss et al., 1986; Matthias et al., 1987). Further regulation may be achieved in particular enhancer or promoter elements by (1) binding sites for negative as well as positive factors; (2) binding sites for two factors, which interact with one another cooperatively (Sawadogo and Roeder, 1985); (3) overlapping, mutually exclusive binding sites for either positive and/or negative factors; or (4) factors which can either activate or repress, depending on the context of the binding site, similar to the case for yeast RAP-1 (Shore and Nasmyth, 1987). Unique regulatory properties of transcription elements may thus be determined by "combinatorial" mechanisms, similar to the generation of antigen–recognition diversity in the immune system by combinatorial joining and association.

Tissue- or developmental stage-specific expression of particular genes could be achieved by varying the level of certain factors in particular tissues. It may not be necessary to vary all required factors, if the presence or absence of one or a few factors alters the activity of the element. This may be the explanation for the observation that most IgH enhancer-binding factors, with the exception of one octamer factor, have a wide tissue distribution. Furthermore, it may be that some enhancers or promoters can function, at different efficiency, with different subsets of factors, allowing different levels of activity in different tissues. This type of modular activity has been suggested by Wasylyk and Wasylyk (1986) for the IgH enhancer, since mutation of different regions of the enhancer has differential effects in fibroblasts and B cells. Levels of transcriptional factors may be regulated transcriptionally, posttranscriptionally, or posttranslationally. The activation of NF-κB in response to LPS or TPA treatment is currently the best-understood example of posttranslational activation of a transcription factor, which appears to occur in a tissue-specific manner (Sen and Baltimore, 1986b).

B. THE OCTAMER AS A TISSUE-SPECIFIC ELEMENT

A variety of evidence indicates that, under some conditions, the octamer is able to activate transcription in a B cell-specific manner. Two groups have demonstrated that the octamer can confer B cell-specific promoter activity in a heterologous system in the absence of additional heavy chain-derived sequences (Wirth et al., 1987; Dreyfus et al., 1987). The octamer is also implicated in the B cell-specific activity of the enhancers in which it occurs (Nomiyama et al., 1987; Gerster et al., 1987; Ondek et al., 1987; Schirm et al., 1987).

In apparent contrast to these studies, other experiments have revealed situations in which the octamer does not confer B cell-specific activity. Octamer sequences are not sufficient for the tissue-specific expression of the *HLA-DR* gene; inclusion of an upstream "class II box" is necessary for proper expression (Sherman et al., 1987). The octamer is a functional element in the promoters of many genes that are active in all cell types. Even within this broad category, the octamer appears to mediate different types of transcriptional control. Octamer-containing sequences are necessary and sufficient for the S-phase induction of H2B transcription (Sive et al., 1986; LaBella et al., 1988) and are also required for normal transcription of genes for U1 and U2 snRNAs (Murphy et al., 1987; Mattaj et al., 1985) and herpes viral tk (Parslow et al., 1987), although transcription of these genes is not subject to cell cycle regulation. In addition to transcription–initiation, the octamer in the U1 and U2 snRNA promoters seems to be involved in proper termination of transcription (Hernandez and Weiner, 1986; Neuman de Vegvar et al., 1986). It has recently been shown that the octamer can also function as a promoter element for the U6 snRNA gene, which is apparently transcribed by RNA polymerase III (Carbon et al., 1987). Finally, in addition to the many transcriptional roles already described for the octamer, this element may also be involved in stimulating DNA replication (Rosenfeld et al., 1987; Pruijn et al., 1987). Thus, it seems that, while the octamer may be sufficient for lymphoid-specific activity under some condition, it can be engaged in a multiplicity of regulatory activities, depending on its context.

The wide range of activities observed for the octamer is partially explained by the recent identification of multiple distinct octamer-binding proteins, all of which seem to have identical sequence specificity, but which may differ in their activities. Gel retardation experiments have demonstrated several nuclear proteins, which differ in electrophoretic mobility, that bind to the octamer motifs of both heavy and light chain promoters (Landolfi et al., 1986, 1987; Staudt et al., 1986). At least one of these has been observed to be restricted to cells of the lymphoid lineage and may contribute to the tissue-specific expression of immunoglobulin genes.

Although different forms of octamer-binding protein may contribute to tissue-specific expression in vivo, it is clear that immunoglobulin octamer sequences can be bound by the ubiquitous factor in vitro. Competition studies indicate that proteins that interact with immunoglobulin promoters are also able to bind octamer motifs in the heavy chain enhancer, the SV40 enhancer, U1, U2, and U6 snRNA promoters, H2B promoter, and the adenovirus replication origin (Pruijn et al., 1987; Bohmann et al., 1987; Sive et al., 1986). Conversely, it seems that octamer sequences, which occur in nontissue-specific promoters, are capable of binding the tissue-specific factor and the ubiquitous factor equally well (Wirth et al., 1987). Regardless of the ability of ubiquitous octamer-binding factors to interact with immunoglobulin octamer sequences in vitro, no such interaction appears to occur in vivo. When protein binding to the heavy chain enhancer is examined in vivo by DMS treatment of nuclei, the octamer is only protected in lymphoid nuclei. Nonlymphoid nuclei, although they presumably contain the ubiquitous factor which binds in vitro, do not, in fact, protect the IgH enhancer octamer (Ephrussi et al., 1985; Church et al., 1985).

Considering the multiplicity of octamer-binding factors and the variable arrangement of octamer sequences with respect to other promoter elements, it is possible to suggest several models, which may explain its B cell-specific activity in immunoglobulin genes.

1. Model 1

Binding to the octamer is unstable (or nonfunctional), unless accompanied by the binding of an adjacent factor. Different octamer-binding proteins have variable domains, which interact with a number of different DNA-binding proteins. The lymphoid-specific octamer factor has an octamer-binding domain and a factor X-binding domain. It will activate transcription only when an octamer sequence occurs next to the target sequence for factor X binding. Note that this model does not require that factor X be restricted to B cells.

In support of this model, many octamer sequences occur next to binding sites for other proteins and are part of larger regions of conservation. *In vivo* DMS protection data show that, in lymphoid nuclei, the IgH enhancer octamer is immediately adjacent to another protein-binding site, μE4 (Ephrussi *et al.*, 1985). In heavy chain promoters, the heptamer sequence immediately upstream of the octamer is necessary for full promoter function (Eaton and Calame, 1987; Ballard and Bothwell, 1986) and may interact with an additional factor (Landolfi *et al.*, 1987; S. Eaton and K. Calame, unpublished). The distal sequence elements (DSEs) in U snRNA gene promoters consist of octamers plus another factor-binding site quite close. For example, in U2, the octamer is next to a SP1 site (Ares *et al.*, 1987). Removal of a class II box sequence upstream of the HLA-DR octamer eliminates tissue-specific expression. The model provides for a great deal of combinatorial diversity; in this way, *n* different sequences could be combined to give *n*! different specificities.

Model 1 is inconsistent with the data of Wirth *et al.* (1987) and Dreyfus *et al.* (1987), which imply that an octamer alone is sufficient for lymphoid-specific transcription. Possibly complicating their data is the fact that the experiments involved heterologous enhancers and the presence of a TATA box. TATA boxes have not been shown to have a function in heavy or light chain promoters.

2. Model 2

The ubiquitous octamer-binding factor(s) are nonfunctional in the absence of other promoter elements, but the tissue-specific factor is capable of activating transcription by itself.

This model is consistent with Wirth *et al.* (1987) and Dreyfus *et al.* (1987) in that an octamer alone would be sufficient for lymphoid-specific expression. It would probably require that, in lymphoid cells, repressors of inappropriate activity or binding exist for genes like H2B, which are cell cycle regulated at the level of transcription (Sive and Roeder, 1986). This model would not predict the existence of the nonoctamer factors, which bind very close to the octamer in immunoglobulin promoters (Scheidereit *et al.*, 1987; Landolfi *et al.*, 1987; S. Eaton and K. Calame, unpublished).

3. Model 3

Although both ubiquitous and tissue-specific octamer factors in nuclear extracts bind to immunoglobulin promoter and enhancer octamers under conditions of probe excess (as in gel retardation), in the cell, the ubiquitous factor cannot bind to Ig octamers either because (1) it is sequestered by the many other genes that bind it or (2) it has a lower binding affinity for Ig genes than for other promoters. Sequences surrounding the octamer in immunoglobulin genes may contribute to an increased affinity for the tissue-specific factor. In support of this model, Ballard and Bothwell (1986) have observed that a heavy chain promoter fragment, containing both the octamer and 12 bp immediately 5' to it (including the heptamer), competes for octamer binding at lower concentrations than a smaller fragment, which does not include the upstream sequence. Other investigators have reported differences in affinity of the octamer-binding protein NF-III (ORP-C) for octamers with different flanking sequences (Pruijn et al., 1987). Also, Falkner et al. (1986) have noted that a fragment of the κ promoter from which the octamer has been deleted can still compete, albeit at a reduced level, with an octamer-containing probe for binding.

4. Model 4

Although both ubiquitous and tissue-specific factors are present in nuclear extracts, they are synthesized at different times during the cell cycle and immediately sequestered. Only the tissue-specific factor is synthesized and available at a time when Ig genes are replicating and capable of binding it. It is interesting to note that the time of replication of V genes is altered after V–D–J joining (Calza et al., 1984).

5. Model 5

The transcriptional activity, but not binding ability, of octamer factors may depend on the context of the binding sites. Perhaps ubiquitous factors act as activators at non-Ig octamer sites, while for Ig octamer sites, they are inactive or repressive. Alternatively, the B cell-specific octamer factors may only be activators for the Ig octamers. This suggestion is based on the recent observation by Shore and Nasmyth (1987) that yeast RAP-1 can act either as a repressor or activator of transcription, depending on the context of its binding site.

C. Immunoglobulin Promoter–Enhancer Interactions

IgH enhancers and promoters display synergy (Garcia et al., 1986), and κ enhancers activate κ promoters better than viral promoters (Queen et al., 1986; Garcia et al., 1986). In addition to IgH promoters and enhancers, there are other examples of promoter–enhancer pairs, which display unusual activity. Parslow et al. (1987) have shown that the ability of the herpes virus tk promoter to be activated by different enhancers depends on the presence or absence of the octamer sequence and on the cell background, i.e., the κ enhancer could activate the tk promoter equally well if it did or did not contain the octamer, while the Moloney sarcoma virus (MSV) enhancer could not activate the tk promoter containing an octamer. These results again suggest that different subsets of factors must be able to interact and ultimately initiate transcription at particular promoters with varying efficiency. Which factors are used and the resulting efficiency of transcription–initiation would be determined by factor levels in particular cells and the configuration of factor-binding sites in relevant promoter and enhancer elements.

Several models for promoter–enhancer interaction can be imagined. These models are not mutually exclusive, and different enhancers may utilize different mechanisms.

1. Model 1

Enhancers may act to increase the local concentration of factors, which eventually combine at the promoter to form a stable transcription complex. This model is consistent with the evidence that enhancers are required for initiation, but not maintenance, of transcription and that there are factors common to both heavy chain promoters and enhancers. It would also explain the inability of certain promoters and enhancers to interact.

2. Model 2

Enhancers may contain binding sites for factors, which modify and activate proteins involved in stable complex formation at the transcription–initiation site, although the enhancer factors themselves are not a part of the initiation complexes. There are examples of this mechanism in prokaryotes and yeast. The *Escherichia coli* glnAp2 promoter is activated by binding sites for factoor NRI, located 100 and 130 bp upstream, which function similar to eukaryotic enhancers. Ninfa *et al.* (1987) have recently shown that phosphorylated NRI facilitates conversion of a promoter–polymerase complex to its open, transcriptionally active form. Model 2 is also consistent with the results of Brent and Ptashne (1985), who were able to activate the yeast Gal 1 promoter by using a hybrid protein, which bound to a *lexA* operator, placed upstream of the Gal 1 promoter, but contained Gal 4 transcription-activating domains. This model is consistent with the observation that enhancers seem to stably activate transcription and with the failure of some promoter–enhancer combinations to be active. It does not require that enhancers and promoters bind common factors.

3. Model 3

The same protein may bind to both an enhancer and a promoter and could bring these elements into close proximity, looping out the DNA between them. Either one protein could have multiple bindings sites for DNA or two protein molecules, each bound to DNA, could interact with one another to form a protein–protein dimer. The resulting close proximity between an enhancer and a promoter could then allow rapid transfer of enhancer factors to the transcription–initiation complex (Model 1 in Section V,C,1) or efficient interaction of enhancer and promoter factors (Model 2 in Section V,C,2). This model predicts the observed binding of common factors to enhancers and promoters, which act synergistically, such as the IgH enhancer and promoter. Evidence for this model comes from cooperative dimerization and DNA looping, which have been demonstrated in *E. coli* for repression in the *lac* (Mossing and Record, 1986) and *deo* (Dandanell *et al.*, 1987) operons and for binding of bacteriophage λ repressor (Hochschild and Ptashne, 1986).

D. Enhancer Independence and Enhancer Dependence of V_H Promoters

If the net result of Models 1, 2, or 3 (see Section V,C) is for enhancers to increase the local concentration of factors required at the promoter, they would be superfluous if the initial concentration of the factor were high enough. This has been shown to be true of NRI, a factor required for glnAp2 promoter activity (Ninfa *et al.*, 1987). When the *in vitro* concentration of NRI is sufficiently high, the upstream NRI-binding sites do not further increase promoter activity. When the concentration is lowered, however, efficient open complex formation is dependent on the presence of NRI-binding sites. The authors suggest that the function of the enhancer may be to increase the local concentration of NRI near the promoter, so that it can alter the transcription complex. Perhaps early in B cell development, nuclear factors, required for heavy chain V gene promoter activity, are present in sufficient amounts to obviate the need for an enhancer. Subsequent down-regulation of these factors as B cells mature would make V gene transcription enhancer dependent. An analogous situation exists for the oocyte and somatic 5 S genes. The relative transcription rate of these genes varies from a ratio of 5:1 (somatic:oocyte) in the ooctye to a ratio of 1000:1 at midblastula transition. Recently, it has been shown that, in activated egg extracts, stable transcription complexes form on somatic genes, whereas oocyte gene complexes are destabilized. Since their associated factors (e.g., TFIIIA) are in equilibrium with free factors, oocyte genes are very sensitive to the concentration of TFIIIA. Somatic genes, however, are relatively insensitive (Wolffe and Brown, 1987). The heavy chain enhancer may act in an analogous fashion to stabilize transcription complexes on rearranged V gene promoters, making them less sensitive to changes in levels of transcription factors.

An alternative model is that expression of germ-line V genes is activated by completely different signals and factors, which function only in pre-B cells. It should be possible to distinguish between these two models by transfection experiments. If there exist sequences that affect germ-line transcription, but not enhancer-dependent transcription, then the latter model would be more likely.

VI. T Cell Receptor Gene Expression

A. GENERAL FEATURES OF TCR GENE EXPRESSION DURING T CELL ONTOGENY

During selection and maturation in the thymus, different functional subsets of T cells are formed (reviewed in Adkins *et al.*, 1987; Mathieson and Fowlkes, 1984). Although the exact ontogeny of each type of T cell is not completely understood (Adkins *et al.*, 1987), several groups have shown that there appears to be an ordered sequence of TCR gene rearrangement, similar to the ordered rearrangement of heavy κ and λ Ig genes (reviewed in Allison and Lanier, 1987 and in Kronenberg *et al.*, 1986; Chien *et al.*, 1987b). When thymus is analyzed at different times in fetal development, it is observed that γ and δ genes rearrange early, followed by β, and later, α gene rearrangement (Chien *et al.*, 1987a,b; Allison and Lanier, 1987; Kronenberg *et al.*, 1986). TCR α- and β- and TCR γ- and δ-expressing cells probably represent distinct mature T cell lineages. α–β Heterodimers are expressed on the surface of MHC-restricted cytotoxic and helper T cells. Recently, it has been shown that suppressor T cells also express α–β heterodimers (Modlin *et al.*, 1987). The exact role of γ–δ heterodimers is not clear, but they are expressed on $CD3^+$ fetal thymocytes, $CD3^+CD4^-CD8^-$ adult thymocytes, and a subset of $CD3^+$ adult peripheral lymphoid organs and represent a lineage that displays non-MHC-restricted cytolysis (Matis *et al.*, 1987).

With the exception of occasional TCR gene rearrangement or expression in some B cells (Calman and Peterlin, 1986), no tissues other than T cells rearrange or express TCR genes. Steady-state mRNA levels for expression of the genes generally follow the same sequence as gene rearrangement (Calman and Peterlin, 1986; Kronenberg et al., 1986). In the case of β mRNA, a "sterile" 1.0-kb transcript, resulting from a D–J joined gene, is often observed early in development before V–D–J joining is complete (Siu et al., 1984). Recent run-on transcription studies have demonstrated that, like immunoglobulin genes, germ-line TCR β chain genes are transcriptionally silent in mature T cells (McDougall and Calame, 1988). These data suggest that rearrangement is required to activate their transcription and imply that transcription enhancers may be brought into functional proximity with TCR promoters as a result of rearrangement.

In all mature T cell types, the number of TCR molecules expressed on the cell surface and the number of mRNAs are probably low. This is in contrast to secreted Ig molecules for which mRNAs are synthesized at extremely high levels (Schibler et al., 1978). The low level of TCR expression suggests that putative transcription signals for TCR genes may be weak. This notion is strengthened by the difficulty several laboratories have experienced in achieving expression of TCR genes following transfection into tissue culture cells (K. Kelly, personal communication; D. Loh, personal communication; S. McDougall and K. Calame, unpublished).

There is evidence for trans-acting negative regulation of TCR gene expression. T–B hybrids (Kozbor et al., 1987) express Ig normally, but TCR β expression is decreased and α expression is abolished. Similarly, when hybrids are formed between T cells expressing and not expressing TCR β mRNAs, β transcription is abolished, suggesting a trans-acting negative factor (MacLeod et al., 1986). The negative factor affects both 1.3 (V–D–J–C) and 1.0 (D–J–C) mRNAs, suggesting that the target for negative regulation may not be the V_β promoter element.

B. Transcriptional Regulatory Elements in T Cell Receptor Genes

Sequences involved in the expression of T cell receptor genes are just beginning to be examined. Luria et al. (1987) have identified promoter and enhancer elements in the human TCR α locus, whose location is analogous to that of their immunoglobulin counterparts. Using *CAT* as a reporter gene, they found that a 600-bp fragment containing sequences 5′ to the ATG functioned as a promoter in T cells, but not B cells or HeLa cells. A 1.1-kb fragment derived from a region just 5′ to C_α contains a duplicated decamer that is homologous to an IgH enhancer core sequence. This fragment enhanced expression from the V_α promoter and the SV40 promoter in T cells or B cells, but not HeLa cells. Thus, the α enhancer is lymphoid specific, although the promoter is strictly T cell specific. Luria et al. have noted that the widely dispersed arrangement of V_α genes seems to suggest that this enhancer is capable of acting over very large distances, similar to the heavy chain enhancer (Wang and Calame, 1986).

β chain transcription is activated by rearrangement (McDougall and Calame, 1988). This implies the presence of an enhancer in the functional vicinity of the β chain constant regions, analogous to that of immunoglobulin heavy and light chain genes. Luria et al. (1987) noted the presence of a duplicated decamer, similar to that in the α enhancer located upstream of $C_\beta 2$. Bier et al. (1985) have reported a T cell-specific DNase I hypersensitive site 5' of $C_\beta 2$ and have suggested that an enhancer might be located in this region. Searches for a functional enhancer in this region, however, have been without success (McDougall and Calame, 1988). Recently, sequences located between $C_\beta 2$ and the downstream variable gene segment $V_\beta 1$ have been examined for the presence of an enhancer. McDougall and Calame (1988) have identified a 1-kb fragment located approximately 8 kb 3' to $C_\beta 2$, which enhances transcription from both a V_β promoter and the SV40 early promoter, when transfected into T cell lines (McDougall and Calame, 1988). The location of this enhancer ensures that it will be retained whether a V gene rearranges to $C_\beta 1$ or $C_\beta 2$. The existence of an enhancer 3' to the constant region is unique among genes of the immunoglobulin super gene family, including the T cell receptor α chain locus. Enhancers, described for the immunoglobulin heavy and light chain genes, and the T cell receptor α chain gene are all located 5' to the constant region. This discovery raises interesting questions about the evolutionary origin of the β chain locus.

The promoter regions of V_β genes have not yet been dissected functionally. Siu et al. (1986) have examined a series of murine V_β upstream regions and identified a 16-bp conserved sequence that generally occurs approximately 100 bp upstream of the ATG. Although it is conserved in the murine and human sequences, its function is as yet undetermined.

Protein binding to a human V_β promoter has been examined by Royer and Reinherz (1987). They have identified four regions within 175 bp of the cap site, which are protected from DNAase I digestion by nuclear extracts. Three of these can be generated by nuclear extracts from a variety of cell types. One seems to be specific to lymphoid cells. Two of the nonrestricted binding sites have limited homology to regions of the immunoglobulin heavy chain and SV40 enhancers, and one protects the conserved sequence identified for V_β genes (Siu et al., 1986). The function of these elements has not yet been examined.

Some V_β genes display a behavior which, thus far, is unique among immunoglobulin and T cell receptor genes. Chou and colleagues have observed variable splicing of the leader exon of one V_β gene to the V–D–J exon of another (Chou et al., 1987a). The V gene, which provides part of the leader region, can also be expressed in its entirety in other T cell lines. Conversely, still other T cell lines express both the leader and the V–D–J exons of the downstream V_β gene. This phenomenon might provide an interesting model system in which to study differential splicing decisions.

VII. Perspectives

A great deal of progress has been made within the past 5 years, since Ig enhancer and promoter elements were first identified. First, cis-acting DNA elements have been defined. Most of the elements of the heavy chain and κ enhancers are probably known; it is also clear that the elements of the Ig promoters are rapidly being defined. Although less is known about TCR cis-acting elements, it appears that technical difficulties with expression systems have been overcome and that we will soon understand the DNA elements responsible for regulated expression of these genes as well. Second, the proteins, which recognize functionally defined DNA elements, are being identified, characterized, and purified. Even though the number and complexity of transcription factors is surprisingly large, this task is well underway, and technology is available to complete it in a few years. Finally, technology is now available for cloning transcription factor genes, either by using affinity chromatography (Kadonaga and Tjian, 1986) to obtain pure protein for sequence analysis and subsequent synthetic probe production or by definition of binding sites and subsequent screening of expression libraries with recognition site probes (Singh et al., 1988).

At present, cloning of transcription factors is just beginning. Completion of this task will provide a significant advance toward our ultimate understanding of the basic questions outlined above. First, by expressing cloned genes, milligram amounts of purified transcription factors should become available for structural and biochemical studies of the transcription–initiation machinery. DNA-binding proteins can also be used to isolate other, non-DNA-binding proteins, which may participate in transcription–initiation. In conjunction with availability of purified transcription factors, it may be possible to develop more dependable and regulation-sensitive *in vitro* transcription systems. Second, by studying the structure, organization, and regulation of genes for transcription factors, we should be able to unravel how the transcription apparatus is modulated in response to inductive, developmental, or tissue-specific signals.

What do we ultimately wish to understand? At minimum, there are three important questions. (1) How are stable transcription–initiation complexes formed, what is their structure and mechanism of action? (2) How is the transcription apparatus modified in response to inductive and developmental signals during the maturation of B and T cells? (3) What events determine tissue-specific expression of Ig and TCR genes? At this time, we do not have complete answers to any of these questions. However, our current understanding of cis-and trans-acting elements involved in transcription regulation provides an excellent springboard for pursuing these questions and obtaining the answers. In this review, we have considered the current data and tried to suggest several models, based on that data, which might answer some of these questions. Present progress and available technology indicate that the next 5 years should be even more exciting than the past 5 years for exploring regulation of these important loci.

REFERENCES

Adams, J., Harris, A., Pinkert, C., Corcoran, L., Alexander, W., Cory, S., Palmiter, R., and Brinster, R. (1985). *Nature (London)* **318,** 533.

Adkins, B., Mueller, C., Okada, C., Reicher, T., Weissman, I., and Spangrude, G. (1987). *Annu. Rev. Immunol.* **5,** 3257.

Aguilera, R., Hope, T., and Sakano, H. (1985). *EMBO J.* **4,** 3689.

Allison, J., and Lanier, L. (1987). *Annu. Rev. Immunol.* **5,** 503.

Alt, F., Blackwell, K., and Yancopoulos, G. (1987). *Science* **238,** 1079.

Angel, P., Imagawa, M., Chiu, R., Stein, B., Imbra, R., Rahmsdorf, H., Jonat, C., Herrlich, P., and Karin, M. (1987). *Cell* **49,** 729.

Ares, M., Chung, J., Giglio, L., and Weiner, A. (1987). *Genes Dev.* **1,** 808.

Atchison, M., and Perry, R. (1986). *Cell* **46,** 253.

Atchison, M., and Perry, R. (1987). *Cell* **48,** 121.

Augereau, P., and Chambon, P. (1986). *EMBO J.* **5,** 1791.

Baldwin, A., and Sharp, P. (1988). *Proc. Natl. Acad. Sci. U.S.A.* In press.

Ballard, D. W., and Bothwell, A. (1986). *Proc. Natl. Acad. Sci. U.S.A.* **83,** 9626.

Banerji, J., Olson, L., and Schaffner, W. (1983). *Cell* **33,** 729.

Bergman, Y., Rice, D., Grosschedl, R., and Baltimore, D. (1984). *Proc. Natl. Acad. Sci. U.S.A.* **81,** 7041.

Bich-Thuy, L., and Queen, C. (1988). *Mol. Cell Biol.* **8,** 511.

Bier, N., Hashimoto, Y., Greene, M., and Maxam, A. (1985). *Science* **229,** 528.

Blackwell, T., Moore, M., Yancopoulos, G., Suh, H., Lutzker, S., Selsing, E., and Alt, F. (1986). *Nature (London)* **324,** 585.

Bohmann, D., Keller, W., Dale, T., Scholer, H., Tebb, G., and Mattaj, I. (1987). *Nature (London)* **325,** 268.

Borst, P., and Greaves, D. (1987). *Science* **235**, 658.
Brent, R., and Ptashne, M. (1985). *Cell* **43**, 729.
Brown, D. (1984). *Cell* **28**, 413.
Calame, K. (1985). *Annu. Rev. Immunol.* **3**, 159.
Calman, A., and Peterlin, B. M. (1986). *J. Exp. Med.* **164**, 1940.
Calza, R., Eckhardt, L., DelGiudice, T., and Schildkraut, C. (1984). *Cell* **36**, 688.
Cambier, J., and Ransom, J. (1987). *Annu. Rev. Immunol.* **5**, 175.
Carbon, P., Murgo, S., Ebel, J. P., Krol, A., Tebb, G., and Mattaj, I. W. (1987). *Cell* **51**, 71.
Chen, W., Tabor, S., and Struhl, K. (1987). *Cell* **50**, 1047.
Chen-Bettecken, U., Wecker, E., and Schimpl, A. (1987). *Immunobiology* **174**, 162.
Chien, Y-H., Iwashima, M., Kaplan, K., Elliot, J., and Davies, M. (1987a). *Nature (London)* **327**, 677.
Chien, Y-H., Iwashima, M., Wettstein, D., Kaplan, K., Elliot, J., Born, W., and Davis, M. (1987b). *Nature (London)* **330**, 722.
Chou, H., Anderson, S., Louie, M., Godambe, S., Pozzi, M., Behlke, M., Huppi, K., and Loh, D. (1987a). *Proc. Natl. Acad. Sci. U.S.A.* **84**, 1992.
Chou, H., Nelson, C., Godambe, S., Chaplin, D., and Loh, D. (1987b). *Science* **238**, 545.
Church, G., Eprussi, A., Gilbert, W., and Tonegawa, S. (1985). *Nature (London)* **313**, 798.
Clarke, C., Berenson, J., Goverman, J., Boyer, P., Crews, S., Siu, G., and Calame, K. (1982) *Nucleic Acids Res.* **10**, 7731.
Cockerill, P., and Garrard, W. (1986). *Cell* **44**, 273.
Cockerill, P., Yuen, M-H., and Garrard, W. (1987). *J. Biol. Chem.* **262**, 5394.
Dandanell, G., Valentin-Hansen, P., Larsen, J., and Hammer, K. (1987). *Nature (London)* **325**, 823.
Davis, M. (1985). *Annu. Rev. Immunol.* **3**, 537.
Davidson, I., Fromental, C., Augereau, P., Wildeman, A., Zenke, M., and Chambon, P. (1986). *Nature (London)* **323**, 544.
Dreyfus, M., Doyen, N., and Rougeon, F. (1987). *EMBO J.* **6**, 1685.
Eaton, S., and Calame, K. (1987). *Proc. Natl. Acad. Sci. U.S.A.* **84**, 7634.
Eckhart, L., and Birshstein, B. (1985). *Mol. Cell. Biol.* **5**, 856.
Ephrussi, A., Church, G., Tonegawa, S., and Gilbert, W. (1985). *Science* **227**, 134.
Falkner, F. G., and Zachau, H. G. (1984). *Nature (London)* **310**, 71.
Falkner, F., Mocikat, R., and Zachau, H. (1986). *Nucleic Acids Res.* **14**, 8819.
Fletcher, C., Heintz, N., and Roeder, R. G. (1987). *Cell* **51**, 773.
Foster, J., Stafford, J., and Queen, C. (1985). *Nature (London)* **315**, 423.
Fried, M., and Crothers, D. (1981). *Nucleic Acids Res.* **9**, 6505.
Garcia, J. V., Thuy, L., Stafford, J., and Queen, C. (1986). *Nature (London)* **322**, 383.
Gerlinger, P., LeMeur, M., Irrmann, C., Renard, P., Wasylyk, C., and Wasylyk, B. (1986). *Nucleic Acids Res.* **14**, 6565.
Gerster, T., Picard, D., and Schaffner, W. (1986). *Cell* **45**, 45.
Gerster, T., Matthias, P., Thali, M., Jiricny, J., and Schaffner, W., (1987). *EMBO J.* **6**, 1323.
Gilman, M., Wilson, R., and Weinberg, R. (1986). *Mol. Cell. Biol.* **6**, 4305.
Gimble, J., and Max, E. (1987). *Mol. Cell. Biol.* **7**, 15.
Goodbourn, S., Burstein, H., and Maniatis, T. (1986). *Cell* **45**, 601.
Gregor, P., Kobrin, B., Milcarek, C., and Morrison, S. (1986). *Immunol. Rev.* **89**, 31.
Grosschedl, R., and Baltimore, D. (1985). *Cell* **41**, 885.
Grosschedl, R., Weaver, D., Baltimore, D., and Constantini, F. (1984). *Cell* **38**, 647.
Gruss, P., Dhar, R., and Khoury, G. (1981). *Proc. Natl. Acad. Sci. U.S.A.* **78**, 943.
Guise, J., Galli, G., Nevins, J., and Tucker, P. (1988). *In* "Immunoglobulin Genes" (T. Honjo, F. Alt, and T. Rabbitts, eds.), in press.

Hamaoka, T., and Ono, S. (1986). *Annu. Rev. Immunol.* **4**, 176.
Hatzopoulos, A., Schlokat, W., and Gruss, P. (1987). *In* "Eukaryotic RNA Synthesis and Processing" (Horizons in Molecular Biology Series) (B. D. Hames and D. M. Glover, eds.), IRL Press, Washington D.C., in press.
Hennighausen, L., Siebenlist, U., Danner, D., Leder, P., Rawlins, D., Rosenfeld, P., and Kelly, T. (1985). *Nature (London)* **314**, 289.
Hernandez, N., and Weiner, A. (1986). *Cell* **47**, 249.
Hochschild, A., and Ptashne, M. (1986). *Cell* **44**, 681.
Hollis, G., Hieter, P., McBride, O., Swan, D., and Leder, P. (1982). *Nature (London)* **296**, 321.
Honjo, T. (1983). *Annu. Rev. Immunol.* **1**, 499.
Hromas, R., Pauli, U., Marcuzzi, A., Lafrenz, D., Nick, H., Stein, J., Stein, G., and Van Ness, B. (1988). *Nucleic Acids Res.* In press.
Hromas, R., and Van Ness, B. (1986). *Nucleic Acids Res.* **14**, 4837.
Imagawa, M., Chiu, R., and Karin, M. (1987). *Cell* **51**, 251.
Imler, J-L., Lemaire, C., Wasylyk, C., and Wasylyk, B. (1987). *Mol. Cell. Biol.* **7**, 2558.
Ishihara, T., Kudo, A., and Watanabe, T. (1984). *J. Exp. Med.* **160**, 1937.
Junker, S., and Pedersen, S. (1985). *Exp. Cell Res.* **158**, 349.
Kadesch, T., Zervos, P., and Ruezinsky, D. (1986). *Nucleic Acids Res.* **14**, 8209.
Kadonaga, J., and Tjian, R. (1986). *Proc. Natl. Acad. Sci. U.S.A.* **83**, 5889.
Kelley, D., and Perry, R. (1986). *Nucleic Acids Res.* **14**, 5431.
Kemp, D., Harris, A., and Adams, J. (1980). *Proc. Natl. Acad. Sci. U.S.A.* **77**, 7400.
Kiledjian, M., Su, L., and Kadesch, T. (1988). *Mol. Cell. Biol.* **8**, 145.
Klein, S., Sablitzky, F., and Radbruch, A. (1984). *EMBO J.* **3**, 2473.
Klein, S., Gerster, T., Picard, D., Radbruch, A., and Schaffner, W. (1985). *Nucleic Acids Res.* **13**, 8901.
Koshland, M. (1985). *Annu. Rev. Immunol.* **3**, 425.
Kozbor, D., Burioni, R., Ar-Rushdi, A., Zmijewski, C., and Croce, C. (1987). *Proc. Natl. Acad. Sci. U.S.A.* **84**, 4969.
Kronenberg, M., Siu, G., Hood, L., and Shastri, N. (1986). *Annu. Rev. Immunol.* **4**, 529.
LaBella *et al.* (1988). Submitted.
Landolfi, N. F., Capra, J. D., and Tucker, P. W. (1986). *Nature (London)* **323**, 548.
Landolfi, N. F., Capra, J. D., and Tucker, P. W. (1987). *Proc. Natl. Acad. Sci. U.S.A.* **84**, 3851.
Lenardo, M., Pierce, J., and Baltimore, D. (1987). *Science* **236**, 1573.
Lennon, G., and Perry, R. (1985). *Nature (London)* **318**, 475.
Livant, D., Blatt, C., and Hood, L. (1986). *Cell* **47**, 461.
Luria, S., Gross, G., Horowitz, M., and Givol, D. (1987). *EMBO J.* **6**, 3307.
MacLeod, C., Minning, L., Gold, D., Terhorst, C., adn Wilkinson, M. (1986). *Proc. Natl. Acad. Sci. U.S.A.* **83**, 6989.
Maeda, H., Kitamura, D., Kudo, A., Kazuo, A., and Watanabe, T. (1986). *Cell* **45**, 25.
Maki, R., Kearney, J., Paige, C., and Tonegawa, S. (1980). *Science* **209**, 1366.
Mason, J. O. Williams, G. T., and Neuberger, M. S. (1985). *Cell* **41**, 479.
Mather, E., and Perry, R. (1981). *Nucleic Acids Res.* **9**, 6855.
Mather, E., Nelson, K., Haimovich, J., and Perry, R. (1984). *Cell* **36**, 329.
Mathieson, B., and Fowlkes, B. J. (1984). *Immunol. Rev.* **89**, 49.
Matis, L., Cron, R., and Bluestone, J. (1987). *Nature (London)* **330**, 262.
Matsuuchi, L., Cann, G. M., and Koshland, M. E. (1986). *Proc. Natl. Acad. Sci. U.S.A.* **83**, 456.

Mattaj, I., Lienhard, S., Jiricny, J., and De Robertis, E. (1985). *Nature (London)* **316**, 163.
Matthias, P., Gerster, T., Bohmann, D., and Schaffner, W. (1987). *Nucleic Acids Res.* **1**, 221.
Max, E., Maizel, J., and Leder, P. (1981). *J. Biol. Chem.* **256**, 5116.
McDougall, S. Peterson, C., and Calame, K. (1988). *Science* In press.
Melchers, F., and Andersson, J. (1986). *Annu. Rev. Immunol.* **4**, 13.
Mercola, M., Wang, X., Olsen, J., and Calame, K. (1983). *Science* **221**, 663.
Mercola, M., Goverman, J., Mirell, C., and Calame, K. (1985). *Science* **227**, 266.
Mocikat, R., Falkner, F., Mertz, R., and Zachau, H. (1986). *Nucleic Acids Res.* **14**, 8829.
Modlin, R., Brenner, M., Krangel, M., Duby, A., and Bloom, B. (1987). *Nature (London)* **329**, 541.
Moreau, P., Hen, R., Wasylyk, B., Everett, R., Gaub, M., and Chambon, P. (1981). *Nucleic Acids Res.* **9**, 6047.
Mossing, M., and Record, M. T. (1986). *Science* **233**, 889.
Murphy, J., Skuzeski, J., Lund, E., Steinberg, T., Burgess, R., and Dahlberg, J. (1987). *J. Biol. Chem.* **262**, 1795.
Nabel, G., and Baltimore, D. (1987). *Nature (London)* **326**, 711.
Neuberger, M. (1983). *EMBO J.* **2**, 1373.
Neuman de Vegvar, H., Lund, E., and Dahlberg, J. (1986). *Cell* **47**, 259.
Ninfa, A., Reitzer, L., and Magasanik, B. (1987). *Cell* **50**, 1039.
Nomiyama, H., Fromental, C., Xiao, J., and Chambon, P. (1987). *Proc. Natl. Acad. Sci. U.S.A.* **84**, 7881.
Ondek, B., Shepard, A., and Herr, W. (1987). *EMBO J.* **6**, 1017.
Palacios, R., Karasuyama, H., and Rolink, A. (1987). *EMBO J.* **6**, 3687.
Parker, C., and Topol, J. (1984). *Cell* **37**, 273.
Parslow, T., and Granner, D. (1982). *Nature (London)* **299**, 449.
Parslow, T. G., Blair, D. L. Murphy, W. J., and Granner, D. K. (1984). *Proc. Natl. Acad. Sci. U.S.A.* **81**, 2650.
Parslow, T., Jones, S., Bond, B., and Yamamoto, K. (1987). *Science* **235**, 1498.
Perry, R., Coleclough, C., and Weigert, M. (1980). *Cold Spring Harbor Symp. Quant. Biol.* **45**, 925.
Perry, R., Kelley, D., Coleclough, C., and Kearney, J. (1981). *Proc. Natl. Acad. Sci. U.S.A.* **78**, 247.
Peterson, C., and Calame, K. (1987). *Mol. Cell. Biol.* **7**, 4194.
Peterson, C., Orth, K., and Calame, K. (1986). *Mol. Cell. Biol.* **6**, 4168.
Peterson, C., Eaton, S., and Calame, K. (1988). Submitted.
Picard, D., and Schaffner, W. (1984). *Nature (London)* **307**, 80.
Picard, D., and Schaffner, W. (1984b). *EMBO J.* **3**, 3031.
Picard, D., and Schaffner, W. (1985). *EMBO J.* **4**, 2831.
Pillai, S., and Baltimore, D. (1987). *Nature (London)* **329**, 172.
Plon, S., and Wang, J. (1986). *Cell* **45**, 575.
Pruijn, G., van Driel, W., van Mittenburg, R., and van der Vliet, P. (1987). *EMBO J.* **6**, 3771.
Ptashne, M. (1986). *Nature (London)* **322**, 697.
Queen, C., and Stafford, J. (1984). *Mol. Cell. Biol.* **4**, 1042.
Queen, C., Foster, J., Stauber, C., and Stafford, J. (1986). *Immunol. Rev.* **89**, 49.
Rabbitts, T., Forster, A., Baer, R., and Hamlyn, P. (1983). *Nature (London)* **306**, 806.
Reik, W., Williams, G., Barton, S., Norris, M., Neuberger, M., and Surani, M. (1987). *Eur. J. Immunol.* **17**, 465.

Reth, M., and Alt, F. (1984). *Nature (London)* **312**, 418.
Rosales, R, Vigneron, M., Macchi, M., Davidson, I., Xiao, J., and Chambon, P. (1987). *EMBO J.* **6**, 3015.
Rosenfeld, P., O'Neill, E., Wides, R., and Kelley, T. J. (1987). *Mol. Cell. Biol.* **7**, 875.
Royer, H., and Reinherz, E. (1987). *Proc. Natl. Acad. Sci. U.S.A.* **84**, 232.
Sawadogo, M., and Roeder, R. G. (1985). *Cell* **43**, 165.
Scheidereit, C., Heguy, A., and Roeder, R. G. (1987). *Cell* **51**, 783.
Schibler, U., Marcu, K., and Perry, R. (1978). *Cell* **15**, 1495.
Schirm, S., Jiricny, J., and Schaffner, W. (1987). *Genes Dev.* **1**, 65.
Schlokat, U., Bohmann, D., Scholer, H., and Gruss, P. (1986). *EMBO J.* **5**, 3251.
Scholer, H., and Gruss, P. (1985). *EMBO J.* **4**, 3005.
Sen, R., and Baltimore, D. (1986a). *Cell* **46**, 705.
Sen, R., and Baltimore, D. (1986b). *Cell* **47**, 921.
Sen, R., and Baltimore, D. (1987). *Mol. Cell. Biol.* **7**, 1989.
Serfling, E., Jasin, M., and Schaffner, W. (1985). *Trends Genet.* **1**, 224.
Sherman, P., Basta, P., and Ting, J. (1987). *Proc. Natl. Acad. Sci. U.S.A.* **84**, 4254.
Shore, D., and Nasmyth, K. (1987). *Cell* **51**, 721.
Short, N. (1987). *Nature (London)* **326**, 740.
Shuey, D., and Parker, C. (1986). *J. Biol. Chem.* **261**, 7934.
Singh, H., LeBowitz, J., Baldwin, A., and Sharp, P. (1988). *Cell* **52**, 415.
Siu, G., Kronenberg, M., Strauss, E., Haars, R., Mak, T., and Hood, L. (1984). *Nature (London)* **311**, 344.
Siu, G., Strauss, E., Lai, E., and Hood, L. (1986). *J. Exp. Med.* **164**, 1600.
Siu, G., Springer, E. A., Huang, H. V., Hood, L. E., and Crews, S. T. (1987). *J. Immunol.* **138**, 4466.
Sive, H. L., and Roeder, R. G. (1986). *Proc. Natl. Acad. Sci. U.S.A.* **83**, 6382.
Sive, H. L., Heintz, N., and Roeder, R. G. (1986). *Mol. Cell. Biol.* **6**, 3329.
Spandidos, D., and Anderson, M. (1984). *FEBS Lett.* **175**, 152.
Staudt, L., Singh, H., Sen, R., Wirth, T., Sharp, P., and Baltimore, D. (1986). *Nature (London)* **323**, 640.
Storb, U. (1987). *Annu. Rev. Immunol.* **5**, 151.
Storb, U., Denis, K., Brinster, R., and Witte, O. (1985). *Nature (London)* **316**, 356.
Suda, Y., Aizawa, S., Hirai, S., Inoue, Y., Furuta, Y., Suzuki, M., Hirohashi, S., and Ikawa, Y. (1987). *EMBO J.* **6**, 4055.
Treisman, R., and Maniatis, T. (1985). *Nature (London)* **315**, 72.
Tsao, B., Peterson, C., Wang, X., and Calame, K. (1988). *Nucl. Acids Res.* **16**, 3239.
Van Ness, B. G., Weigert, M., Coleclough, C., Mather, E. L., Kelleu, D. E., and Perry, R. P. (1981). *Cell* **27**, 593.
Voss, S., Schlokat, U., and Gruss, P. (1986). *Trends Biochem. Sci. Pers. Ed.* **11**, 287.
Wabl, M., and Burrows, P. (1984). *Proc. Natl. Acad. Sci. U.S.A.* **81**, 2452.
Wall, R., Briskin, M., Carter, C., Govan, H., Taylor, A., and Kincade, P. (1986). *Proc. Natl. Acad. Sci. U.S.A.* **83**, 295.
Wang, J., Nishiyama, K., Araki, K., Kitamura, D., and Watanabe, T. (1987). *Nucleic Acids Res.* **15**, 10105.
Wang, X., and Calame, K. (1985). *Cell* **43**, 659.
Wang, X., and Calame, K. (1986). *Cell* **47**, 241.
Wasylyk, B. (1986). In "Promoter Elements and Transacting Factors" (W. Reznikoff, and L. Gold, eds.), pp. 79–99. Butterworth, London.
Wasylyk, C., and Wasylyk, B. (1986). *EMBO J.* **5**, 553.
Weber, F., and Schaffner, W. (1985). *Nature (London)* **315**, 75.

Weiher, H., Konig, M., and Gruss, P. (1983). *Science* **219,** 626.
Weinberger, J., Baltimore, D., and Sharp, D. (1986). *Nature (London)* **322,** 846.
Wirth, T., Staudt, T., and Baltimore, D. (1987). *Nature (London)* **329,** 174.
Wolffe, A., and Brown, D. (1987). *Cell* **51,** 733.
Wu, C. (1985). *Nature (London)* **317,** 84.
Xiao, J., Davidson, I., Ferrandon, D., Rosales, R., Vigneron, M., Macchi, M., Ruffenach, F., and Chambon, P. (1987). *EMBO J.* **6,** 3005.
Yamamura, K. Kikutani, H., Takahashi, N., Taga, T., Akira, S., Kawai, K., Fukuchi, K., Kumahara, Y., Honjo, T., and Kishimoto, T. (1984). *J. Biochem.* **96,** 357.
Yancopoulos, G., and Alt, F. (1985). *Cell,* **40,** 271.
Yancopoulos, G., and Alt, F. (1986). *Annu. Rev. Immunol.* **4,** 339.
Yuan, D., and Tucker, P. (1984). *J. Exp. Med.* **160,** 564.
Zaller, D., and Eckhardt, L. (1985). *Proc. Natl. Acad. Sci. U.S.A.* **82,** 5088.

Molecular Aspects of Receptors and Binding Factors for IgE

HENRY METZGER

*Section on Chemical Immunology,
Arthritis and Rheumatism Branch,
National Institute of Arthritis
and Musculoskeletal and Skin Diseases,
National Institutes of Health,
Bethesda, Maryland 20892*

I. Introduction

Immunoglobulin E (IgE) is notorious principally for its role in mediating allergic disorders—disorders which plague some 35 million in the United States alone (Kaplan, 1985). Elevated synthesis of IgE also regularly accompanies parasitic infestations in animals and in humans. It is reasonably assumed that IgEs role in combatting the latter infections, rather than its nuisance value in the former disorders, provided the advantage that led to its evolution (Capron et al., 1986a).

Like immunoglobulins in general, IgE by itself can do nothing except bind to the antigen it recognizes. However, through its interaction with cell surface receptors that bind the Fc domains of antibodies (Fc receptors), IgE, like other immunoglobulins, can muster a formidable army of protective mechanisms: phagocytosis, cell-mediated killing, and release of potent mediators. Only the complement systems do not seem to be importantly mobilized by IgE–antigen complexes.

Fc receptors or Fc-"binding factors" or both are also used by immunoglobulins to regulate their biosynthesis (Fridman et al., 1987), and IgE is no exception (Ishizaka, 1984). This role for IgE receptors, particularly for those on lymphocytes, is the most recently recognized aspect of IgE metabolism and is currently one of the most actively pursued subjects by those interested in IgE.

In this review, I summarize the present knowledge about the proteins through which IgE mediates its actions. Substantial progress has been made recently in clarifying the structure of several of these proteins. There is also considerable new information about the biological function of these IgE-binding proteins. The largest gap in our information is about how the one leads to the other and about the molecular mechanisms by which these proteins induce cellular responses.

II. Sites on IgE Interacting with Receptors

A. OVERVIEW

Virtually all studies that have attempted to define the site(s) on IgE, that interact with receptors or binding factors, have focused on the binding sites for the high-affinity receptor. Three complementary approaches have been (or could be) employed for such studies: (1) "protection" experiments, which probe the accessibility of the ε heavy chain of receptor-bound IgE; (2) inhibition experiments, which employ chemically or biosynthetically synthesized segments of the ε chain and test for their capacity to inhibit the binding of IgE; and (3) site-specific modifications of IgE or of inhibitory peptides [see (2) above], in modern times produced by genetic manipulation, and testing of their effect on binding.

Probing the exposure of the ε chain may be performed by chemical modification (probe MW < 500) with proteolytic enzymes (probe MW 20,000) or with antibodies [probe MW 20,000 (Fv fragments) and probe MW 150,000 (IgG)]. It is clear that the smaller the probe, the higher its resolving power. That is, there is an improved likelihood that smaller probes will reveal those changes in accessibility that are specific, rather than those that are due to steric inhibition by adjoining regions. However, in science, as in other areas of life, strategies are chosen not simply on the basis of efficacy, but also on the basis of expediency. This likely accounts for the reporting of a moderate number of studies employing intact antibodies, but only a few with Fab fragments and none with Fv and only a couple with proteolytic enzymes and none with chemical probes.

Because the binding of small peptides can be anticipated to be relatively weak and the concentration of receptors that can be practically achieved is low, direct binding studies are unlikely to be fruitful. Since the affinity of intact IgE is so high, inhibition assays performed at equilibrium will reveal only tightly binding peptides, whereas kinetic studies could uncover much more weakly binding inhibitors. Only during the more or less linear rate of binding of the IgE will the observed inhibition approach the maximum possible; [(inhibitor)/(indicator)]k_{rel}, where k_{rel} is the forward rate constant for the inhibitor relative to that of the indicator (e.g., labeled IgE). Review of the literature on inhibition of binding of IgE suggests that either this aspect is not appreciated or is not used because the available assays [e.g., skin tests (Helm *et al.*, 1988)] preclude kinetic analysis.

B. STUDIES WITH ANTI-IgE ANTIBODIES

Several new probing types of studies have been reported recently. Rousseaux-Prevost *et al.* (1987) described a *monoclonal anti-rat IgE* ("MARE-1"), which appeared to be directed against the $C_\varepsilon 3$ domain(s), since it reacted with intact IgE, but neither with $F(ab')_2$ (which, in IgE, includes $C_\varepsilon 2$) nor with $C_\varepsilon 4$. The antibody inhibited the binding of rat IgE to peritoneal mast cells, but *only partially* (maximum $\sim 55\%$), reminiscent of similar results reported by Baniyash and Eshhar (1984). Rousseaux-Prevost *et al.* neither employed Fab fragments nor studied the binding of the antibody to bound IgE nor assessed the amount of antibody bound per IgE quantitatively. It is not possible, therefore, to know whether the alternative explanations, proposed to account for the earlier results (Metzger *et al.*, 1986), would be applicable in this instance. Hook *et al.* (1987) tested 12 monoclonal antibodies to human IgE and analyzed the distinctive epitopes with which they appeared to interact. Notably, antibodies that failed to react with bound IgE mapped to several discrete areas. Whether this reflects a broad zone of interaction or nonspecific steric inhibition cannot be distinguished. Robertson and Liu (1988) produced polyclonal antibodies to six peptides based on murine sequences in $C_\varepsilon 2$, $C_\varepsilon 3$, and $C_\varepsilon 4$. These were chosen for their overall hydropathy, for high homology to both rat and human IgE, and, where possible, to contain a cysteine (to simplify conjugation of the peptide to the hemocyanin carrier protein).

The positions in the ε chain of the peptide immunogens they used are shown in Fig. 1. All the antipeptide antiserums reacted with unbound IgE, but one of these failed to react with cell-bound IgE. Based on the latter result, it appears that of the regions selected for study, that in the NH_2-terminal one-half of the $C_\varepsilon 3$ domain was closest to a site on IgE that interacts with the high-affinity receptor.

Very similar experiments were reported by Burt *et al.* (1987) using polyclonal antibodies to peptides based on sequences of rat IgE (Fig. 1). In these studies, the principle used for selecting segments was that they should correspond to portions of the ε chain that can be expected to be well exposed on the ε dimer. Antiserums to all of the segments inhibited the binding of IgE to mast cells to a variable extent. Nevertheless, each was able to react with cell-bound IgE (as assessed by degranulation studies) *with an effectiveness that paralleled their ability to interact with unbound IgE*. The authors propose that these results suggest that only

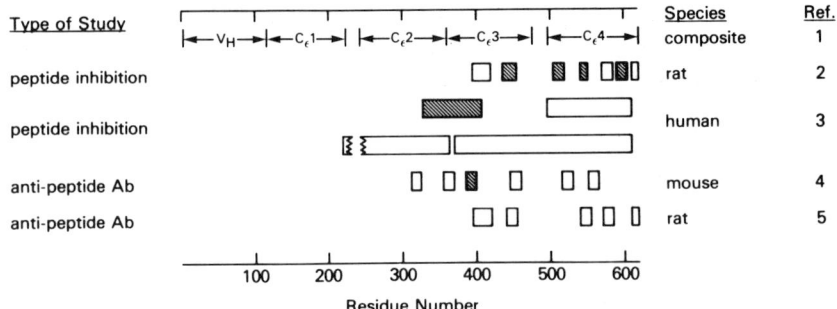

FIG. 1. Localization of IgEs binding site for the high-affinity receptor. The figure schematically illustrates the segments of the ε chain implicated as interacting with the receptor on the basis of two types of study: inhibition of binding of IgE to cells by a peptide or inhibition of binding of antipeptide antibody (Ab) to IgE in the presence of a receptor. *Negative* results are shown as open boxes; *positive* results are shown as hatched boxes. The peptides have been positioned relative to the consensus numbering scheme used in Ref. 1. References: 1, Kabat *et al.* (1987); 2, Burt and Stanworth (1987); 3, Helm *et al.* (1988); 4, Robertson and Liu (1988); 5, Burt *et al.* (1987).

one of the two ε chains interact with the receptor (see below). Quantitative binding studies with Fab fragments of the antibodies would be useful for assessing the correctness of this proposal.

A novel approach has been used by Baniyash and Eshhar (1987). After prolonged immunization of guinea pigs with mouse monoclonal anti-IgE antibodies, the authors were able to detect antiidiotypic antibodies that bound to mast cell receptors for IgE. If the specificity of the original anti-IgE antibodies could be more precisely defined, one might, in this way, clarify the actual contact sites between IgE and the high-affinity receptor.

No new studies have appeared employing proteases to probe the sites on ε chains protected by their interaction with receptors. Such experiments earlier suggested the importance of the $C_\varepsilon 2:C_\varepsilon 3$ interdomain area in particular (Perez-Montfort and Metzger, 1982). However, Rousseaux-Prevost *et al.* (1987), in the course of documenting once again the inability to obtain Fc fragments from rat IgE, have detected an interesting intermediate fragment—an IgE missing the $C_\varepsilon 3$ and $C_\varepsilon 4$ domain from one of the ε chains. Regrettably, they were unable to purify this fragment free of intact IgE in order to check its bindability. This would be interesting to test because of the observations by Burt *et al.* (1987) (above) and by Helm *et al.* (1988) (below), which suggest that a single ε chain may be sufficient to engage the receptor. One way by which the bindability of this novel fragment could be tested would be to use a digestion mixture prepared from labeled IgE, incubate it with an *excess* of cell-bound receptors, remove the unbound IgE, and examine the acid-eluted material on gels to see if the "one-legged" IgE had been bound.

C. Inhibition Studies

Burt and Stanworth (1987) reported inhibition of binding of rat IgE to rat mast cells by seven synthetic IgE peptides (Fig. 1). Even the most active was 1000 times less active than native rat IgE. Four of the seven peptides, three of which were based on sequences from the rat $C_\varepsilon 4$ domain, were reported to be active.

There are several disturbing aspects to the results reported by Burt and Stanwork (1987). First, it can be calculated from the results, quoted in Fig. 1 and Table 3 of their article, that, even in the absence of inhibitor, the amount of radioactive IgE that was bound accounts for only 2–3% saturation of the expected number of receptors present on the cells. Second, with unlabeled IgE, they observed only 50% inhibition using a 5-fold excess. Third, the inhibition with the peptides was unaffected by washing the cells prior to the addition of the labeled IgE. It can be calculated from the concentrations of peptide (e.g., #1290) required to obtain ~50% inhibition that the K_A for the peptide is $2 \times 10^5 \, M^{-1}$, so that it would be expected to have dissociated in less than a second! Finally, combinations of certain peptides gave synergistic inhibitions that are difficult to explain. All these quantitative anomalies suggest that these results must be treated with great caution.

Helm *et al.* (1988) reported on the use of biosynthetically engineered peptides based on cDNA constructs of the human ε chain. Two substantial peptides were inactive: one that contains most of the $C_\varepsilon 3$ region and all of the $C_\varepsilon 4$ region and another that contains largely $C_\varepsilon 2$. However, a peptide of 77 residues that bridges the carboxy-terminal portion of $C_\varepsilon 2$ and the amino-terminal portion of $C_\varepsilon 3$ had considerable activity and so did two other constructs (not shown in Fig. 1), which included the 77-residue segment. In intact IgE, the region encompassed by the active peptide would include the cleft between $C_\varepsilon 2$ and $C_\varepsilon 3$ (Fig. 2). As the authors suggest, the inactivity of the two larger peptides that independently include the residues comprising the cleft suggests that the cleft (with or without the connecting residues) interacts with the receptor.

Molecular-weight studies indicate that the active peptide likely exists as a monomer. This, at least, raises the possibility that the high-affinity receptor interacts with only one of the two ε chains in IgE, a proposal consistent with the model proposed by Burt *et al.* (1987)

(above). Prior studies on the proteolysis of receptor-bound IgE had shown that digestion of both ε chains was retarded (Perez-Montfort and Metzger, 1982). Although the simplest interpretation of this result is that both chains interact with the receptor, an alternative explanation cannot be ruled out, namely, that, through stabilization of one ε chain by its interaction with the α chain of the receptor, the other ε chain is also stabilized, albeit indirectly. As noted above, it would be interesting to know whether the unusual fragment of IgE, reported by Rousseaux-Prevost et al. (1987), was capable of binding to mast cells.

Helm et al. (1988) used skin tests on the forearms of one of the coauthors to assay the peptides. This is because there are no cell lines that are practical to use for assaying the binding of human IgE to the high-affinity receptor. The cDNA for the IgE-binding α chain of the human receptor has recently been cloned (see below). Once expression is accomplished, safer, less-cumbersome assays will become possible to study the binding of human IgE.

III. The Mast Cell-Specific Receptor

Mast cells and basophils have a surface protein that binds monomeric IgE with very high affinity. Often referred to as $Fc_\varepsilon RI$, this receptor is found on these cells exclusively and can trigger their degranulation—a phenomenon that served as the original hallmark of reaginic (i.e., IgE-mediated) activity.

The detailed analysis of this protein became practical when a unique "rat basophilic leukemia" (RBL) tumor (Eccleston et al., 1973) was discovered to have ample numbers of these receptors (Kulczycki et al., 1974). Our original assumption (faith) that the receptors would be normal, despite their being derived from cells with abnormal growth characteristics, has been confirmed wherever tested. Molecular genetic techniques are providing additional evidence in this regard.

FIG. 2. Three-dimensional model of the site on human IgE that presumptively binds to the receptor with high affinity for IgE on mast cells. On the left, the peptide, found to be active by Helm et al. (1988), is shown in white, superimposed on the presumed backbone configuration (Padlan and Davies, 1986) of the ε chain in blue. The $C_\varepsilon 2$, $C_\varepsilon 3$, and $C_\varepsilon 4$ domains of the right ε chain are shown in red, green, and magenta, respectively. The carbohydrate between the $C_\varepsilon 3$ domains is shown in yellow. Asparagine residues bearing the carbohydrate in human Fc_ε (Asn-265 and -271) are shown in yellow. The beginning and end of the inhibitory peptide (Gln-301–Arg-376) and two of the three residues that separate the inactive peptides (see Fig. 1) are numbered. [From Helm et al. (1988), reproduced with permission.]

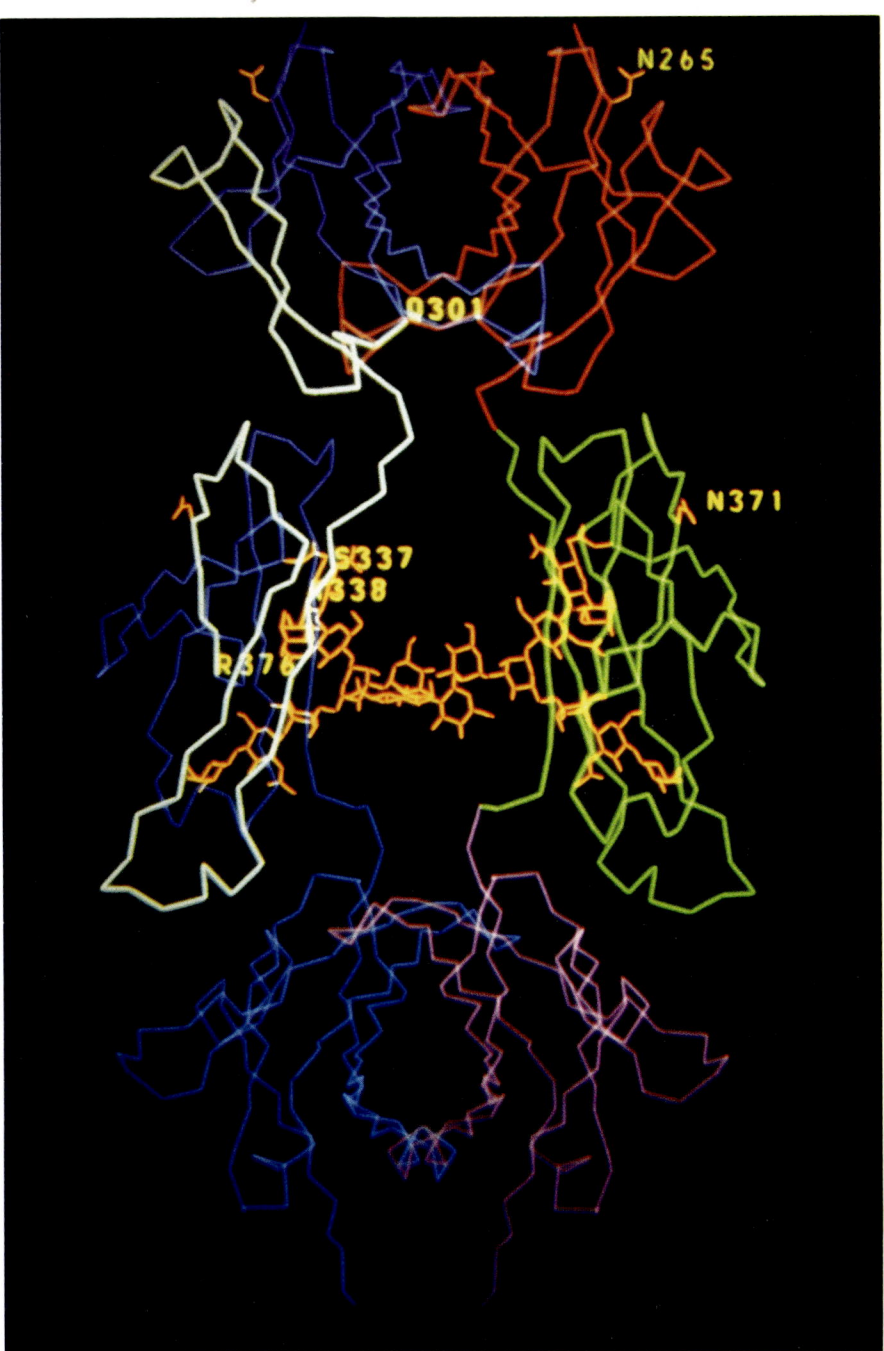

Fig. 2

The evolving information about this protein was reviewed in depth by our group relatively recently (Metzger et al., 1986). Therefore, after a brief summary of the salient features, I will concentrate on the major new findings.

A. Principal Characteristics

Several hundred thousand receptors are present on most surviving isolates of RBL cells and on normal rat peritoneal mast cells; a similar number of receptors are present on cultured human cells (Ishizaka et al., 1985). The receptors are univalent and bind monomeric IgE with an affinity variously estimated as $10^9 - > 10^{10} M^{-1}$ (Metzger and Bach, 1978; Froese, 1980a). They are diffusely distributed on the cell surface, where they rotate and translate in the plane of the membrane with kinetics similar to most surface proteins (Schlessinger et al., 1976; Zidovetski et al., 1986).

The IgE binds to a glycosylated polypeptide "α," which is associated with a 30- to 34-kDa β and two 7- to 6-kDa γ chains, the last apparently disulfide linked to each other. An unusual feature of this tetrameric complex is its tendency to dissociate in the micellar detergent required to solubilize it (Rivnay et al., 1982; Kinet et al., 1985).

Little is so far known about how the different subunits interact with each other, but the availability of a potent monoclonal anti-β antibody has recently provided some further information. Rivera et al. (1988) have shown that, under conditions in which the dissociation of the receptor occurs more slowly, e.g., in submicellar detergent, β and γ_2 dissociate as a unit, a finding that explains the earlier observation that β and γ_2 dissociate in unison (Kinet et al., 1985). These results are also consistent with another earlier finding that, under certain conditions, β and γ_2 become disulfide linked to each other (Kinet et al., 1983). Together, these results indicate that β and γ_2 must be in contact with each other in the native structure.

B. Gene Cloning Studies

It may seem surprising that it took over 10 years from the time that the α chain of the receptor was first defined (Conrad and Froese, 1976; Kulczycki et al. 1976; Newman et al., 1977; Rossi et al., 1977) to the time that a cDNA, which encodes its amino acid sequence, was first identified (Kinet et al., 1987). Similarly, the β chain was first recognized in 1980 (Holowka et al., 1980), and the γ chains were first recognized 3 years later (Perez-Montfort et al., 1983a); yet, the cDNA coding for the former has just now been isolated and sequenced (Kinet et al., 1988), and work on the cDNA for γ chains has been only recently initiated.

Several factors contributed to this delay, as far as our own group was concerned. We opted to invest our efforts in developing a procedure for purifying the receptor that would (1) selectively purify the high-affinity receptor free of an additional low-affinity receptor known to be present on RBL cells (Froese, 1980b); (2) simultaneously purify α, β, and γ_2, despite the instability of the receptor; and (3) yield receptors that had not been exposed to denaturants in order to optimize the likelihood that any intrinsic activity would be retained. All three subunits appear to have a blocked amino-terminus. It required considerable material to verify this and then to isolate internal peptides to obtain the necessary sequence data for creating suitable probes. Nevertheless, these long-term investments appear to be paying off, and it can be anticipated that the cDNA for α, β, and γ will soon be isolated and sequenced (see below).

1. Alpha Chain

The cDNA for rat α was isolated by our group in 1986, but submission (and acceptance for publication) of the report about it was delayed because it could not be expressed in transfection experiments (see below). Nevertheless, the amino acid sequence, predicted by the cDNA, is so consistent with the structural information on α (Kinet et al., 1987) that there is no reason to doubt that the correct cDNA has been isolated. Two new types of data have since provided additional support; (1) antibodies to peptides, synthesized on the basis of the sequence predicted by the cDNA, react with α chains (Ra et al., 1988) and (2) the genomic locus, identified with the cDNA probe, appears to be in a chromosomal region previously determined to contain the genes coding for the homologous Fc_γ receptors (Huppi et al., 1988).

The cDNA predicts an amino acid sequence with the following characteristics.

1. An amino-terminal stretch consists of 23 amino acids that likely represent a leader sequence with an expected clevage site for a signal peptidase.

2. This sequence is followed by a stretch of approximately 180 amino acids, which we interpreted as representing the extracellular portion of the α chain. It contains seven potential N-glycosylation sites (five, if those triads of Asn-X-Ser/Thr, where X is Asp, are excluded) and five cysteines. Four of the latter are surrounded by sequences reminiscent of those associated with disulfide loops in immunoglobulins and related proteins (Table I). The extracellular immunoglobulin-binding portion of the α chain of the IgE receptor is highly homologous to a corresponding portion of Fc_γ receptors (Ravetch et al., 1986). In the latter, it has

TABLE I
RELATIONSHIP BETWEEN THE CYSTEINE LOOPS IN VARIOUS IMMUNOGLOBULIN DOMAINS[a]

Domain	Length of S–S loop	Consensus loop sequence							Reference
		13 or 14	14 or 15	22 or 23	−13	−7	−6	−2	
		W, F, Y	W, F, Y	K, R, H	h(I)	D, N, E, Q	D	Y	
V_κ	66	W	+	+	I	+	+	+	1
C_κ	60	W	−	−	h	+	+	+	1
$Fc_{poly(Ig)}R$ (1–5)	61–71	+ (0.6)	−	−	h (0.4I)	+	+	+	2
$Fc_\varepsilon R(1)$	43	W	+ (0.75)	+ (0.75)	I (0.5)	+	+	+	3–6
$Fc_\varepsilon R(2)$	45	+	+	+ (0.5)	I	+ (0.5)	+ (0.5)	+	3–6

[a] Consensus residues are numbered starting with either the amino-terminal cysteine (N) or the carboxy-terminal cysteine (C). Conventions used in this table include the following: +, all sequences show the indicated residue; −, no sequences show the indicated residue. Where only some of the sequences show a particular type of residue the fraction is given in parentheses. Abbreviations: D, E, F, H, I, K, N, Q, R, W, and Y, one-letter code for amino acids (Dayhoff et al., 1978); h, hydrophobic amino acid (I, L, V, or M); V_κ, variable domain of murine κ light chains; C_κ, constant domain of murine light chains; $Fc_{poly(Ig)}R$, receptor for polymeric immunoglobulins; $Fc_\varepsilon R$, Fc receptor for IgG or IgE. References: 1, Kabat et al. (1987); 2, Mostov et al. (1984); 3, Ravetch et al. (1986); 4, Kinet et al. (1987); 5, Stuart et al. (1987); 6, Kochan et al. (1988).

recently been demonstrated that the corresponding four cysteines are indeed disulfide linked (Hibbs et al., 1988; J. Unkeless, personal communication). It is likely, therefore, that those in the α chain of the Fc_ε receptor are similarly linked. Notably, these putative disulfide loops are about 20 residues shorter than the homologous loops in immunoglobulins—sufficiently shorter to make it difficult to model the secondary and tertiary folding on the basis of known immunoglobulin structures (E. Padlan, personal communication).

The predicted sequence shows two homologous domains (residues 65–102 and 148–182), having 26% of their residues identical and showing even greater similarity (average 34% identity) with the corresponding portions coded by the Fc_γ receptor α genes.

The new data on the corresponding human receptors (Kochan et al., 1988; Stuart et al., 1987) confirm the original assessment that the two immunoglobulin-like domains were duplicated and began to diverge prior to the differentiation into genes coding for isotype-specific receptors.

3. The putative extracellular portion is followed by a sequence of 20 amino acids that likely forms a transmembrane segment. It contains two hydroxyamino acids (Ser and Thr) and one charged residue (Asp), but the remaining residues are all hydrophobic or aromatic. Eight of the amino acids surrounding the more or less central Asp are identical to those predicted by the α cDNA of the mouse Fc_γ receptor.

4. The carboxy-terminal segment is rich in basic amino acids and is the only portion of the sequence that shows no resemblance to Fc_γ receptors (only 1 out of ~ 25 residues is identical).

The homologous cDNA that codes for the human α chain (Kochan et al., 1988) is strikingly similar. Although the putative transmembrane segment encompasses the longest stretch of identical residues (10), the extracellular portion contains runs of identical residues almost as long and contains several segments in which only highly homologous residues interrupt runs of identical side chains. Two stretches show more extensive differences: the putative signal sequence and the carboxy-terminus, which shows, as its only common feature, a concentration of basic residues.

Shimizu et al. (1988) have now also reported the isolation and sequencing of cDNAs, presumptively coding for the α chain, from "libraries" of RBL cells and human mast cells. Their rat sequence differs by several nucleotides (largely in the untranslated regions) from that reported by Kinet et al. (1987). Further analysis by J.-P. Kinet and J. Kochan (personal communication) of their orginal gels, as well as of new clones, indicates that, with one exception, their data are in complete agreement with the results of Shimizu et al. (1988). Even the one exception pertains only to the original clone in which 10 adenines were found (and reconfirmed) near the 3' end of the open reading frame (Kinet et al., 1987). All subsequent clones and those published by Shimizu et al. have only nine, thereby leading to a polypeptide sequence five amino acids shorter than originally thought. The data from Kochan et al. (1988) and Shimizu et al. (1988) for the human sequences agree completely.

The α chain has not yet been successfully expressed in transfected cells. As noted in our original report (Kinet et al., 1987), it is interesting that the cDNA that codes for the Fc_γ receptor (α), whose transmembrane domain most resembles that of the Fc_ε receptor, also has not been successfully expressed. We, therefore, surmised that the β or γ subunits (or both) may be required for expression, a hypothesis which should soon be testable. It is interesting, in this respect, that α does not become modified (whereas β and γ do), when the receptor is reacted with an intramembrance probe either *in situ* (Holowka et al., 1981; Perez-Montfort et al., 1983a) or after solubilization (B. Grasberger and H. Metzger, unpublished observations). This suggests that the presumed transmembrane segment of α is substantially surrounded by β or γ_2 or both.

2. β and γ Chains

A so-far unique aspect of the mast cell receptor for IgE relative to other Fc receptors is its multisubunit structure. The experimental findings that led to the recognition of the β and γ chains as constituting part of this receptor, as well as some of their characteristics, were recently reviewed (Metzger et al., 1986). It seems likely that major new progress in clarifying the part they play in the structure and function of the receptor will result from isolating the genes that code for β and γ.

Several cDNAs that code for the β subunit have been isolated from a cDNA library derived from RBL cells and have been cloned and sequenced (Kinet et al., 1988). The cDNA predicts a sequence of 243 amino acids, and the correspondingly predicted composition is in reasonable agreement with direct analyses, as well as with biosynthetic labeling studies (Alcaraz et al., 1987). It also correctly predicts several peptides isolated in substantial yields from tryptic digests of the β subunit. Although the cDNA predicts a MW of around 26,000, it yields a protein product on *in vitro* synthesis with a molecular weight similar to newly synthesized β chains, i.e., ~32,000 MW, when analyzed on polyacrylamide gels (Kinet et al., 1988). Furthermore, when transcribed and then translated in an *in vitro* translation system, the protein product reacts with two distinctive monoclonal anti-β antibodies (Rivera *et al.*, 1988). Finally, a polyclonal antibody to the protein, synthesized by transfected *Escherichia coli*, reacts with native IgE–receptor complexes. There is, therefore, convincing evidence that the correct coding sequence has been identified.

The predicted sequence for β shows no NH_2-terminal signal sequence. Instead, there is a stretch of ~60 residues, which contains many basic and acidic residues, and a section in which over 40% of the amino acids are prolines. This N-terminal hydrophilic segment is followed by four hydrophobic sections, each containing at least 20 residues and separated from each other by short stretches of hydrophilic amino acids. The carboxy-terminal sequence of ~40 amino acids again shows many basic and acidic amino acids. Together, these findings suggest that the protein has four transmembrane sections and that the termini are both on the same side of the plasma membrane (Fig. 3). Detailed analysis of the sequence and of other features make it somewhat more likely that the ends of the polypeptide are on the cytoplasmic side, but experimental evidence will be more convincing. For example, antibodies to peptides, predicted to be in the hydrophilic sections, should be helpful for obtaining the necessary information.

One notable feature of the four putative transmembrane segments of the β chain is that none includes a charged residue. Since the two γ

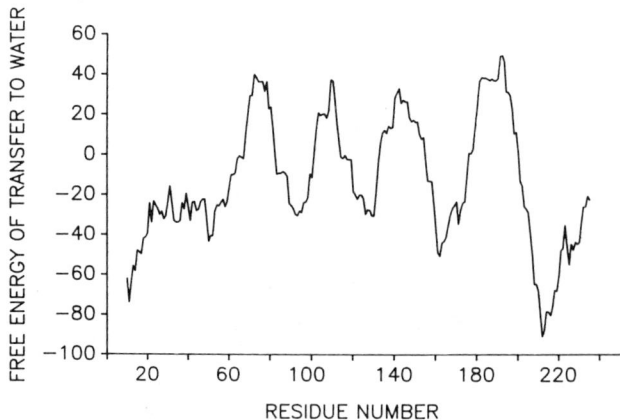

FIG. 3. Hydropathicity profile of the β subunit of the receptor with high affinity for IgE. The hydropathicity scale used is that described by Engleman *et al.* (1986). A 20-residue window was used, and the scores were plotted over the tenth residue in the window.

chains appear to be identical, we anticipated that a single basic residue might have been found in a transmembrane segment of β to balance the aspartic acid residue in the putative transmembrane of the α chain. It will be interesting to see what the sequence of the γ chains will reveal, and whether the sequence-predicted topology of the chains can be experimentally verified.

As already noted, the α chain cannot be expressed on the surface of cells transfected with the cDNA for α alone (Kinet et al., 1987). Major efforts to express β or to coexpress α and β have not yet been reported. Our own strategy is to first clone the cDNA coding for γ and to then attempt cotransfection and expression of all three polypeptides. Attempts to clone cDNA coding for the γ chain is currently underway.

C. MECHANISMS OF ACTION

1. Overview

The initial step in stimulation of cells by the IgE receptors is the aggregation of receptors. Earlier studies showed that dimerization of the receptors is sufficient to initiate a cascade of cellular events, although higher aggregates appear to have increased activity (Fewtrell and Metzger, 1980; Metzger et al., 1986). The precise molecular events are difficult to explore experimentally, until one of the most fundamental aspects of the receptor's action is clarified. That is, it is still unclear whether aggregation of the receptors results in some *intrinsic* activity in the associated receptors themselves or whether the aggregates simply activate a more distal component (*extrinsic* mechanism). There continues to be few published experiments that speak to this point. Mazurek et al. failed to observe electrogenic channel activity in bilayers, when reconstituted receptors from RBL cells were aggregated (Mazurek et al., 1984; Pecht et al., 1986). Quarto and Metzger (1986) investigated the tyrosine kinase activity they observed in preparations of "purified" receptors and concluded that the activity was not due to the receptor itself. (Whether the activity was due simply to an enzyme that artifactually contaminated the preparations or was due to an unstable, but meaningful, association of enzyme with the receptor was not determined.) As emphasized previously, there are too many trivial reasons why intrinsic activities of the receptor might be missed to make such negative results meaningful.

One way by which possible intrinsic activities might become recognizable is from the sequences of the receptor subunits themselves. In several instances, distinctive sequences are associated with particular activities of proteins. No such regions have been detected in the sequences of the α and β subunits, so far; whether they exist in the γ subunits should soon be known.

In the absence of direct evidence, alternative approaches are required. One approach is to ask whether the requirement for aggregation of receptors means that they actually have to touch each other. If not, then the initial biochemical event cannot be due to an intrinsic activity arising in the receptors directly because of their interaction with each other. Kane et al. (1988) have attempted to test this explicitly, by constructing a rigid bivalent ligand with a distance between the epitopes sufficiently large so that the receptors would be unlikely to be able to contact each other when their bound IgE reacted with the antigen. Kane et al. (1988) found that such ligands are still capable of triggering degranulation. Although their experiments cannot provide absolute proof that receptors in fact do not have to come in van der Waals contact with each other in order to activate the cells, the data are highly suggestive.

Another approach consists of defining early, quantitatively meaningful, biochemical perturbations stimulated by the receptors and then of attempting to define the molecular interactions that produce them. A large number of such perturbations have been and are continuing to be explored (Metzger et al. 1986; Beaven and Cunha-Melo, 1988). Some of the earlier studies have been difficult to extend, and the molecular basis for the phenomena they described have not been further clarified. I shall, therefore, limit the discussion to the two reactions that have been explored most intensively for the past few years: the hydrolysis of phosphoinositides and the rise in intracellular Ca^{2+}.

2. Phosphoinositide Hydrolysis

Many receptors activate cellular responses by stimulating the hydrolysis of phosphoinositides (Berridge, 1987). Upon binding ligand, the receptor is thought to activate a still-unidentified guanosine 5'-triphosphate (GTP)-binding protein, "Gp" (Cockcroft and Gomperts, 1985), analogous to the activation of the better-characterized GTP-binding proteins, Gs, Gi, and Gt (transducin) by receptors that stimulate adenyl cyclase or cyclic guanosine 5'-monophosphate (GMP)-specific phosphodiesterase (Gilman, 1987). The activation of Gp is thought to activate one or more types of phospholipase C, which releases inositol phosphates from the phosphoinositides. One of these inositol phosphates, in particular, inositol 1,4,5-phosphate, is thought to interact with receptors on the endoplasmic reticulum, leading to the release of Ca^{2+}. Thus, these initial reactions are thought to be the cause of a rise in cytoplasmic Ca^{2+}, rather than a consequence of it. The increase in cytoplasmic free Ca^{2+} is then thought to stimulate a variety of other Ca^{2+}-dependent proteins (Carafoli, 1987).

The hydrolysis of the phosphoinositide leaves stoichiometric amounts of diacylglycerol in membrane. The diacylglycerol, in turn, can act synergistically with the released Ca^{2+} to stimulate protein kinase C, which modulates still other proteins by phosphorylating serine or threonine residues (Nishizuka, 1986).

These pathways have been extensively studied in mast cells and, in particular, in RBL cells. A scholarly review of these studies has recently been completed by Beaven and Cunha-Melo (1988). I shall, therefore, provide only a brief summary in this review of the principal findings emphasizing, in particular, the uncertainties that remain.

Under physiological conditions, i.e., with millimolar Ca^{2+} in the extracellular medium, aggregation of IgE receptors leads to substantial release of counts from cells preincubated with radioactive inositol. The release is proportional to the extent of aggregation of the receptors and continues as long, and only as long, as the aggregation of the receptors is maintained. It had been thought earlier that, unlike in other systems (see above), hydrolysis was strictly dependent on extracellular Ca^{2+} (Beaven et al., 1984a), but more recent studies indicate that this dependence is neither absolute in RBL cells (Pribluda and Metzger, 1987; Cunha-Melo et al., 1987) nor in other mast cells (Ishizaka, 1985; Musch and Seigel, 1986; Baranes et al., 1986; Ishizaka et al., 1987). It is curious that, although extracellular Ca^{2+} may markedly enhance the receptor-induced breakdown of phosphoinositides, it is effective even in the presence of La^{3+}, which blocks the influx of extracellular Ca^{2+} (Beaven et al., 1984b, 1987). If this apparent anomaly does not have a simple quantitative explanation (i.e., that the blockage of the influx of Ca^{2+} by La^{3+} is incomplete), it suggests a still-undefined additional role for extracellular Ca^{2+}. Further evidence for such a role comes from the observation that a variant of RBL cells exhibits virtually no influx of Ca^{2+} on aggregation of the IgE receptors, yet the receptor-stimulated hydrolysis of phosphoinositide is dependent on extracellular Ca^{2+} (WoldeMussie et al. 1987).

a. Role of G-Proteins. Bacterial toxins, such as those from *Vibrio cholera* and *Bordetella pertussis*, can with adenosine 5'-diphosphate (ADP) ribosylate the α subunit of certain G-proteins, thereby inhibiting its action. The inhibition by toxin has been an effective analytical tool for implicating the role of G-proteins in the receptor-mediated activation in several systems (Gilman, 1987). Initial reports by Nakamura and Ui (1983, 1985) suggested that pertussis toxin could inhibit receptor-mediated activation of rat mast cells, but this could not be replicated with RBL cells (Metzger et al., 1986) or other mast cells (Saito et al., 1987), even though activation of these cells by other stimulants was inhibited. That GTPγS, a nonhydrolyzable activator of G-proteins, can stimulate breakdown of phosphoinositide when introduced into transiently permeabilized mast cells (Cockcroft and Gomperts, 1985) or permanently permeabilized RBL cells (Ali et al., 1988) or lysed cytoplasts from RBL cells (S. Dreskin and H. Metzger, unpublished), is consistent with a role for G-proteins in *some* of the pathways that stimulate phosphoinositide breakdown, but not necessarily those that are initiated by the receptor for IgE. In this last case, more persuasive evidence stems from the inhibition of receptor-mediated hydrolysis of phosphoinositides by GDPβS (S. Dreskin and H. Ali, personal communication), but more experimental verification of this inhibition is required. Direct demonstration of receptor-induced GTPase activity would provide even more convincing evidence. Detailed analysis will, of course, require reconstitution of the initial pathway with purified components.

   b. Role of Inositol Phosphates. The pathways of phosphoinositide metabolism are complex (Beaven and Cunha-Melo, 1988), and nine discrete inositol phosphates, more, if cyclic isomers are included, can be generated. Two of these, inositol 1,4,5-phosphate and inositol 1,3,4,5-tetrakisphosphate, are implicated in promoting the rise of cytoplasmic Ca^{2+} (see below), but the possible role of others cannot yet be excluded.

   It is notable that substantial receptor-mediated degranulation can occur in the absence of appreciable breakdown of phosphoinositides or a rise in cytoplasmic Ca^{2+}, when cells are reacted with phorbol esters (which by themselves do not stimulate secretion) (Sagi-Eisenberg and Pecht, 1984; Beaven et al., 1987). Although this does not rule out a physiological role in secretion for these signals under normal conditions, it suggests that alternative or additional signals exist (Beaven et al., 1987).

   c. Role of Protein Kinase C. As already noted, breakdown of phosphoinositides to inositol phosphates *pari passu* generates diacylglycerol. The latter may be further metabolized, will participate in the regeneration of phosphoinositides (or other phospholipids), and can itself acts as a "second messenger" by stimulating protein kinase C. The phorbol esters (presumptively acting as nonmetabolizable analogs of diacylglycerol) can stimulate phosphorylation of what appear to be similar proteins, as those phosphorylated during IgE receptor-mediated activation (Katakami et al., 1984; Teshima et al., 1986), and in combination with ionophore can stimulate degranulation (Katakami et al., 1984; Sagi-Eisenberg and Pecht, 1984; Beaven et al., 1987) provides circumstantial evidence only, the logical basis for implicating the kinase being similar to the logic for implicating G-proteins, because GTPγS can synergize with Ca^{2+} to mediate hydrolysis of phosphoinositides (see above).

There is some direct evidence for activation of the kinase, however, White et al. (1985) described increased kinase activity in membranes prepared from receptor-stimulated mast cells. An unusual feature was that this was *not* accompanied by depletion of the enzyme from the cytoplasm, as is usually seen in other systems. White and Metzger (1988) have observed the more common pattern, when RBL cells were examined. On aggregation of the IgE receptors, there was recruitment of kinase activity to a membrane fraction with equivalent loss of activity from the cytosolic fraction of the sonicated cells. The translocation reached maximal values (about 30%) in ~30 seconds, with subsequent rapid return to near control values within 2 minutes. After that, only a few percent of the activity remained membrane associated relative to control cells, but only so long as aggregation of the receptors was maintained. The receptor-induced translocation was ablated if the cells had been exposed to phorbol ester for prolonged periods. However, whereas such cells were now refractive to stimulation by phorbol ester plus ionophore, their exocytotic response to aggregation of IgE receptors was only reduced and not eliminated. This suggests that the translocation of protein kinase, observed after aggregation of the receptors, may be an accompanying event, rather than an essential step in the exocytotic pathway. It is possible that the mobilization of protein kinase C is somehow related to inactivation of the stimulatory pathway, but there are no firm data in this regard (Beaven and Cunha-Melo, 1988).

3. Rise in Cytoplasmic Calcium

Aggregation of the receptors leads to a rise in cytoplasmic free Ca^{2+}, a rise that is at least partially maintained as long, and only as long, as the receptors remain aggregated. The Ca^{2+}, whose cytoplasmic concentration is increased, is derived from two sources: intracellular stores and the extracellular fluids.

a. *Mobilization of Intracellular Stores.* i. *Storage sites.* Maintenance and utilization of the intracellular stores of Ca^{2+} require ATP (Mohr and Fewtrell, 1988). That the receptor-induced release is largely unaffected by mitochondrial inhibitors (providing ATP is maintained by the glycolytic pathway) makes it likely that the endoplasmic reticulum, rather than the mitochondria, is the principal site for storage of the intracellular Ca^{2+}.

The extent to which these intracellular stores are mobilized appears to vary with the stimulus used to aggregate the receptors. For example, if high doses of a multivalent hapten conjugate such as dinitrophenylated protein are added to cells bearing anti-dinitrophenyl IgE, the stores are recruited efficiently, lower doses giving proportionately slower responses. On the other hand, if aggregated ovalbumin (which is 1000-fold less potent in stimulating secretion) is added to cells bearing anti-ovalbumin IgE, the stores are mobilized hardly at all (Beaven et al., 1984b); anti-IgE produces intermediate responses (Millard et al., 1988a). It is reasonable to assume that the number and type of receptor aggregates which are formed, influence the rate and extent of formation of some second messenger that is required to mobilize the stores.

The rise in cytoplasmic Ca^{2+}, in general, and even that portion of it which is attributable to mobilization of intracellular stores, in particular, can be studied at the single cell level by digital video fluorescence microscopy (Millard et al., 1988a,b). The earliest changes are observed even when Ca^{2+} influx is prevented, for example, by (brief) depletion of external free Ca^{2+} or by addition of La^{3+} (Millard et al., 1988a) prior to the addition of a stimulant. Therefore, under these conditions, the changes must reflect mobilization of intracellular stores. On addition of a stimulant, there is a lag time before any changes are observed. With an antigen such as the multivalent hapten conjugate as the stimulant, the lag time varies somewhat from cell to cell, but is quite short, averaging ~ 25 seconds. Since, above a certain optimal dose of such an antigen, there is no further shortening of the lag time, the delay cannot principally reflect simply the binding of antigen to antibody. Given other considerations, the subsequent formation of receptor aggregates is also unlikely to be the time-dependent step (Dembo et al., 1979). (With other stimulants, such as aggregated ovalbumin, anti-IgE, or preformed oligomers of IgE, the kinetics of binding or aggregation of receptors or both may contribute to the delay.) Since, with an ionophore, such as ionomycin, the lag times may be as short as 2 seconds (Millard et al., 1988a), release of the stored Ca^{2+} may also not be the limiting step. It appears most likely that the principal rate-determining event is the generation of the postulated second messenger. Studies on the kinetics of receptor-induced breakdown of phosphoinositides, as determined either by measuring the generation of inositol phosphates (Pribluda and Metzger, 1987) or by the recuitment of protein kinase C to the (plasma?) membrane (White and Metzger, 1988), are consistent with such a model.

ii. Mechanism of mobilization. There is evidence from other systems that inositol 1,4,5-trisphosphate, generated by the hydrolysis of plasma membrane polyphosphoinositide, interacts with a protein in the membrane of the endoplasmic reticulum and, thereby, opens a pathway through which stored Ca^{2+} is released. There are no data from mast cells or basophils inconsistent with such a mechanism, but there is no direct demonstration of it either. In some systems, an alternative second messenger has been implicated—arachadonic acid released by the action of a phospholipase A_2 or by neutral lipases (Wolf *et al.*, 1986). The possible influence of this metabolite on the intracellular stores in mast cells deserves study.

iii. Consequences of mobilization. A discussion of whether the intracellular stores can provide sufficient Ca^{2+} to drive the cascade of reactions that lead to degranulation (under the artificial conditions in which influx of extracellular Ca^{2+} is prevented) falls outside the focus of this review. Suffice it to say that the prestored Ca^{2+} may be sufficient to support secretion in certain instances (Musch and Siegel, 1986; Baranes *et al.*, 1986; Mohr and Fewtrell, 1987a), but in most cases, it is not. Whether mobilization of these stores is *necessary* for degranulation, when the other mechanism of raising cytoplasmic Ca^{2+}—influx from the outside—is functional, is unclear.

b. *Influx of Extracellular Ca^{2+}.* Although the initial rise in cytoplasmic Ca^{2+} is derived from intracellular stores, the subsequent continued elevation and increase in total cellular Ca^{2+} are due to influx of extracellular Ca^{2+}. The characteristics of this pathway are gradually becoming better defined, but several fundamental questions, such as the molecular basis for the transport and the proximal stimulus that initiates it, remain unanswered.

i. Functional characteristics. Using 3 mM of intracellular quin 2 as an infinite "sink" for intracellular Ca^{2+}, Fewtrell and Sherman (1987) measured the true, unidirectional influx of Ca^{2+} by examining the uptake of $^{45}Ca^{2+}$ by RBL cells stimulated by aggregation of the receptors. A maximum rate of $\sim 10^7$ Ca^{2+} ions/cell/second was observed, the rate being $\frac{1}{2}$ maximal at an extracellular concentration of 0.6 mM. Once activated, the permeability pathway was maintained for over an hour, providing the aggregates of receptors were not disrupted.

The potential across the plasma membrane of these cells is -70 mV (Kanner and Metzger, 1983; Sagi-Eisenberg and Pecht, 1983) and is required for the influx of Ca^{2+}. Depolarizing the cells by high concentrations of K^+ prevents the influx of Ca^{2+} (Kanner and Metzger, 1984; Mohr and Fewtrell, 1987b). In turn, new studies with bis-oxonol

(Mohr and Fewtrell, 1987c)—a better probe for estimating the potential across the plasma membrane, than the tetraphenyl phosphonium ion used in previous studies on RBL cells—confirm that, under normal conditions, influx of Ca^{2+} collapses the plasma membrane potential. In the absence of extracellular Ca^{2+}, aggregation of the receptor depolarizes the membrane through stimulated influx of Na^+, presumably by the same pathway. Provided that intracellular ATP levels are maintained via glycolysis and that a suitable sink for the intracellular Ca^{2+} is provided (Mohr and Fewtrell, 1988). the influx of Ca^{2+} can be maintained, even in the absence of functional mitochondria. Apparently, the mitochondria serve as the physiological buffer or sink during stimulated influx of Ca^{2+} (see above).

ii. Molecular basis of influx pathway. It is apparent from the above that the Ca^{2+} permeability pathway is not mediated by a conventional voltage-regulated calcium channel; correspondingly, Ca^{2+} transport is not inhibited by conventional antagonists of such channels (Ritchie *et al.*, 1984). Indeed, it remains to be proved that the mechanism of entry of Ca^{2+} is by way of a simple channel, since patch–clamp studies have failed to reveal receptor-induced electrogenic movements of Ca^{2+} (Lindau and Fernandez, 1986a,b).

Could the pathway result simply by aggregation of the receptors themselves? The experiments of Maeyama *et al.* (1986) indicated that the maximum early increase in the concentration of cytoplasmic Ca^{2+} was progressively greater, as successively larger numbers of receptors were aggregated. The proportionality closely paralleled the stimulated hydrolysis of phosphoinositides assessed on the same samples. Since it now appears that, even in the presence of extracellular Ca^{2+}, the early rise in cytoplasmic Ca^{2+} is importantly contributed to by Ca^{2+} released from intracellular stores (see above), the data of Maeyama *et al.* may support a correlation between aggregation of receptors and mobilization of Ca^{2+} from storage sites, rather than with influx of extracellular Ca^{2+}. Indeed, when studied explicitly using $^{45}Ca^{2+}$, influx appeared to saturate, when only about 10% of the receptors could have been aggregated (Fewtrell and Sherman, 1987). Together, these results, in turn, suggest that the direct molecular mediator of the influx of extracellular Ca^{2+} is not the aggregated unmodified receptors themselves.

This conclusion agrees with the findings of Pecht and colleagues (1986), who have failed to observe electrogenic Ca^{2+} channels, when putatively intact receptors, reconstituted in artificial bilayers, were aggregated (Mazurek *et al.*, 1984). Of course, such negative results by themselves could be due to a variety of other causes. This same group has proposed that a cromolyn-binding protein, isolatable from RBL cells, constitutes the physical entity that mediates stimulated influx of Ca^{2+} by direct interaction with aggregated receptors. Their studies have been extensively critiqued previously (Gomperts and Fewtrell, 1985; Metzger et al., 1986), and there is little new to add. From a functional perspective, the principal dilemmas are that, in vivo, the influx pathway appears to require ATP and a substantial membrane potential, whereas the phenomenon, described by Pecht and colleagues, shows no such dependence. From a structural perspective, the evidence is incomplete that the 60-kDa unreduced/67-kDa reduced entity, they isolated (Mazurek et al., 1982), accounts for the functional observations.

iii. Stimulation of influx pathway. That stimulated influx of extracellular Ca^{2+} requires ATP raises the possibility that the mechanism involves more than a passive direct interaction between unmodified aggregated receptors and a permeability-inducing component. Possibly the receptors must first be phosphorylated. Although initial studies provide scant evidence for such a model (Perez-Montfort et al., 1983b), this possibility has not been exhaustively tested (see above). Alternatively, the influx pathway may require stimulation by the products of ATP-requiring hydrolysis of phosphoinositides (Irvine and Moore, 1986; Kuno and Gardner, 1987) or, even more indirectly, may depend on the initial rise in cytoplasmic Ca^{2+} (Tscharner et al., 1986) released from intracellular stores by newly formed inositol polyphosphates.

iv. Consequences of influx. Since the dependence of secretion on extracellular Ca^{2+} was demonstrated three decades ago (Mongar and Schild, 1958), it has been assumed that the consequent rise in intracellular Ca^{2+} was the essential stimulant. That a variety of maneuvers, which artificially raise cytoplasmic Ca^{2+}, could induce degranulation provided strong support for this model, but not proof. One needs to show, in addition, that the artifactually raised intracellular Ca^{2+} is not incidentally producing a perturbation normally induced by aggregation of receptors, irrespective of the rise in Ca^{2+}. Stimulated hydrolysis of phosphoinositides is a prime candidate, and the interesting studies, by Beaven and colleagues, provide experimental support for such a possibility (Lo et al., 1987). This group raised cytoplasmic Ca^{2+} by exposing RBL cells to progressive doses of ionophore. Doses that led to increases in Ca^{2+}, equivalent to those induced by aggregation of receptors, failed to stimulate degranulation. Secretion was observed only when doses of ionophore were used that induced breakdown of phosphoinositides, equivalent to the hydrolysis induced by IgE-mediated stimulation. It remains to be proved that, in the other studies in which cytoplasmic Ca^{2+} was increased, the Ca^{2+} per se, rather than some other metabolite, was responsible for the stimulated secretion. It is noteworthy, in this regard, that phorbol esters [which by themselves cannot induce degranulation (Sagi-Eisenberg et al., 1985; Beaven et al., 1987)] have little (Sagi-Eisenberg et al., 1985) or no effect (Beaven et al., 1987) on receptor-induced degranulation. Yet, though the secretion requires extracellular Ca^{2+}, essentially no rise in cytoplasmic Ca^{2+} above the baseline is observed under these conditions! At the very least, it seems fair to say that *substantial* rises in cytoplasmic Ca^{2+}, although normally a consequence of receptor-mediated stimulation, may not be a critical element in the pathway that leads to secretion. As already noted above, studies on variants of RBL cells lead to a similar conclusion (WoldeMussie et al., 1987).

IV. Immunoregulatory Fc_ε Receptors and IgE-Binding Factors

A. STRUCTURAL PROPERTIES

1. Fc_ε Receptors on Human B Lymphocytes

Considerable progress has been made in the molecular definition of, at least, the principal receptor for IgE on human B cells. This has resulted from the use of several discrete monoclonal antibodies to the receptor and, in particular, from the successful isolation of a cDNA coding for the receptor. The latter was accomplished independently and virtually contemporaneously by three groups, Kikutani et al. (1986), Ludin et al. (1987), and Ikuta et al. (1987), using the human lymphoblastoid line RPMI 8866. Their results show the following.

1. There is an open reading frame of 963 nucleotides which, therefore, predicts a polypeptide of 321 amino acids. At positions 14–20 in the amino-terminal region, there is a strongly basic cluster [(R-R-R-C-C-R-R), R = Arg, C = Cys], which is followed (at positions 24–45) by a segment containing largely hydrophobic residues and interspersed with only two threonines, but no other hydrophilic amino acids. The subsequent sequence shows no further extensive runs of hydrophobic residues. It contains a single potential site for N-glycosylation about 20 residues distal to the hydrophobic segment.

2. Starting at positions 148 and 150, a sequence is found to be identical to the alternative NH_2-terminal sequences found on a 25-kDa fragment secreted into the medium by the RPMI 8866 cells. Further, toward the COOH-terminus, additional sequences are found to be

identical to peptides isolated from this fragment and on the basis of which the oligonucleotide probes were synthesized and used to originally identify the cDNA.

3. The sequence shows extensive homology with that of the integral membrane protein chicken hepatic lectin (Drickamer, 1981). There are 44 identities, allowing for 2 gaps in residues 150–285, and almost perfect alignment of 6 tryptophans and 7 cysteines in that region. Like the lectin, the protein is, therefore, likely to be disposed relative to the plasma membrane, such that the NH_2-terminus is on the cytoplasmic side and the 250-residue carboxy-terminal portion is on the exterior.

4. Messenger RNA transcribed from the cDNA, when injected into frog oocytes, led to material in the lysates that reacted with monoclonal antibodies to the receptor (Kikutani et al., 1986), and transfected COS cells expressed IgE binding activity and reactivity with antireceptor antibodies.

5. Northern blot analysis revealed a single 1.7-kb mRNA in RPMI 8866 and in normal B cells, as well as in the human macrophage line U937, but not in normal T lymphocytes.

Together, these results support extensive other data, showing the close relationship between the cell surface receptor with low affinity for IgE on B cells and the IgE-binding factor released by such cells (e.g., Nakajima et al., 1987). The common differences (6–7 kDa) in molecular masses between those predicted by the amino acid sequences for the intact receptor and binding factor (36.1 and 19.4 kDa) and those observed on polyacrylamide gels for the proteins themselves (43 and 25 kDa) provides further supporting evidence and is likely accounted for by glycosylation. However, since the 25-kDa fragment does not contain the single putative N-glycosylation site (see above), the carbohydrate must be attached otherwise.

A substantial number of monoclonal antibodies have been generated to the Fc_ε receptor and have been used to study it and related components (Nakajima and Delespesse, 1986; Delespesse et al., 1986; Suemura et al., 1986; Noro et al., 1986; Bonnefoy et al., 1987; Kisaki et al., 1987). The general picture that is emerging is as follows.

1. The antigen previously identified as CD23 is the Fc_ε receptor.
2. The 65- to 95-kDa component, irregularly seen in preparations prepared by affinity chromatography using IgE, shares epitopes and is likely an aggregate of the principal 43-kDa component material, as originally suggested by Peterson and Conrad (1985). The 31-kDa component (37 kDa after reduction), as well as those with apparent

MWs of 25,000 (see above) and 16,000, are likely proteolytic breakdown products of the 43-kDa component.

3. Some of the monoclonal antibodies appear to be more effective in detecting Fc_ε receptors than is IgE. With these antibodies, 30–50% of B lymphocytes may score positive. The antigen is regularly observed on certain macrophages, but only on small numbers of T lymphocytes and natural killer (NK) cells.

Bonnefoy et al. (1988) have observed recently that two monoclonal antibodies, which inhibit the binding of IgE to RPMI 8866, in fact, are reactive with the α and β chains of HLA-DR. These and other experiments suggest that at least some of the Fc_ε receptors are associated with class II histocompatibility molecules, much as has been proposed for Fc_γ receptors on murine lymphocytes (Dickler and Sachs, 1974). The functional significance of this apparent association remains to be explored.

2. IgE-Binding Factor from Human B Cells

As already noted, a fragment, having an apparent MW of 25,000 (on polyacrylamide gels), is released into the medium by RPMI 8866 cells and appears to be a proteolytic fragment of the 45-kDa Fc_ε receptor. Kisaki et al. (1987) have also reported on such IgE-binding factors released by the same cells. They chromatographed the material through Sephadex G-100 and assessed the molecular weights of the components that had the capacity to inhibit rosette formation between RPMI 8866 cells and ox red blood cells (RBC) coated with human IgE. They observed three active components with apparent MWs of 60,000, 30,000, and 15,000. These molecular weights are virtually identical to those Ishizaka and colleagues have regularly observed for IgE-binding factors from human and rodent T cells (see below). It must be remembered that, although molecular weights estimated from polyacrylamide gels run in the strong denaturant sodium dodecyl sulfate can also be inaccurate, this is even more so with respect to molecular weights estimated from gel filtration in nondenaturing solvents. How then, can we rationalize these discrepant findings? I suggest that the "60,000"-MW fragment may be an aggregate of the 25,000-MW (or sightly larger) binding factor, that the "30,000"-MW component is the monomeric form of this factor, and that the "15,000"-MW component is a proteolytic breakdown fragment of the latter. Such 16,000-MW fragments are occasionally seen in preparations of binding factors from these cells (see discussion in Sarfati et al., 1987) and may be the same as the papain fragment noted by Peterson and Conrad (1985) and the 16,000-MW binding factor observed in colostrum (Sarfati et al., 1986). Using the monoclonal antibody H107 prepared against Fc_ε receptors on RPMI 8866 cells (Noro et al., 1986) Kisaki et al. (1987) were able to absorb out the activity of the 60- and 30-kDa factors, but not the 15-kDa factor.

3. Fc_ε Receptors on Human T Cells

The presence of Fc_ε receptors on human T cells has been controversial. Nutman et al. (1987) transformed with human T cell leukemia lymphoma virus the putatively Fc_ε receptor-positive T cells from two patients with hyper-IgE syndrome and tropical pulmonary eosinophilia, respectively. Soon after infection, they were able to isolate T cell clones that bound a monoclonal antibody ("135"), which recognizes Fc_ε receptors on human macrophages, T cells, and B cells (Sarfati et al., 1986). Seventeen discrete clones also produced IgE-binding factor, i.e., material in the culture supernatant that either reacted with two monoclonal antibodies (176 and 135) directed to distinct epitopes on Fc_ε receptors, as assessed in a solid-phase plate assay, or with IgE by a blot assay on nitrocellulose paper.

One of the positive clones was more extensively characterized by Sarfati et al. (1987). Northern blot analysis showed that it contained a mRNA identical in size to that in RPMI 8866 cells, when probed with the cDNA derived from mRNA of the latter (see above).

Kisaki et al. (1987) have analyzed human T cell hybridomas for Fc_ε receptors with the monoclonal antibody to B cell Fc_ε receptors and with monoclonal antibodies to the IgE-binding factors released by the hybridoma (see below). Both types of antibody failed to stain the cells by fluorescence microscopic techniques, presumably because the cells have too few surface receptors (Suemura et al., 1986), but *both* were able to induce release of IgE-binding factors, apparently by reacting with the few Fc_ε receptors that the cells do possess. Young et al. (1984) have also reported human T cell lines bearing Fc_ε receptors.

4. IgE-Binding Factors from Human T Cells

Sarfati et al. (1987) found that the virally transformed human T cells they examined released IgE-binding factors into the medium that were indistinguishable from the 25-kDa IgE-binding factors released by human B cells. A 16-kDa breakdown product was also occasionally observed. On the other hand, Kisaki et al. (1987) found distinguishing characteristics on the binding factors released by the human T cell hybridomas they studied. These cells released binding factors having molecular weights identical to those determined by the same investigators for the factors released by the B cells. Whereas an antibody to the B cell Fc_ε receptor reacted with all except the smallest T cell-derived binding factors, the antibodies induced by the latter failed entirely to react with the B cell Fc_ε receptor or the factors released by the B cells. These findings suggest, at least, some difference between the T cell and B cell IgE receptors and binding factors. As we shall see below, there is analogous evidence in the murine system.

5. Comparative Studies on Rodent Cells

Murine B cells, as well as B cell hybridomas, have Fc_ε receptors, whose structural characteristics closely resemble their counterparts on human B cells (Conrad and Peterson, 1984; Lee and Conrad, 1986). There is evidence that this receptor binds more than one IgE; however, cross-linking experiments suggest that this may be due to associations between two or more receptors, rather than because of multivalency of the 45- to 49-kDa component (Lee and Conrad, 1984, 1985). Such cross-linking studies, either on murine or human B cells, have failed to reveal any associated polypeptides, such as the β and γ components associated with the IgE-binding α chain of the high-affinity receptor. Interestingly, these same experiments by Lee and Conrad (1985) did suggest that some of the Fc_ε receptors are in close association with the membrane IgM or IgD on the same cells.

Biosynthetic studies indicate that the receptor goes through a still-undefined "processing" step prior to acquiring IgE-binding activity, but that N-glycosylation of the processed form, if anything, reduces the affinity for IgE (Keegan and Conrad, 1987). The Fc_ε receptor that becomes situated on the plasma membrane has a relatively short half-life, but its degradation is slowed if IgE binds to it (Lee et al., 1987). This accounts for the "up-regulation" (i.e., increase in number of Fc receptors per cell) seen when cells are cultured in the presence of IgE. A similar phenomenon has been observed for the high-affinity receptor on rat basophilic leukemia cells (Furuichi et al., 1985; Quarto et al., 1985).

The degradation of the Fc_ε receptor leads to release into the medium of fragments having molecular weights (estimated on polyacrylamide gels) of 38,000, 35,000, 28,000, and 25,000 (Lee et al., 1987). Only the two larger forms appear to contain N-linked carbohydrate. If this receptor is analogous to the human receptor, the extracellular portion could contain up to about 275-amino acid residues with an estimated peptide MW of ~30,000. Given the estimated mass of carbohydrate associated with the human Fc_ε receptor (6–7 kDa; see above), this would give a molecular weight consistent with the observations of Lee et al. (1987). Notably, none of the fragments released by the murine B cells or B cell hybridoma bind to IgE (Lee et al., 1987).

Most investigators have failed to detect substantial Fc_ε receptors on T cells in normal or even parasite-infected mice (c.f. Chen et al., 1981). However, especially in mice bearing an IgE-secreting hybridoma, considerable Fc_ε receptors can be demonstrated on L3T4−Lyt2$^+$ T cells (Mathur et al., 1986). A goat antibody to the Fc_ε receptor of murine B cells inhibited IgE binding to the T cell Fc_ε receptor and immunoprecipitated a surface-iodinated 49-kDa component. On the other hand, a rabbit polyclonal antibody and a monoclonal rat anti-mouse B cell Fc_ε receptor failed to inhibit IgE binding to the T cell Fc_ε receptor or to precipitate a surface component from the same cells. These results (Mathur et al., 1988) indicate that the T cell Fc_ε receptor shares structural features with the receptor on B cells, but may not be identical.

In the presence of high levels of IgE, the percentage of T cells expressing Fc_ε receptors increases substantially, and biosynthesis of IgE is suppressed, apparently by selective loss of expression of ε mRNA (Mathur et al., 1987). Whether this occurs via a T cell-derived, IgE-binding suppressive factor, as the data from studies with rat and human cells suggest (see below), is unknown.

In rats, much of the work on IgE-binding proteins has been directed toward clarifying IgE-binding factors, rather than cell surface receptors. These factors released by T lymphocytes are capable of selectively enhancing or suppressing IgE responses, the nature of their regulatory activity being correlated with indirectly determined changes in their state of glycosylation. The molecular weights for these factors (as estimated by gel filtration) has revealed species of ∼15,000, 30,000, and 60,000 MW (Ishizaka, 1984). The factors share antigenic epitopes with each other and with low-affinity Fc_ε receptors on B and T cells (Yodoi et al., 1982).

Huff et al. (1982) fused AKR mouse thymoma cells with mesenteric lymph node cells from rats and secured a hybridoma, which releases such IgE-binding factors. The 23B6 hybridoma has been used in an attempt to obtain a cDNA that codes for the regulatory factors. Using subtractive hybridization and hybrid selection, Martens et al. (1985) identified cDNA clones that appeared to code for binding factors by several criteria. When injected into oocytes, mRNA hybrid selected with the cDNA-induced IgE-binding material. Similarly, when expressed in COS cells, several cDNA clones that hybridized with a 1200-bp probe, prepared from the candidate clone, induced the production of material that was adsorbed by rat IgE, but neither by human IgE nor by rat or mouse IgG.

When a 3.1-kb restriction fragment of one of the candidate clones was cloned in both orientations in a yeast expression vector, only that in the correct orientation led to yeast mRNA, which yielded binding factor when injected into oocytes. Three clones, when transfected into COS cells, yielded binding factor with no IgE-regulatory activity, two others (one of which produced both a 60- and an 11-kDa factor) had potentiating activity.

The sequences of the cDNAs that appear to code for these factors were unexpected. The first one analyzed (from the clone that led to the 60- and 11-kDa potentiating factors) showed an open reading frame of 1668 bp (556 amino acids, ~62 kDa) (Martens et al., 1985). There was a putative leader sequence and two potential glycosylation sites, but no obvious transmembrane segment. (Presumably, if this protein were to serve also as a surface Fc receptor, it would have to be anchored in the membrane with a lipid "tail.") Sequence, as well as other studies, showed that this clone is a member of an endogenous *mouse* retrovirus-like gene family (Moore et al., 1986). Compared to a reference intracisternal A particle (IAP) sequence, it showed a 3.4-kb deletion, leading to a fusion of the coding sequence for the *gag* and *pol* regions. Another clone (one of which, presumptively, coded for the 60-kDa binding factor without regulatory activity) had an open reading frame over twice as long, similarly representing a *gag–fol* fusion protein. Where the two genes were similar, the homology was ~85%. The three other clones, though not fully sequenced, were similarly found to be related to the viral genes, the common feature among all of them being in the *gag* regions.

Independent evidence for the relationship between the IgE-binding factors and retroviral proteins has been obtained with antibodies. Polyclonal antibodies to IAP core proteins reacted with the IgE-binding factors, although variably with binding factors from different sources. Also, a monoclonal antibody directed against a peptide constructed on the basis of the sequence from the smaller clone (see above) was said to react with IgE-binding factors induced by that clone, those produced by the parent hybridoma from which the cDNA was originally isolated, and those produced by normal mouse cells (Moore et al., 1986).

Jardieu et al. (1985) have attempted to relate the various factors obtained from the 23B6 hybridoma and from the transfected COS cells. The principal finding was that, in part, the higher molecular-weight factors gave rise to the smaller ones on reduction and alkylation. This would be explainable if, in some of the larger peptides, cleavage within intrachain disulfide bridges had occurred. Nevertheless, some findings require more complex explanations. For example, the 60-kDa binding factor (produced by the hybridoma) fails to bind to peanut agglutinin, whereas its putative 30-kDa breakdown product does; the 30-kDa factor is inactive, whereas the parent 60-kDa factor is partially active, and the 14-kDa product substantially so. The 60-kDa factor reacts with an anti-Ia monoclonal antibody [and, therefore, may be similar to the 60-kDa suppressor factor originally described by Suemura et al. (1981)], yet the sequence, which led to its production in transfected COS cells, shows no homology with Ia antigen sequences. New data suggest that the cross-reactivity may be based on carbohydrate epitopes (Martens et al., 1987).

The major discrepancy in the findings raised by the gene cloning studies of human and rat IgE-binding factors and receptors cannot now be resolved. However, clarification should soon be forthcoming. One can anticipate, from the studies on the high-affinity receptors, that there should be sufficient similarity between the species to allow for probing of the one by genetic probes developed in the other. It should be possible to test whether mouse and rat cells contain mRNA that codes for the analog of the human Fc_ε receptor on human B and T cells. If so, it can be sequenced (via cDNA), antibodies can be made to suitable peptides predicted by the sequence, and the reactivity of the binding factors with such antibodies can be tested. Similar studies can be performed in the reverse direction. It seems futile to speculate further, until the results of some of these comparative analyses are available.

B. Functional Characteristics

A number of isotype-specific immunoregulatory processes appear to implicate the Fc receptors or their secreted products (Fridman et al., 1987). For IgE in particular, there have been numerous studies which demonstrate that IgE-binding factors, derived from T lymphocytes, can modify IgE responses *in vitro* and are likely to have similar effects *in vivo* (Ishizaka, 1984; Ishizaka, 1987; Marcelleti and Katz, 1987). So far, there is almost nothing known about how the factors work mechanistically. Whether the suppression (or stimulation) of IgE production is always related to changes in levels of mRNA coding for the ε heavy chains (Mathur et al., 1987) and how such changes are induced are still virtually unexplored.

V. Surface Receptors on Monocytes, Eosinophils, and Platelets

A. Monocytes

Receptors, with an affinity for IgE lower than that possessed by the mast cell-specific receptor for IgE, have been observed on monocytes and macrophages (reviewed in Spiegelberg, 1984). As already noted, an mRNA, having a size identical to the one that codes for the low-affinity receptor on human B lymphocytes, has been observed in the human macrophage line U937, using the cDNA that codes for the lymphocyte receptor. Furthermore, a goat antilymphocyte Fc_ε receptor (Meinke et al., 1978), which prevents binding of IgE to lymphocytes, will also prevent binding of IgE to macrophages (Spiegelberg, 1984; Capron et al., 1986b). On polyacrylamide gels, immunoprecipitates of extracts from surface-iodinated U937 cells have given patterns closely similar to those observed with extracts from lymphocytes (Melewicz et al., 1982). However, a monoclonal antibody (BB10) has recently been described, which appears to distinguish between the macrophage receptor and its analog on lymphocytes (Capron et al., 1986b), so that the receptors may not be identical. Although this could simply reflect minor differences in glycosylation, the diversity of sequences, among the highly homologous family of proteins that code for Fc_γ receptors (Ravetch et al., 1986; Hibbs et al., 1988), indicates that one must be prepared for a greater complexity among these proteins than might have been earlier anticipated.

In studies on the macrophage receptor for IgE in rats, Finbloom and Metzger (1983) described a 33,000-MW protein that copurified with the ~50-kDa IgE-binding protein, when chemical cross-linking reagents were employed. Its molecular weight appeared to be the same as that for the β chain of the high-affinity receptor, although the protease digestion pattern was different. Recent studies with a potent monoclonal anti-β chain antibody [mAbβ (JRK)] failed to reveal cross-rective material in such macrophages by immunoblotting (Rivera et al., 1988), and Northern blotting studies (J. P. Kinet and C. Ra, personal communication) have, so far, also failed to reveal a mRNA closely homologous to that for β chains in a mouse macrophage line. Therefore, there really are no convincing data as yet that the protein, observed by Finbloom and Metzger, represents a homolog to the mast cell β subunit of the high-affinity receptor.

B. Eosinophils and Platelets

The polyclonal anti-Fc_ε receptor of lymphocytes reacts also with eosinophils and platelets (Spiegelberg, 1984; Capron et al., 1986b), cells earlier shown to have low-affinity receptors for IgE (Capron et al., 1981; Joseph et al., 1984). The monoclonal antibody BB10 reacts with the IgE receptors on both types of cells and immunoprecipitates, from the extracts of surface-iodinated platelets, components similar to those precipitated by it from monocytes (Capron et al., 1986b).

That the BB10 antibody fails to react with mast cells indicates not only that the epitope it recognizes is not on the high-affinity receptor for

IgE, but also shows that the epitope is not on the low-affinity receptor for IgE, found on mast cells and RBL cells (Froese, 1980b). The use of existing molecular genetic probes should make it soon possible to trace the relationships between these receptors.

Lichtenstein and colleagues have recently explored human "histamine-releasing factors" that appear to mediate their action via IgE (Schulman *et al.*, 1985; Warner *et al.*, 1986; MacDonald *et al.*, 1987). These factors bind to some, but not all, IgE—a curiosity the molecular basis for which remains to be defined—and are released by various inflammatory cells. The apparent molecular-weight range of these factors, 15,000–30,000 MW, would be consistent with their being the proteolytic fragments of low-affinity Fc_ε receptors, as MacDonald *et al.* suggest. If these fragments are multivalent (or form multivalent aggregates) like the murine lymphocyte receptor, as Lee and Conrad (1984) propose, then they might well be capable of aggregating IgE-loaded mast cell receptors. Whether this plausible model is correct, must await further molecular definition of these factors.

VI. Other IgE-Binding Proteins

An unusual IgE-binding protein has been described by F.-T. Liu and colleagues. They had observed that oocytes, injected with mRNA from RBL cells, expressed a 31-kDa protein, which appeared to have specific IgE-binding activity (Liu and Orida, 1984). They were able to clone a cDNA that coded for it from a library prepared from RBL cells (Liu *et al.*, 1985). Although the initial cDNA was only a partial one, they were able to use it as a probe and have now reported a full-length sequence (Albrandt *et al.*, 1987). The cDNA predicts a protein ("εBP") of 262 residues. It contains no apparent leader sequence, and starting at position 35, there is a nonapeptide consensus sequence YPGXXXPGA, which, with small variations, is repeated 10 times (through position 131).

The carboxy-terminal 120 amino acids of εBP contains 2 regions of 26 and 52 residues, respectively, that Albrandt *et al.* (1987) propose are homologous to sequences in the extracellular portion of the mouse Fc_γ receptor. The homology is, however, not very impressive, even though it is statistically significant. For one thing, the candidate sequences are contiguous in Fc_γ receptors, but not in εBP. Second, three gaps must be introduced to optimize the homology. Third, highly conserved residues, such as cysteine and tryptophan, are not among the similar residues, and fourth, the homologies between Fc_ε receptor and εBP are not in the same residues, which are shared by Fc_γ receptors and the α chain of the high-affinity receptor. Finally, εBP has no cysteine loop, a hallmark of immunoglobulin-like domains. The predicted sequence for the εBP protein shows neither a potential site for N-glycosylation nor a potential transmembrane segment.

An mRNA was synthesized *in vitro* from the full-length cDNA and was translated *in vitro*. The synthesized protein had an apparent MW of 28,500 and bound to IgE, but not an IgG1 mouse protein. It remains to be determined where this protein normally acts, which cells produce it, and what its function is. Its relationship to other proteins, having MWs of around 30,000, that have irregularly been seen in RBL cells (Kulczycki and Parker, 1979; Holowka and Baird, 1984) is also unknown. εBP clearly is unrelated to the β chain of the high-affinity receptor (see above).

REFERENCES

Albrandt, K., Orida, N. K., and Liu F.-T. (1987) *Proc. Natl. Acad. Sci. U.S.A.* **84,** 6859.
Alcaraz, G., Kinet, J.-P., Liu, T. Y., and Metzger, H. (1987). *Biochemistry* **26,** 2569.
Ali, H., Cunha-Melo, J. R., and Beaven, M. A. (1988). *Biochim. Biophs. Acta,* in press.
Baniyash, M., and Eshhar, Z. (1984). *Eur. J. Immunol.* **14,** 799.
Baniyash, M., and Eshhar, Z. (1987). *Eur. J. Immunol.* **17,** 1337.
Baranes, D., Liu, F.-T., and Razin, E. (1986). *FEBS Lett.* **206,** 64.
Beaven, M. A., and Cunha-Melo, J. R. (1988). *Prog. Allergy* **42,** in press.
Beaven, M. A., Moore, J. P., Smith, G. A., Hesketh, T. R., and Metcalfe, J. C. (1984a). *J. Biol. Chem.* **259,** 7129.
Beaven, M. A., Rogers, J., Moore, J. P., Hesketh, T. R., Smith, G. A., and Metcalfe, J. C. (1984b). *J. Biol. Chem.* **259,** 7129.
Beaven, M. A., Guthrie, D. F., Moore, J. P., Smith, G. A., Hesketh, T. R., and Metcalfe, J. C. (1987). *J. Cell. Biol.* **105,** 1129.
Berridge, M. J. (1987). *Annu. Rev. Biochem.* **56,** 159.
Bonnefoy, J.-Y., Aubry, J.-P., Peronne, C., Wijdeness, J., and Banchereau, J. (1987). *J. Immunol.* **138,** 2970.
Bonnefoy, J.-Y., Guillot, O., Spits, H., Blanchard, D., Ishizaka, K., and Banchereau, J. (1988). *J. Exp. Med.* **167,** 57.
Burt, D. S., and Stanworth, D. R. (1987). *Eur. J. Immunol.* **17,** 437.
Burt, D. S., Hastings, G. Z., Healy, J., and Stanworth, D. R. (1987). *Mol. Immunol.* **24,** 379.
Capron, M., Capron, A., Dessaint, J. P., Torpier, G., Johansson, S. G. D., and Prin, L. (1981). *J. Immunol.* **126,** 2087.
Capron, A., Dessaint, J.-P., Capron, M., Joseph, M., Ameisen, J. C., and Tonnel, A. B. (1986a). *Immunol. Today* **7,** 15.
Capron, M., Jouault, T., Prin, L., Ameisen, J. C., Butterworth, A. E., Papin, J. P., Kusnierz, J. P., and Capron, A. (1986b). *J. Exp. Med.* **164,** 72.
Carafoli, E. (1987). *Annu. Rev. Biochem.* **56,** 395.
Chen, S. S., Bohn, J. W., Liu, F.-T., and Katz, D. H. (1981). *J. Immunol.* **127,** 166.
Cockcroft, S. and Gomperts, B. D. (1985). *Nature (London)* **314,** 534.
Conrad, D. H., and Froese, A. (1976). *J. Immunol.* **116,** 319.
Conrad, D. H., and Peterson, L. H. (1984). *J. Immunol.* **132,** 796.
Cunha-Melo, J. R., Dean, N. M., Moyer, J. D., Maeyama, K. and Beaven, M. A. (1987). *J. Biol. Chem.* **262,** 11455.

Dayhoff, M. O., Hunt, L. T., and Hurst-Calderone, S. (1978). "Atlas of Protein Sequence and Structure" (M. O. Dayhoff, ed.) vol. V. (Suppl 3). pp. 363–373. National Biomedical Research Foundation Washington, D.C.
Dembo, M., Goldstein, B., Sobotka, A. K., and Lichtenstein, L. M. (1979). *J. Immunol.* **123,** 1864.
Delespesse, G., Sarfati, M., Rubio-Trujillo, M. and Wolowiec, T. (1986). *Eur. J. Immunol.* **16,** 815.
Dickler, H., and Sachs, D. (1974). *J. Exp. Med.* **140,** 779.
Drickamer, K. (1981). *J. Biol. Chem.* **256,** 5827.
Eccleston, E., Leonard, B. J., Lowe, J. S., and Welford, H. J. (1973). *Nature (London) New Biol.* **244,** 73.
Engelman, D. M., Steitz, T. A., and Goldman, A. (1986). *Annu. Rev. Biophys. Biophys. Chem.* **15,** 321.
Fewtrell, C., and Metzger, H. (1980). *J. Immunol.* **125,** 701.
Fewtrell, C., and Sherman, E. (1987). *Biochemistry* **26,** 6995.
Finbloom, D. S., and Metzger, H. (1983). *J. Immunol.* **130,** 1489.
Fridman, W. H., Teillaud, J.-L., Amigorena, S., Daeron, M., Blank, U., and Neauport-Sautes, C. (1987). *Int. Rev. Immunol.* **2,** 221.
Froese, A. (1980a). *CRC Crit. Rev. Immunol.* **1,** 79.
Froese, A. (1980b). *J. Immunol.* **125,** 981.
Furuichi, K., Rivera, J., and Isersky, C. (1985). *Proc. Natl. Acad. Sci. U.S.A.* **82,** 1522.
Gilman, A. G. (1987). *Annu. Rev. Biochem.* **56,** 615.
Gomperts, B. D., and Fewtrell, C. M. S. (1985). *In* "Molecular Mechanisms of Transmembrane Signalling" (P. Cohen and M. D. Houslay, eds.), Elsevier, New York.
Grasberger, B., Minton, A. P., Delisi, C., and Metzger, H. (1986). *Proc. Natl. Acad. Sci. U.S.A.* **83,** 6258.
Helm, B., Marsh, P., Vercelli, D., Padlan, E., Gould, H., and Geha, R. (1988). *Nature (London)* **331,** 180.
Hibbs, M. L., Classon, B. J., Walker, I. D., McKenzie, I. F. C., and Hogarth, P. M. (1988). *J. Immunol.* **140,** 544.
Holowka, D., and Baird, B. (1984). *J. Biol. Chem.* **259,** 3720.
Holowka, D., Hartmann, H., Kanellopoulos, J., and Metzger, H. (1980). *J. Recept. Res.* **1,** 41.
Holowka, D., Gitler, C., Bercovici, T., and Metzger, H. (1981). *Nature (London)* **289,** 806.
Hook, W. A., Berenstein, E. H., and Siraganian, R. P. (1987). *Fed. Proc. Fed. Am. Soc. Exp. Biol.* **46,** 1346 (Abstr. 6008).
Huff, T. F., Uede, T., and Ishizaka, K. (1982). *J. Immunol.* **129,** 509.
Huppi, K. Mack, B. A., Hilgers, J., Kochan, J., and Kinet, J.-P. (1988). Submitted.
Ikuta, K., Takami, M., Kim, C. W., Honjo, T., Miyoshi, T. Tagaya, Y., Kawabe, T., and Yodoi, J. (1987). *Proc. Natl. Acad. Sci. U.S.A.* **84,** 819.
Irvine, R. F., and Moore, R. M. (1986). *Biochem. J.* **240,** 917.
Ishizaka, K. (1984). *Annu. Rev. Immunol.* **2,** 159.
Ishizaka, K. (1987). *Int. Rev. Immunol.* **2,** 75.
Ishizaka, T. (1985). *In* "Advances in the Biosciences Frontiers in Histamine Research" (C. R. Ganeillin and J. C. Schwartz, eds.), pp. 401–409, Pergamon Press, New York.
Ishizaka, T., Conrad, D. H., Huff, T. F., Metcalfe, D. C., Stevens, R. L., and Lewis, R. A. (1985). *Int. Arch. Allergy Appl. Immunol.* **77,** 137.
Ishizaka, T., White, J. R., and Saito, H. (1987). *Int. Arch. Allergy Appl. Immunol.* **82,** 327.
Jardieu, P., Moore, K., Martens, C., and Ishizaka, K. (1985). *J. Immunol.* **135,** 2727.

Joseph, M., Ameisen, J. C., Kusnierz, J. P., Pancre, V., Capron, M., and Capron, A. (1984). *C. R. Hebd. Seances Acad. Sci. Ser. D*: **298**, 55–60.
Kabat, E. A., Wu, T. T., Reid-Miller, J., Perry, H. M., and Gottesman, K. S. (1987). "Sequences of Protein of Immunological Interest." 4th ed. U.S. Dept. Health and Human Services, Bethesda, Maryland.
Kane, P. M., Holowka, D., and Baird, B. (1988). *J. Cell. Biol.*, in press.
Kanner, B. I., and Metzger, H. (1983). *Proc. Natl. Acad. Sci. U.S.A.* **80**, 5744.
Kanner, B. I., and Metzger, H. (1984). *J. Biol. Chem.* **259**, 10188.
Kaplan, A. P. (1985). "*Allergy.*" Churchill Livingstone, New York.
Katakami, Y., Kaibuchi, K., Sawamura, M., Takai, Y., Nishizuka, Y. (1984). *Biochem. Biophys. Res. Commun.* **121**, 573.
Keegan, A. D., and Conrad, D. H. (1987). *J. Immunol.* **139**, 1199.
Kikutani, H.,Inui, S., Sato, R., Barsumian, E. L., Owaki, H., Yamasaki, K., Kaisho, T., Uchibayashi, N., Hardy, R. R., Hirano, T., Tsunasawa, S., Sakiyama, F., Suemura, M., and Kishimoto, T. (1986). *Cell* **47**, 657.
Kinet, J.-P., Perez-Montfort, R., and Metzger, H. (1983). *Biochemistry* **22**, 5729.
Kinet, J.-P., Alcaraz, G., Leonard, A., Wank, S., and Metzger, H. (1985). *Biochemistry* **24**, 4117.
Kinet, J.-P., Metzger, H., Hakimi, J., and Kochan, J. (1987). *Biochemistry* **26**, 4605.
Kinet, J.-P., Blank, U., Ra., C., Metzger, H., and Kochan, J. (1988). *Proc. Natl. Acad. Sci. U.S.A.*, in press.
Kisaki, T., Huff, T. F., Conrad, D. H., Yodoi, J., and Ishizaka, K., (1987). *J. Immunol.* **139**, 3345.
Kochan, J., Pettine, L. F., Hakimi, J., Kishi, K., and Kinet, J.-P., (1988). *Nucl. Acid Res.* **16**, 3584.
Kulczycki, A., Jr., and Parker, D. W. (1979). *J. Biol. Chem.* **254**, 3187.
Kulczycki, A., Jr., Isersky, C., and Metzger, H. (1974). *J. Exp. Med.* **139**, 600.
Kulczycki, A., Jr., McNearney, T. A., and Parker, C. W. (1976). *J. Immunol.* **117**, 661.
Kuno, M., and Gardner, P. (1987). *Nature (London)* **326**, 301.
Lee, W. T., and Conrad, D. H. (1984). *J. Exp. Med.* **159**, 1790.
Lee, W. T., and Conrad, D. H. (1985). *J. Immunol.* **134**, 518.
Lee, W. T., and Conrad, D. H. (1986). *J. Immunol.* **136**, 4573.
Lee, W. T., Rao, M., and Conrad, D. H. (1987). *J. Immunol.* **139**, 1191.
Lindau, M., and Fernandez, J. M. (1986a). *Nature (London)* **319**, 150.
Lindau, M., and Fernandez, J. M. (1986b). *J. Gen. Physiol.* **88**, 349.
Liu, F.-T., and Orida, N. (1984). *J. Biol. Chem.* **259**, 10649.
Liu, F.-T., Albrandt, K., Mendel, E., Kulczycki, A., Jr., and Orida, N. K. (1985). *Proc. Natl. Acad. Sci. U.S.A.* **82**, 4100.
Lo, T. N., Saul, W. F., and Beaven, M. A. (1987). *J. Biol. Chem.* **262**, 4141.
Ludin, C., Hofstetter, H., Sarfati, M., Levy, C. A., Suter, U., Alaimo, D., Kilchherr, E., Frost, H., and Delespesse, G. (1987). *EMBO J.* **6**, 109.
MacDonald, S. M., Lichtenstein, L. M., Proud, D., Plaut, M., Naclerio, R. M., MacGlashan, D. W., and Kagey-Sobotka, A. (1987). *J. Immunol.* **139**, 506.
Maeyama, K., Hohman, R. J., Metzger, H., and Beaven, M. A. (1986). *J. Biol. Chem.* **261**, 2583.
Marcelletti, J. F. and Katz, D. H. (1987). *Int. Rev. Immunol.* **2**, 43.
Martens, C. L., Huff, T. F., Jardieu, P., Trounstine, M. L., Coffman, R. L., Ishizaka, K., and Moore, K. W. (1985). *Proc. Natl. Acad. Sci. U.S.A.* **82**, 2460.
Martens, C. L., Jardieu, P., Trounstine, M. L., Stuart, S. G., Ishizaka, K., and Moore, K. W. (1987). *Proc. Natl. Acad. Sci. U.S.A.* **84**, 809.

Mathur, A., Maekawa, S., Ovary, Z. and Lynch, R. G.(1986). *Mol. Immunol.* **23,** 1193.
Mathur, A., Kamat, D. M., VanNess, B. G., and Lynch, R. G. (1987). *J. Immunol.* **139,** 2865.
Mathur, A., Conrad, D. H., and Lynch, R. G. (1988). Submitted.
Mazurek, N., Bashkin, P., and Pecht, I. (1982). *EMBO. J.* **1,** 585.
Mazurek, N., Schindler, H., Schürholz, T. H., and Pecht, I. (1984). *Proc. Natl. Acad. Sci. U.S.A.* **81,** 6841.
Meinke, G. C., Magro, A. M., Lawrence, D. A., Spiegelberg, H. L. (1978). *J. Immunol.* **121,** 1321.
Melewicz, F. M., Plummer, J. M., Spiegelberg, H. L. (1982). *J. Immunol.* **129,** 563.
Metzger, H., and Bach, M. K. (1978). *In* "Modern Concepts and Developments in Immediate Hypersensitivity" (M. K. Bach, ed.) pp. 561–588. Dekker, New York.
Metzger, H., Alcaraz, G., Hohman, R., Kinet, J.-P., Pribluda, V., and Quarto, R. (1986). *Annu. Rev. Immunol.* **4,** 419.
Millard, P. J., Gross, D., Webb, W. W., and Fewtrell, C. (1988a). Submitted.
Millard, P. J., Gross, D., Webb, W. W., and Fewtrell, C. (1988b). *Proc. Natl. Acad. Sci. U.S.A.* **85,** 1854.
Mohr, F. C., and Fewtrell, C. (1987a). *J. Biol. Chem.* **262,** 10638.
Mohr, F. C., and Fewtrell, C. (1987b) *J. Cell. Biol.* **104,** 783.
Mohr, F. C., and Fewtrell, C. (1987c) *J. Immunol.* **138,** 1564.
Mohr, F. C., and Fewtrell C. (1988). Submitted.
Mongar, J. L., and Schild, H. O. (1958). *J. Physiol. (London)* **140,** 272.
Moore, K. W., Jardieu, P., Mietz, J. A., Trounstine, M. L., Kuff, E. L., Ishizaka, K., and Martens, C. L. (1986). *J. Immunol.* **136,** 4283.
Mostov, K. E., Friedlander, M., and Blobel, G. (1984). *Nature (London)* **308,** 37.
Musch, M. W., and Seigel, M. I. (1986). *Biochem. J.* **234,** 205.
Nakajima, T., and Delespesse, G. (1986). *Eur. J. Immunol.* **16,** 809.
Nakajima, T., Sarfati, M., and Delespesse, G. (1987). *J. Immunol.* **139,** 848.
Nakamura, T., and Ui, M. (1983). *FEBS Lett.* **173,** 414.
Nakamura, T., and Ui, M. (1985). *J. Biol. Chem.* **260,** 3584.
Newman, S. A., Rossi, G., and Metzger, H. (1977). *Proc. Natl. Acad. Sci. U.S.A.* **74,** 869.
Nishizuka, Y. (1986). *Science* **233,** 305.
Noro, N., Yoshioka, A., Adachi, M., Yasuda, K., Masuda, T., and Yodoi, J. (1986). *J. Immunol.* **137,** 1258.
Nutman, T. B., Delespesse, G., Sarfati, M., and Volkman, D. J. (1987). *J. Immunol.* **139,** 4049.
Padlan, E. A., and Davies, D. R. (1986). *Mol. Immunol.* **23,** 1063.
Pecht, I., Dulic, V., Rivnay, B., Coria, A. (1986). *In* "Mast Cell Differentiation and Heterogeneity" (A. D. Befus, J. Denburg, and J. Bienenstock, eds.), pp. 301–312. Raven Press, New York.
Perez-Montfort, R., and Metzger, H. (1982). *Mol. Immunol.* **19,** 1113.
Perez-Montfort, R., Kinet, J.-P., and Metzger, H. (1983a). *Biochemistry* **22,** 5722.
Perez-Montfort, R., Fewtrell, C., and Metzger, H. (1983b). *Biochemistry* **22,** 5733.
Peterson, L. H., and Conrad, D. H. (1985). *J. Immunol.* **135,** 2654.
Pribluda, V. S., and Metzger, H. (1987). *J. Biol. Chem.* **262,** 11449.
Quarto, R., and Metzger, H. (1986). *Mol. Immunol.* **23,** 1215.
Quarto, R., Kinet, J.-P., and Metzger, H. (1985). *Mol. Immunol.* **22,** 1045.
Ra., C., Kochan, J., Hakimi, J., Danho, W., Metzger, H., and Kinet, J.-P. (1988). *Fed. Proc. Fed. Am. Soc. Exp. Biol.* **47,** A1248.
Ravetch, J. V., Luster, A. D., Weinshank, R., Kochan, J., Pavlovec, A., Portnoy, D. A., Hulmes, J., Pan, Y.-C. E., Unkeless, J. C. (1986). *Science* **234,** 718.

Ritchie, D. M., Sierchio, J. N., Bishop, C. M., Hedli, C. C., Levinson, S. L., Capetola, R. J. (1984). *J. Pharmacol. Exp. Ther.* **229,** 690.
Rivera, J., Kinet, J.-P., Kim, J., Pucillo, C., and Metzger, H. (1988). *Mol. Immunol.*, in press.
Rivnay, B., Wank, S. A., Poy, G., and Metzger, H. (1982). *Biochemistry* **21,** 6922.
Robertson, M. W., and Liu, F.-T., (1988). *Mol. Immunol.*, in press.
Rossi, G., Newman, S. A., and Metzger, H. (1977). *J. Biol. Chem.* **252,** 704.
Rousseaux-Prevost, R., Rousseaux, J., and Bazin, H. (1987). *Mol. Immunol.* **24,** 187.
Sagi-Eisenberg, R., and Pecht, I. (1983). *J. Membr. Biol.* **75,** 97.
Sagi-Eisenberg, R., and Pecht, I. (1984). *Immunol. Lett.* **8,** 237.
Sagi-Eisenberg, R., Lieman, H., and Pecht, I. (1985). *Nature (London)* **313,** 59.
Saito, H., Okajima, F., Molski, T. F., Sha'afi, R. I., Ui, M., and Ishizaka, T. (1987). *J. Immunol.* **138,** 3927.
Sarfati, M., Vanderbeeken, Y., Rubio-Trujillo, M., Duncan, D., and Delespesse, G. (1986). *Eur. J. Immunol.* **16,** 1005.
Sarfati, M., Nutman, T. B., Suter, U., Hostetter, H., and Delespesse, G. (1987). *J. Immunol.* **139,** 4055.
Schlessinger, J., Webb, W. W., Elson, E. L., and Metzger, H. (1976). *Nature (London)* **264,** 550.
Schulman, E. S., Liu, M. C., Proud, D., MacGlashan, D. W., Jr., Lichtenstein, L. M., and Plaut, M. (1985). *Am. Rev. Respir. Dis.* **131,** 230.
Shimizu, A., Tepler, I., Benfey, P. N., Berenstein, E. H., Siraganian, R. P., and Leder, P. (1988). *Proc. Natl. Acad. Sci. U.S.A.* **85,** 1907.
Spiegelberg, H. L. (1984). *Adv. Immunol.* **35,** 61.
Stuart, S. C., Trounstine, M. L., Vaux, D. J., Koch, T., Martens, C. L., Mellman, I., and Moore, K. W. (1987) *J. Exp. Med.* **166,** 1668.
Suemura, M., Shiho, O., Deguchi, H., Yamamura, Y., Bottcher, I., and Kishimoto, T. (1981). *J. Immunol.* **127,** 465.
Suemura, M., Kikutani, H., Barsumian, E. L., Hattori, Y., Kishimoto, S., Sato, R., Maeda, A., Nakamura, H., Owaki, H., Hardy, R. R., and Kishimoto, T. (1986). *J. Immunol.* **137,** 1214.
Teshima, R., Suzuki, K., Ikebuchi, H., Terao, T. (1986). *Mol. Immunol.* **23,** 279.
Tscharner, V. von, Prod-Hom, B., Baggiolini, M., Reuter, H. (1986) *Nature (London)* **324,** 369.
Warner, J. A., Pinekowski, M. M., Plaut, M., Norman, P. S., and Lichtenstein, L. M. (1986). *J. Immunol.* **136,** 2583.
White, J. R., Pluznik, D. H., Ishizaka, K., and Ishizaka, T. (1985). *Proc. Natl. Acad. Sci. U.S.A.* **82,** 8193.
White, K., and Metzger, H. (1988). *Fed. Proc. Fed. Am. Soc. Exp. Biol.* **47,** in press.
WoldeMussie, E., Ali, H., Takaishi, T., Siraganian, R. P., and Beaven, M. A. (1987). *J. Immunol.* **139,** 2431.
Wolf, B. A., Turk, J., Sherman, W. R., and McDaniel, M. L. (1986). *J. Biol. Chem.* **261,** 3501.
Yodoi, J., Hirashima, B., and Ishizaka, K. (1982). *J. Immunol.* **128,** 289.
Young, M. C., Leung, D. Y. M., and Geha, R. S. (1984). *Eur. J. Immunol.* **14,** 871.
Zidovetzki, R., Bartholdi, M., Arndt-Jovin, D., and Jovin, T. M. (1986). *Biochemistry* **25,** 4397.

Index

A

Activation, T cell receptor $\gamma\delta$ lymphocytes, 171-172
α chain, 282, 284-287
α enhancer, 268
α-helix, amphipathic, 204
Amino acids
  antigenic equivalence, 68-72
  conservation, 115, 221-222
  contact residue, 119
  critical residue, 67-68
  critical to binding energy, 24
  energy penalty of voids, 72
  epitope, 65-67
  human C_γ sequences, 143
  human J_γ sequences, 142
  human V_γ sequences, 141-142
  intrinsic hydrophilicity parameters, 49
  MHC contact, 210
  propensity factors, sequential epitopes, 66-67
  replaceability, 69-71, 77
  sequence, 111, 113
  sequential epitope role, 39
  T cell receptor, 157-159
  V_L and V_H domains, 109, 125
Amphipathicity, T cell receptor, 204
Antibody
  antigen binding, 105
  catalysis, 1-2
  combining site, structure variability, 14-15
  conformation, 17, 124-125
  crystallographic studies, 1
  diversity, 133
  Dob structure, 8
  electrostatic recognition with antigen, 12-13
  enzymatic activity, 81
  Fab Kol, crystal packing, 19-20
  Fab region, 8-9
  Fab structures, 17
  Fc region, 8-9
  induced complementarity, 16-20
  immunological, biochemical and crystallographic data, 16-18
  mobility data, 20
  variable domain structure superposition, 18-20
  raised against native protein, 3
  sequence variation, 13-16
  structure
    conformational changes, 113-114
    domain-domain interactions, 110
    domain structure, 107-109
    Fab fragments, 107
    H and L chain CDR, 108-109
    module structure, 110-113
    polypeptide loops, 107
    sheet-sheet interactions, 110
    V and C domains, 109
    variable domains, 109
    V_L-V_H interface, 110, 112, 124
    V_L-V_L pairing, 112
    V module interface, 111
  X-ray analysis, 121
Antibody-antigen interaction, 99-128, *see also* Antigen recognition
  affinities to small molecules, 82
  analysis of structure, 22
  buried surface, 82
  chemistry and binding energy, 72-84
  concerted shifts, 22
  contact residue, 26
  criteria for systematic studies, 5-8
  critical residues, 27
  crystallographic approaches, 4
  degree of fidelity, 60
  dominant epitopes and collective response, 46-49
  electrostatic attraction, 80
  evaluation of predictive methods
    criteria for assessing prediction success, 50-59
    sequence data, 49-50
  immunological cross-reactivity, 64
  interface adaptor hypothesis, 125-128
  local antibody conformation effect, 17-18

314 INDEX

lock-and-key fit, 78, 128
MHr, 42
mobile structural elements, 80
nature of immune response, 83
recognition and binding components, 79
reconciling crystallographic and peptide mapping data, 23–27
surface mobility role, 16–17
synthesis of structural description, 26
three-dimensional structures, 119–120
  complementarity, 122
  conformational changes, 123–125
  epitope and paratope size, 120–121
  epitope character, 121–122
X-ray crystallography, 8–11
X-ray structures, 73
Antibody complexes, crystallographic structures, 4
Antibody-hapten complexes, 119
Antigen, 117–119
changes in amino acid side chain, 24
competition, 208–211
complementarity-determining regions, 9–10
conformational changes, 123–124
electrostatic recognition with antibody, 12–13
epitopes, 106
equivalence, 68–72
flexibility, 118–119
fragment properties, 204
global structure, binding process, 22
induced complementarity, 20–23
interaction energy, 23–24
levels of conformational hierarchy, 74
microassemblies, 42
mutants, 117–118
processing, T cells, 196–197
receptor structure, 105–107, see also Antibody, structure
replaceability, amino acids, 69–70
residues, definition, 23
sites, hydrophilicity, 76
structural characteristics, antibody-binding sites, 118
surface accessibility, 119
three-dimensional structures, 117
Antigen–antibody binding, first stage, 79
Antigenicity
chemistry, 2
epitope prediction, 58
flexibility, 105, 118
relationship with local mobility, 61
secondary structure, 76
in terms of recognition frequency, 29–32
value, 50
Antigen–MHC determinant, formation
antigen competition, 208–211
binding of antigen and MHC molecules in solution, 211–213
immune response gene control, 206–208
Antigen–MHC interactions, 219
binding rate constants, 212
correlation with immune-response phenotype, 212
Antigen recognition, 60, 223–225
altered self model, 202
antigen dose–response curve, 202
chemistry, 6
amino acid residues, 27
antigenicity in terms of recognition frequency, 29–32
reactive site stereochemistry, 32–39
systematic data base, 27–29
Class II MHC molecules, 203–205
difference between T and B cells, 134
dual recognition, 202
hydrogen bonds and electrostatic forces, 83
hypothesis for origin, 224
mechanisms, 39
  critical residues, 39–42
  induced complementarity, 42–46
MHC-restricted, 202–203
T cell, 194
T cell receptor γδ, 176
Anti-IgE antibodies, 279–281
Antipeptide antibodies, 61
binding, 118
design, 59–65
  affinities of monoclonal anti-MHr antibodies, 61–62
  local mobility and antigenicity, 61
Antiprotein antibody, binding, 42
Apo-MHr, monoclonal anti-MHr antibody affinity, 61–62
Atomic solvation parameter, 101
Autoimmunity, 26

B

BB10 antibody, 306–307
B cell, 133
antigen recognition, 223
C_μ-only transcription, 237–238

development
  heavy chain and ϰ gene rearrangement and expression, 236–237
  immunoglobulin, gene expression, 236–240
  terminally differentiated plasma cells, 239
  IgE-binding factor, 300–301
  immunoglobulin, enhancer preference, 242
  recognition, 224
  V-D-J joining, 239
  V_H-only transcription, 236–238
B cell epitope, 78
B cell lymphoma A20-2J, 196–197
B cell peptide epitope vaccine, 78
Bence-Jones proteins, structural analyses, 112
β bulges, 111
β chain, see also T cell receptor
  mast cell-specific receptor, gene cloning studies, 287–289
  primary structure, 220
β sheet, 102, 104, 111
β strand, 108–110
B lymphocytes
  antigen recognition, 223
  Fc_ϵ receptors, 298–300
  receptor, 133
Buried surfaces, protein–protein interactions, 100–101

C

Calcium, see Cytoplasmic calcium
CD3 molecule, 135
  expression, 166, 169
CD4 accessory molecule, 218
CD8 accessory molecule, 218
cDNA
  α chain, 284–286
  biosynthetically engineered peptides based constructs, 281
  clone
    characterization, 148
    T cell receptor δ genes, 154–155
    T cell-specific, 135
  coding, 287, 304
  T cell receptor γ, 145
  T cell-specific, 135, 220
CDR, see Complementarity-determining regions

C-helix, peptide homologs, 63
C_H gene, 239
Chromosomal translocation, T cell receptor δ genes, 160–161
Complementarity
  antibody-antigen complexes, 122
  induced
    antigen, 20–23
    differential contributions, 73
    evidence, 42–46
    protein folding, 83
    protein–protein interactions, 101–103
Complementarity-determining regions, 9–10, 107
  buried surface, 111
  conformational changes, 18–20, 124
  diversity, 222
  forming antigen binding site, 13
  framework residues, 16
  length, 222
  loop
    flexibility, 107
    structures, 20
  mobility, 20
  models, 10
  modulating V_L-V_H pairing, 125
  movement from canonical positions, 127
  sequences, 125
  size, 120–121
  spatial arrangement, 109
  structural variability, 73
  structures, 107–108
  variable domain, 79
  V module interface modification, 113, 126
Contact residue, 26, 72, 119
Critical residue, 27, 72
  amino acids, 67–68
  antigen binding, 61, 63
  antiprotein antibodies, 42
  buried, 42–43
  identification, 39–42
  peptide mapping, 74
  properties, 40–41
  sequential epitope, 64
  stereochemical relationships, 43
Cromolyn-binding protein, 297
Crystallographic temperature factors, mobility data from, 20
Crystallography, reconciling data, 23–27
Crystal packing, Fab Kol, 19–20
CTL, see T cell, cytotoxic

Cysteine
  relationship between loops,
    immunoglobulin domains, 285
  residues, 137, 156
Cytochrome c
  acetimidylated, 200
  as antigen, 198–200
  beef, 199
  cross-reactive, 202
  immune response, 224
  moth, 203
  mouse, 199
  pigeon, 199–200
  secondary structure, 204
  sequence differences in, 4
Cytolysis, T cell receptor γδ lymphocytes,
    172–174
Cytoplasmic calcium, 293–298
  influx of extracellular, 295–298
  influx pathway, 296–297
  intracellular stores mobilization, 293–295
  ionophore dose effect, 297
  lag times, 294
  phosphoinositide hydrolysis, 291
Cytotoxicity, 173

D

D1.3, 20–21, 222
Determinant selection model, MHC
    restriction, 207
DNA, promoters and enhancers, 239
DNase I, T cell-specific hypersensitive site,
    268
Dob antibody structure, 8
Domain–domain interactions, 110

E

Electrostatic complementarity, 12–13
Electrostatic forces, binding role, 13
Electrostatic potential, MHr, 35
Electrostatic recognition, between antigen
    and antibody, 12–13
Eosinophils, IgE, 306–307
Epitope, see also Sequential epitope
  amino acids, 65–67
    binding energy contribution, 24
    character, 121–122

  conformational, sites mapped by peptides
    cluster into, 38–39
  crystallographically defined, 24–25
  definition, 25–26, 72
  dominant, 46–49
  epicenter, 27
  flexibility, 63, 119
  mobility, 121
  observed and predicted, 52–55
  overlapping, fine specificity studies, 25
  prediction, 56–58
    methods, 75–76
  sequence variability, 77
  size, 120–121
  three-dimensional structure, 76–77
  as vaccines, 59–60
εBP, 307–308
ε chain, 278
  biosynthetically engineered peptides based
    cDNA constructs, 281
  digestion, 282
  peptide immunogens, 279–280
  sites protected by receptor interaction,
    280
  stabilization, 282
Equilibrium dissociation constant, 101

F

Fab-lysozyme complexes, 121
Fc-binding factors, 277
Fc receptors, 277, 305
Fc$_\epsilon$ receptors, 282
  degradation, 302
  human B lymphocytes, 298–300
  murine B cells, 302
  T cells, 301, 303
Fibroblasts, trans-acting negative factors,
    248
Fine specificity, 3
  overlapping epitopes, 25
  HyHEL5, 26
Flexibility
  antigenicity, 105, 118
  epitope, 57, 63, 119
  loop, CDR, 107
  main-chain parameters, 50
Footprinting studies, 2–3
Free energy, protein–protein interactions,
    100

G

γ chains, mast cell-specific receptor, gene cloning studies, 288-289
GDPβS, 292
Gene cloning studies, mast cell-specific receptor, 283-289
Genomic organization, TCR δ gene segments, 155-156
Gp, activation, 290
G-proteins, role in phosphoinositide hydrolysis, 291-292
GTPγS, 292
Guanosine 5'-triphosphate-binding protein, 290

H

Hapten, binding, 17
Heavy chain gene expression, sequences upstream of Ig promoter, 253-254
Helix, movement, 104
Histamine-releasing factors, 307
HLA-A2, structure, 215
Hybridoma technology, 1
Hydropathicity, mast cell-specific receptor, β subunit profile, 288
Hydrophilicity, 76
   epitope prediction, 56
   intrinsic parameters, 49
   MHr, 35
   profiles, 14
Hydrophobic regions, 76
HyHEL5
   binding, 20-21
   blind peptide mapping study, 26
   fine specificity study, 26
   lysozyme binding, 12-13, 122
   model, 121
HyHEL10, blind peptide mapping study, 26

I

Idiotope-determining region, 14, 16
IgE, *see also* ε chain, mast cell-specific receptor
   α chain, 282
   anti-dinitrophenol, 294
   anti-ovalbumin, 294
   binding site localization, 279-280
   biosynthesis, 303
   eosinophils, 306-307
   ε chain, 278
   Fc receptors, 277
   functional characteristics, 305
   high-affinity receptor, 281
   histamine-releasing factors, 307
   60-kDa factor, 304-305
   macrophage receptor, 306
   monoclonal anti-β antibody, 283
   monocytes, 305-306
   platelets, 306-307
   protein-binding sites, 281-282
   role, 277
   sites interacting with receptors, 278-279
      anti-IgE antibody studies, 279-281
      inhibition studies, 281-282
   structural properties
      comparative studies on rodent cells, 302-305
      Fc_ϵ receptors, 298-301, 303
      IgE-binding factor, 300-302
   synthesis, 277
IgE-binding factor
   cDNA coding, 304
   functional characteristics, 305
   human B cells, 300-301
   retroviral protein, 304
   T cells, 301-303
IgE-binding proteins, 307-308
Ig fold, 221
Immune response
   affinity maturation, 126-127
   assessment, 6
   diversity in, antigen receptors, 106
   gene control, 206-208
   phenotype, correlation with antigen-MHC interactions, 212
Immunogenicity, 75
Immunogenic peptide, conformation, 63
Immunoglobulin, 235, *see also* IgE
   characterization, 7
   C_H switching rearrangement, 239
   domains, cysteine loop relationships, 285
   fold, 108
   gene expression
      B cell-specific, 253
      during B cell development, 236-240
      heavy chain and κ chain, 236-237
      tissue specific, 235

Immunoglobulin (cont.)
  mRNA, 239
  µ chains, 238–239
  promoter elements, 269
    B cell preference, 258
    enhancer interactions, 263–265
    heavy chain, common factors with enhancers, 256–257
    interaction between multiple proteins and conserved octanucleotide, 254–255
    ϰ activity, required octamer, 257–258
    ϰ V gene promoters, 258
    light chain, sequence conservation, 257
    nuclear factor interaction with conserved heptamer, 255
    sequence conservation, heavy chain V gene promoters, 252
    sequences required for heavy chain gene expression, 253–254
    tissue and stage specificity, 251, 255–256
    V_H promoter activity, 252–253
  regulating transcription, 240
  sequence variability, 11, 14
  transcriptional enhancer elements, see also Protein-binding sites
    activation and V-D-J joining, 241–242
    B cell preference, 258
    common factors with heavy chain promoters, 256–257
    containing bind sites, 264
    dependent transcription, 246
    EcoRI site, 249
    IgH enhancer-binding proteins, 247–248
    increase local concentration of factors, 264
    in vivo protein-binding sites, 243–244
    ϰ enhancer, 249
    light chain enhancer elements, 249–250
    location, 242
    mechanism, 240–241
    negative regions, 248–249
    promoter interactions, 263–265
    tissue specificity, 242
    transcription initiation, 242–243
  transcriptional regulation mechanisms, 258–259
    enhancer independence and dependence of V_H promoter, 265–266

  lymphoid-specific octamer factor, 261–262
  octamer as tissue-specific element, 260–263
  octamer-binding protein forms, 261
  promoter-enhancer interactions, 263–265
  tissue and stage specificity, 259–260
  ubiquitous factor, 263
  V-D-J joining, 237–239, 241–242
Immunological antigen receptors, 133–136
Inhibition studies, IgE, 281–282
Inositol phosphates, role in phosphoinositide hydrolysis, 292
Inositol 1,4,5-trisphosphate, 295
Insulin, hexamers, 103–104
Interface adaptor hypothesis, 125–128
Intracisternal A particle core protein, polyclonal antibody, 304
Ir gene, 207, 212

K

ϰ enhancer, 249–251
ϰ gene, 238, 240
ϰ light chain promoter, 257
ϰ promoter, 257–258
KOL, crystal packing, 113

L

λ light chain promoter, 257
λ phage repressor protein, 209–210
Lock-and-key fit, 78
Lysozyme
  antibody complex, 22–23
  antigenic structure, 118
  hen egg, T cell response, 201
  HyHEL5 binding, 12–13, 122
  mobility, 21
  side-chain movement, 22

M

Main-chain flexibility parameters, 50
MARE-1, 279
Mast cell-specific receptor, 282–283
  β subunit, hydropathicity profile, 288
  characteristics, 283

gene cloning studies, 283–289
  α chain, 284–287
  β and γ chains, 287–289
  mechanisms of action, 289–298
    aggregation of receptors, 289–291
    cytoplasmic calcium rise, 293–298
    intrinsic and extrinsic, 289
    phosphoinositide hydrolysis, 290–293
μ chains, 238–239
McPC603
  binding site, 9, 19
  electrostatic complementarity with phosphorylcholine, 10
  variable domains, 18–19
Membrane, proximal and distal domains, 214, 221
MHC, 134
  allele, 219
  antigen
    binding in solution, 211–213
    conformation, 203–205
    class I, 213–214
      binding process, 214
      determinant nature, 205–206
    contact amino acids, 210
    electron density in cleft, 214
    membrane proximal and distal domains, 214
    recognition, 194–196, 224
    restriction, determinant selection model, 207
    self-encoded, 213
    three-dimensional structure, 213–215
MHr
  antibody-antigen interaction, 42
  antigenic response, 29–31, 47
  antigenic sites, 80
  critical residues, stereochemical relationships, 43
  crystallographic structure, 21
  electrostatic potential, 35
  epitopes, 51, 53
  features, 28
  hexapeptide homologs, 32
  hydrophilicity, 35
  immune response, 28–29, 44–46, 75
  mobility, 33–35
  molecular surface chemistry and immunological reactivity, 33–34
  monoclonal anti-MHr antibody affinity, 61–62

packing density, 33–35
peptides, monoclonal anti-MHr antibody affinity, 61–62
protein fold, 28
reactive sites, 29
secondary structure, 38
sequential epitopes, 36, 38–40
shape accessibility and exposed surface area, 35–36
side-chain contributions, 36–38
site 4–9, 37, 40, 43, 45
site 90–95, 45–46
Mobility, 33–35
  crystallographic temperature factors, 20
  functional significance, 73–74
  interacting structural elements, 80
  local, relationship with antigenicity, 61
  secondary structure, 76
  T cell receptor γδ proteins, 150
Monoclonal antibody, Fc_ϵ receptors, 299–300
Monoclonal anti-β antibody, 283
Monoclonal anti-MHr antibodies, protein and peptide-induced, 61–62
Monocytes, IgE, 305–306
mRNA
  immunoglobulin, 239
  T cell receptor γ, 145
Myoglobin
  antigenic structure, 118
  epitopes, 51–52, 55

N

Naturally occurring proteins, T cell specificity for liner sequences, 198–201
NC41, 21, 123–125
NF-κB, 250–251
NRI, concentration, 265
Nuclear factors, protein-binding sites, 255
Nuclease, sequence, 209
Nucleotides, human T cell receptor γ V-J junction sequences, 145

O

Oligomeric proteins, buried surfaces, 101
Ontogeny, T cell receptor γδ, 165–171, 177
  gene expression, 266

P

Packing density, 33–35
Paratope, see Complementarity-determining regions
Peptide
  α-helical conformation, 203–204
  epitope vaccines, 77–78
  MHC interaction, rate constants for binding reaction, 211
  monoclonal anti-MHr antibody affinity, 61–62
  synthesis, 77
  antigens, see T cell, specificity for peptide antigens
Peptide mapping, 3
  blind study, 26
  critical residue, 74
  reconciling data, 23–27
  sites mapped into conformational epitopes, 38–39
  T cell receptor $\gamma\delta$ subunits, 150
Pertussin toxin, 291–292
Phosphoinositide hydrolysis, mast cell-specific receptor, 290–293
  G-protein role, 291–292
  inositol phosphate role, 292
  protein kinase role, 292–293
Phosphorylcholine, electrostatic complementarity with antibody McPC603, 10
Platelets, IgE, 306–307
Polyclonal anti-MHr antisera, 29
Pre-B cells, exogenous recombination substrates, 237
Pre-pre B cells, 237
Protein
  epitopes, 2–4
  folding, 83, 100
  immunogenicity, 2
  interaction with Ig conserved octanucleotide, 254–255
  molecule, 118
  naturally occurring, T cell specificity, 198–201
  oligomeric, buried surfaces, 101
  retroviral, IgE-binding factor, 304
Protein-binding sites
  B cell-specific, 256
  enhancers containing, 264
  fibroblast-specific site, 256
  heavy and $\varkappa$ chain enhancers, 243–244
  IgE, 281–282
  in vivo, 243–244
  multiple
    in vitro mapping, 244–245
    $\varkappa$ enhancer, 250–251
    required, 245–247
  NF-$\varkappa$B, 250–251
  nuclear factors, 255
  protein purification, 247
  sequences, 246
  site A, 248
  site B/μE1, 247
  site D, 248
  site E, 247
  V_β promoter, 268–269
Protein kinase C, role in phosphoinositide hydrolysis, 292–293
Protein-protein interactions, 81, 265
  buried surfaces, 100–101
  complementarity, 101–103
  conformational changes, 103–105
  free energy, 100
  structural complementarity, 104–105
  thermodynamics, 100

R

Rat basophilic leukemia tumor, 282–283
Reactive site
  association with electrostatic regions and surface grooves, 79
  MHr, 29
  stereochemistry, 32–39
    electrostatic potential, 35
    hydrophilicity, 35
    mobility and packing density, 33–35
    secondary structure, 38
    shape accessibility and exposed surface area, 35–36
    side-chain contributions, 36–38
    sites mapped by peptides cluster into conformational epitopes, 38–39
  superassemblies, 74–75
Retroviral protein, IgE-binding factor, 304
Rhe, 112–113
RPMI 8866, 298–300

S

SDS-PAGE, T cell receptor, 150
Secondary structure, 38, 76
Sequential epitope, 25, 72
  amino acid, 39, 66–67
  collective responses of different species, 48
  critical residue, 64
  DFLEKI, 63–64
  EVVPH, 63–64
  individual side chain role, 40
  replaceability matrix, 77–78
  superassemblies, 36, 38
Shape accessibility, MHr, 35–36
Sheet-sheet interactions, 102–103, 110
Sheet-sheet packing, schematic, 102–103
Side chain
  contribution to solvent-exposed surface area, 36–38
  hydrophobic, 42, 80–81
  sequential epitope role, 40
Solvation energy, estimating, 101
Structural complementarity, protein-protein interactions, 104–105
Subtilisin, site-directed mutants, 23
Subtractive hybridization, 135
Surface area, exposed, MHr, 35–36
Surface variability analysis, 11, 14
SV40 early promoter, 268
Switch peptides, 113
Synthetic peptide technology, applications, 59

T

T cell, 134
  antigen
    conformation, 203
    recognition, 134
  clone, 134
    activation, 216
    antigen dose–response curves, 202
    lines, 196
  cross-reactivity, 200
  cytotoxic, 195
    MHC restriction, 214–215
    response to foreign antigens, 205
    response to virus infection, 205
  epitope, 4, 60, 73, 78

Fc_ϵ receptors, 301, 303
high-affinity self-reactive, deletion, 219
IgE-binding factors, 301–303
λ phage repressor protein response, 209–210
MHC recognition, 224
ontogeny, T cell receptor gene expression, 266
selection, 217
self-versus nonself-discrimination, 223
specific for foreign antigens, 219–220
specificity for peptide antigens, 197–206
  approaches, 197–198
  cytochrome c, 198–200
  hen egg lysozyme, 201
  liner sequences, naturally occurring proteins, 198–201
  minimum fragment size, 197
suppressor, 266
T cell receptor, 105, 114–117, see also Antigen-MHC determinant
  $\alpha\beta$, 115
  antigen receptor dimer, 116–117
  CD3 complex, 135
  chain alignment, 221
  compared to T cell receptor $\gamma\delta$, 175
  expression, 146–147, 165
  genes, 136
  heterodimers, 221, 266
  loci, 140
  lymphocytes, ontogeny, 169
  molecules, 134
  mRNA, 146–147
  protein complex, 134–135
  transcripts, 168
  variable region sequences, 221
α enhancer, 268
amino acid conservation, 221–222
amphipathicity, 204
antigen
  conformation and recognition, 194, 223–225
  MHC interaction model, 201
  processing, 196–197
  specificity, 193–194
β chain, 136
  germ-line genes, 267
  primary structure, 220
  transcription, 268
C_α domain, 221

CD3 complex, 171
C_δ locus deletion, 170
CDR, diversity, 222
class I and class II-specific, 205
combining site
  antigens bound to MHC, 215–220
  selections based on receptor specificity, 218–219
  shape, 222
  T cell receptor structure, 220–222
δ gene
  breakpoints, 160
  C_δ-C_α locus organization, 155–156
  C_δ gene segment, 156–157
  chromosomal translocation, 160–161
  genomic organization, 155–156
  identification, 153–155
  IDP2 transcript organization, 154
  rearrangement and diversity, 158–160, 167
δ subunit, 150
$\gamma\delta$, 175–176
  heterodimers, 266
  importance, 177–178
  ontogeny, 165–171
  protein structure, 174–175
  thymic ontogeny, 177
$\gamma\delta$ lymphocytes
  activation, 171–172
  allogenic target cell recognition, 173–174
  anti-CD3 mAb, 172
  cell surface expression, 162
  cytolysis, 172–174
  functional studies, 174
  peripheral blood and thymic, 161–164
  skin, 164–165
$\gamma\delta$ proteins, 146
  CD3 polypeptides, 150–152
  cytofluorographic analysis, 146–147
  forms, 148–149
  human subunit structure, 148–152
  identification, 146–148
  mobility, 150
  murine subunit structure, 152–153
  peptide mapping, 150
  SDS-PAGE, 150
γ-γ homodimers, 150
γ gene
  disulfide-linked form, 148
  human, 140–143
  IDP2 transcript organization, 154
  mRNA transcripts, 167
  murine genes, 136–140
  puzzle, 144–145
  rearrangement and diversity, 143–144
  role in fetal life, 144
gene expression, 135
  during T cell ontogeny, 266–267
  putative transcription signals, 267
  tissue specific, 235
  trans-acting negative regulation, 267
  transcriptional regulatory elements, 267–269
Ig fold, 221
interaction with antigen bound to MHC, 222
J gene segment, 115
mAb βF1, 161
MHC
  molecule recognition, 194–196
  polymorphism recognition, 214–215
  restricted recognition, 202–203, 216
  restriction, 195–196
  specific, 193
  structure, 220–222
  V and C domains, 115
  V-V interface edge strands, 115–116
  V-V pairings, 116
TCR1, 169
TFIIIA, concentration, 266
Thermodynamics, protein-protein interactions, 100
Thy-1$^+$ dendritic epidermal cells, 164–165
Thymocytes, precursor, 166
T lymphocytes, antigen recognition, 223
T3 membrane proteins, 116
TMV coat protein, 54, 76
Transcription, activation, Ig enhancers, 241–242
Transcription factors, cloning, 269
Transfection studies, κ V gene promoter, 258
Trypsin, site-directed mutants, 23
Tumor cells, lysis, 173
Two-color cytofluorographic analysis, T cell receptor $\gamma\delta$ proteins, 146–147

U

U snRNA gene promoters, 260, 262

V

Variable domain structures, superposition, 18–20
V gene, 252, 255
V_β gene, 268–269
V_H genes, 237–238
    promoter, 238
        activation, 243
        conserved octanucleotide and activity, 252–253
        enhancer independence and dependence, 265–266
V_x promoters, 238

V_L-V_H interface, 103, 110, 112
    CDR, 125
V_L-V_L pairing, 112

W

WT31, CD3, 161–163

X

X-ray analysis, antibody, 121
X-ray crystallography, 2–3, 8–11
X-ray structures, antibody-antigen interaction, 73

CONTENTS OF RECENT VOLUMES

Volume 33

The CBA/N Mouse Strain: An Experimental Model Illustrating the Influence of the X-Chromosome on Immunity
  IRWIN SCHER

The Biology of Monoclonal Lymphokines Secreted by T Cell Lines and Hybridomas
  AMNON ALTMAN AND DAVID H. KATZ

Autoantibodies to Nuclear Antigens (ANA): Their Immunobiology and Medicine
  ENG M. TAN

The Biochemistry and Pathophysiology of the Contact System of Plasma
  CHARLES G. COCHRANE AND JOHN H. GRIFFIN

Binding of Bacteria to Lymphocyte Subpopulations
  MARIUS TEODORESCU AND EUGENE P. MAYER

INDEX

Volume 34

T Cell Alloantigens Encoded by the IgT-C Region of Chromosome 12 in the Mouse
  F. L. OWEN

Heterogeneity of *H-2D* Region Associated Genes and Gene Products
  TED H. HANSEN, KEIKO OZATO, AND DAVID H. SACHS

Human Ir Genes: Structure and Function
  THOMAS A. GONWA, B. MATIJA PETERLIN, AND JOHN D. STOBO

Interferons with Special Emphasis on the Immune System
  ROBERT M. FRIEDMAN AND STEFANIE N. VOGEL

Acute Phase Proteins with Special Reference to C-Reactive Protein and Related Proteins (Pentaxins) and Serum Amyloid A Protein
  M. B. PEPYS AND MARILYN L. BALTZ

Lectin Receptors as Lymphocyte Surface Markers
  NATHAN SHARON

INDEX

Volume 35

The Generation of Diversity in Phosphorylcholine-Binding Antibodies
  ROGER M. PERLMUTTER, STEPHEN T. CREWS, RICHARD DOUGLAS, GREG SORENSEN, NELSON JOHNSON, NADINE NIVERA, PATRICIA J. GEARHART, AND LEROY HOOD

Immunoglobulin RNA Rearrangements in B Lymphocyte Differentiation
  JOHN ROGERS AND RANDOLPH WALL

Structure and Function of Fc Receptors for IgE on Lymphocytes, Monocytes, and Macrophages
  HANS L. SPIEGELBERG

The Murine Antitumor Immune
Response and Its Therapeutic
Manipulation
  ROBERT J. NORTH

Immunologic Regulation of Fetal–
Maternal Balance
  DAVID R. JACOBY, LARS B. OLDING,
  AND MICHAEL B. A. OLDSTONE

The Influence of Histamine on Immune
and Inflammatory Responses
  DENNIS J. BEER, STEVEN M. MATLOFF,
  AND ROSS E. ROCKLIN

INDEX

Volume 36

Antibodies of Predetermined Specificity
in Biology and Medicine
  RICHARD ALAN LERNER

A Molecular Analysis of the Cytolytic
Lymphocyte Response
  STEVEN J. BURAKOFF, OFRA WEINBERGER,
  ALAN M. KRENSKY, AND CAROL S. REISS

The Human Thymic Microenvironment
  BARTON F. HAYNES

Aging, Idiotype Repertoire Shifts,
and Compartmentalization of the
Mucosal-Associated Lymphoid System
  ANDREW W. WADE AND
  MYRON R. SZEWCZUK

A Major Role of the Macrophage
in Quantitative Genetic Regulation
of Immunoresponsiveness
and Antiinfectious Immunity
  GUIDO BIOZZI, DENISE MOUTON, CLAUDE
  STIFFEL, AND YOLANDE BOUTHILLIER

INDEX

Volume 37

Structure, Function, and Genetics
of Human Class II Molecules
  ROBERT C. GILES AND J. DONALD CAPRA

The Complexity of Virus—Cell Interactions in Abelson Virus Infection
of Lymphoid and Other Hematopoietic
Cells
  CHERYL A. WHITLOCK AND
  OWEN N. WITTE

Epstein–Barr Virus Infection
and Immunoregulation in Man
  GIOVANNA TOSATO AND
  R. MICHAEL BLAESE

The Classical Complement Pathway:
Activation and Regulation of the First
Complement Component
  NEIL R. COOPER

Membrane Complement Receptors
Specific for Bound Fragments of C3
  GORDON D. ROSS AND
  M. EDWARD MEDOFF

Murine Models of Systemic Lupus
Erythematosus
  ARGYRIOS N. THEOFILOPOULOS
  AND FRANK J. DIXON

INDEX

Volume 38

The Antigen-Specific, Major
Histocompatibility Complex-Restricted
Receptor on T Cells
  PHILIPPA MARRACK AND JOHN KAPPLER

Immune Response (Ir) Genes
of the Murine Major Histocompatibility
Complex
  RONALD H. SCHWARTZ

The Molecular Genetics of Components of Complement
R. D. CAMPBELL, M. C. CARROLL, AND R. R. PORTER

Molecular Genetics of Human B Cell Neoplasia
CARLO M. CROCE AND PETER C. NOWELL

Human Lymphocyte Hybridomas and Monoclonal Antibodies
DENNIS A. CARSON AND BRUCE D. FREIMARK

Maternally Transmitted Antigen
JOHN R. RODGERS, ROGER SMITH III, MARILYN M. HUSTON, AND ROBERT R. RICH

Phagocytosis of Particulate Activators of the Alternative Complement Pathway: Effects of Fibronectin
JOYCE K. CZOP

INDEX

Volume 39

Immunological Regulation of Hematopoietic/Lymphoid Stem Cell Differentiation by Interleukin 3
JAMES N. IHLE AND YACOB WEINSTEIN

Antigen Presentation by B Cells and Its Significance in T–B Interactions
ROBERT W. CHESNUT AND HOWARD M. GREY

Ligand–Receptor Dynamics and Signal Amplification in the Neutrophil
LARRY A. SKLAR

Arachidonic Acid Metabolism by the 5-Lipoxygenase Pathway, and the Effects of Alternative Dietary Fatty Acids
TAK H. LEE AND K. FRANK AUSTEN

The Eosinophilic Leukocyte: Structure and Function
GERALD J. GLEICH AND CHERYL R. ADOLPHSON

Idiotypic Interactions in the Treatment of Human Diseases
RAIF S. GEHA

Neuroimmunology
DONALD G. PAYAN, JOSEPH P. MCGILLIS, AND EDWARD J. GOETZL

INDEX

Volume 40

Regulation of Human B Lymphocyte Activation, Proliferation, and Differentiation
DIANE F. JELINEK AND PETER E. LIPSKY

Biological Activities Residing in the Fc Region of Immunoglobulin
EDWARD L. MORGAN AND WILLIAM O. WEIGLE

Immunoglobulin-Specific Suppressor T Cells
RICHARD G. LYNCH

Immunoglobulin A (IgA): Molecular and Cellular Interactions Involved in IgA Biosynthesis and Immune Response
JIRI MESTECKY AND JERRY R. MCGHEE

The Arrangement of Immunoglobulin and T Cell Receptor Genes in Human Lymphoproliferative Disorders
THOMAS A. WALDMANN

Human Tumors Antigens
RALPH A. REISFELD AND DAVID A. CHERESH

Human Marrow Transplantation: An Immunological Perspective
PAUL J. MARTIN, JOHN A. HANSEN, RAINER STORB, AND E. DONNALL THOMAS

INDEX

Volume 41

Cell Surface Molecules and Early Events Involved in Human T Lymphocyte Activation
ARTHUR WEISS AND JOHN B. IMBODEN

Function and Specificity of T Cell Subsets in the Mouse
JONATHAN SPRENT AND SUSAN R. WEBB

Determinants on Major Histocompatibility Complex Class I Molecules Recognized by Cytotoxic T Lymphocytes
JAMES FORMAN

Experimental Models for Understanding B Lymphocyte Formation
PAUL W. KINCADE

Cellular and Humoral Mechanisms of Cytotoxicity: Structural and Functional Analogies
JOHN DING-E YOUNG AND ZANVIL A. COHN

Biology and Genetics of Hybrid Resistance
MICHAEL BENNETT

INDEX

Volume 42

The Clonotype Repertoire of B Cell Subpopulations
NORMAN R. KLINMAN AND PHYLLIS-JEAN LINTON

The Molecular Genetics of the Arsonate Idiotypic System of A/J Mice
GARY RATHBUN, INAKI SANZ, KATHERYN MEEK, PHILIP TUCKER, AND J. DONALD CAPRA

The Interleukin 2 Receptor
KENDALL A. SMITH

Characterization of Functional Surface Structures on Human Natural Killer Cells
JEROME RITZ, REINHOLD E. SCHMIDT, JEAN MICHON, THIERRY HERCEND, AND STUART F. SCHLOSSMAN

The Common Mediator of Shock, Cachexia, and Tumor Necrosis
B. BEUTLER AND A. CERAMI

Myasthenia Gravis
JON LINDSTROM, DIANE SHELTON, AND YOSHITAKA FUJII

Alterations of the Immune System in Ulcerative Colitis and Crohn's Disease
RICHARD P. MACDERMOTT AND WILLIAM F. STENSON